OCR Gateway
GCSE Chemistry
Student Book

Author:
Nigel Saunders

Series Editor:
Philippa Gardom Hulme

OXFORD
UNIVERSITY PRESS

Contents

How to use this book

Welcome to your *OCR Gateway GCSE Chemistry* Student Book. This introduction shows you all the different features *OCR Gateway GCSE Chemistry* has to support you on your journey through GCSE Chemistry.

Being a scientist is great fun. As you work through this Student Book, you'll learn how to work like a scientist, and get answers to questions that science can answer.

This book is packed full of questions well as plenty of activities to help build your confidence and skills in science.

Higher Tier
If you are sitting the Higher Tier exam, you will need to learn everything on these pages. If you will be sitting Foundation Tier, you can miss out these pages. The same applies to boxed content on other pages.

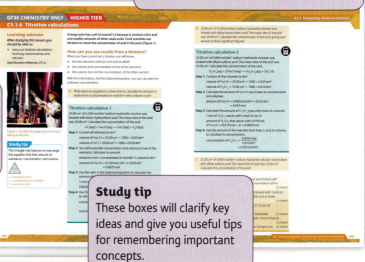

GCSE Chemistry only
This content is only needed by students studying for GCSE Chemistry. If you are studying Chemistry for Combined Science, you can miss out these pages. The same applies to boxed content on other pages.

Learning outcomes
These are statements describing what you should be able to do at the end of the lesson. You can use them to help you with revision.

Study tip
These boxes will clarify key ideas and give you useful tips for remembering important concepts.

Using Maths
Together with the *Maths for GCSE Chemistry* chapter, you can use these feature boxes to help you to learn and practise the mathematical knowledge and skills you need.

Literacy
Literacy boxes help you develop literacy skills so that you are able to demonstrate your knowledge clearly in the exam.

Go further
This feature shows interesting ways that you can explore a topic further. These ideas go beyond the GCSE Chemistry specification.

Key words
These are emboldened in the text, to highlight them to you as you read. You can look them up in the Glossary if you are not sure what they mean.

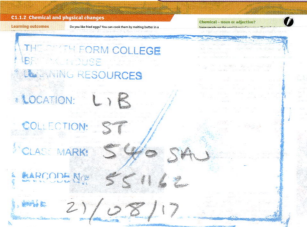

Synoptic link
This feature shows links between the lesson and content in other parts of the course, as well as any links back to what you learned at Key Stage 3.

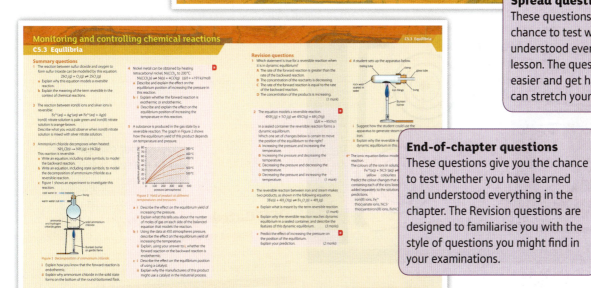

Kerboodle

This book is also supported by Kerboodle, offering unrivalled digital support for building your practical, maths, and literacy skills.

If your school subscribes to Kerboodle, you will also find a wealth of additional resources to help you with your studies and with revision:

- animations, videos, and revision podcasts
- webquests
- maths and literacy skills activities and worksheets
- on your marks activities to help you achieve your best
- practicals and follow up activities
- interactive quizzes that give question-by-question feedback
- self-assessment checklists.

Watch interesting animations on the trickiest topics, and answer questions afterward to check your understanding.

Check your own progress with the self-assessment checklists.

If you are a teacher reading this, Kerboodle also has plenty of practical support, assessment resources, answers to the questions in the book, and a digital markbook along with full teacher support for practicals and the worksheets, which include suggestions on how to support and stretch your students. All of the resources that you need are pulled together into ready-to-use lesson presentations.

Assessment objectives and key ideas

There are three Assessment Objectives in OCR GCSE (9–1) Chemistry A (Gateway Science). These are shown in the table below.

Assessment Objectives		Weighting	
		Higher	Foundation
AO1	Demonstrate knowledge and understanding of: • scientific ideas • scientific techniques and procedures.	40	40
AO2	Apply knowledge and understanding of: • scientific ideas • scientific enquiry, techniques, and procedures.	40	40
AO3	Analyse information and ideas to: • interpret and evaluate • make judgements and draw conclusions • develop and improve experimental procedures.	20	20

Studying science at GCSE helps us to understand the world around us. It is important to understand the essential aspects of the knowledge, methods, processes, and uses of science. There are a number of key ideas that underpin the complex and diverse phenomena of the natural world. These key ideas are shown in the table below.

GCSE Chemistry (9–1) Key Ideas
Conceptual models and theories are used to make sense of the observed diversity of natural phenomena.
It is assumed that every effect has one or more causes.
Change is driven by differences between different objects and systems when they interact.
Many interactions occur over a distance and over time without direct contact.
Science progresses through a cycle of hypothesis, practical experimentation, observation, theory development, and review.
Quantitative analysis is a central element of many theories and of scientific methods of inquiry.

Working Scientifically

WS1 The power of science

Learning outcomes

After studying this lesson you should be able to:

- describe some applications of science
- evaluate the implications of some applications of science.

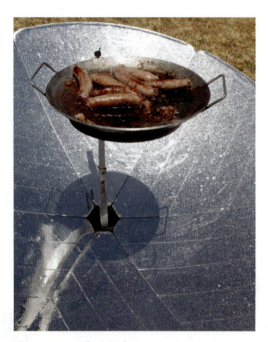

Figure 1 *A solar cooker.*

In a solar cooker (Figure 1), mirrors direct heat from the Sun onto the frying pan, cooking the sausages. Could you use a solar cooker at home?

What are applications of science?

The solar cooker is an example of an application of science. Its designer used knowledge of reflection to work out the best shape for the mirror, and where to place the pan holder. **Technology** is the application of science for practical purposes.

Scientists are developing new ways of using energy transferred from the Sun to generate electricity. In a solar thermal power plant (Figure 2), mirrors focus radiation from the Sun onto boilers at the top of high towers.

Figure 2 *The mirrors in this solar thermal power plant focus light to the top of the tower, which is so hot that it glows.*

Water in the towers boils, producing steam. The steam makes a machine called a turbine turn. The turbine is connected to another machine called a generator, which generates an electrical potential difference, which makes a current flow.

A Write down an application of science that has improved human health.

How can we evaluate science applications?

Any application of science brings drawbacks as well as benefits. When you evaluate an application of science, think about its personal, social, economic, and environmental implications. You will also need to consider **ethical issues**. An ethical issue is a problem where a choice has to be made concerning what is right and what is wrong.

Consider for example the Ivanpah solar thermal power plant in the USA. Its construction and operation has created jobs, so it has personal and social benefits. Building the power plant was expensive, but selling its electricity will make money. These are examples of **economic impacts**.

The power plant was built on the habitat of the desert tortoise (Figure 3). Its glowing towers attract insects, which in turn attract birds, but few birds survive close to the hot towers (Figure 4). Is it morally acceptable to kill birds in order to generate electricity?

The power plant generates electricity without producing carbon dioxide, which is a greenhouse gas. This is an **environmental** benefit. However, carbon dioxide was produced in making and transporting the mirrors and towers.

B Describe an ethical issue arising from the Ivanpah solar thermal plant.

People are concerned about hazards linked to the power plant, such as aeroplane pilots being dazzled by glare from the mirrors. A **hazard** is anything that threatens life, health or the environment. Managers at the power plant work hard to reduce the **risks** linked to these hazards. They do this by reducing the probability of a hazard occurring, as well as by trying to reduce the consequences if it does occur.

C Suggest how to reduce the risk caused by dazzling.

How can we make decisions about applications of science?

Government officials had to decide whether or not to allow the Ivanpah power plant to be built. It is often difficult to make decisions about the applications of science, as most have both benefits and drawbacks. The government had to weigh up these benefits and drawbacks before making its decision.

Other applications of science

There are many other applications of science (Figure 5). Scientists use their knowledge of viruses and antibodies to create new vaccines. They use knowledge of microwaves to improve mobile phone technology, and an understanding of properties of materials to develop scratch-free cars and chip-free nail varnish.

Figure 3 *The Ivanpah solar thermal power plant was built on the habitat of the desert tortoise.*

Figure 4 *This scientist is looking for birds flying close to the solar towers.*

Key words

As you read this chapter, make a glossary by writing down all the key words and their meanings.

1 Write down what the word *technology* means. *(1 mark)*

2 Suggest two hazards of using a solar cooker. *(1 mark)*

3 ✏ Write down one impact of mobile phones in each of these four categories: personal, social, environmental, and economic. State whether each impact is a benefit or a drawback. *(5 marks)*

4 ✏ Evaluate the impacts of building a solar thermal power plant. *(6 marks)*

Figure 5 *This scientist is using an advanced electron microscope to research and control the structure and behaviour of new nanomaterials. Nanomaterials have many potential uses, including in medical treatments.*

WS2 Methods, models, and communication

The MRI scan in Figure 1 shows the brain of a person with alcoholic dementia. The enlarged gap between the brain and skull shows that his brain has shrunk.

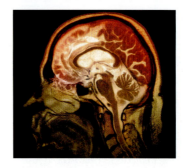
Figure 1 *The brain of a person with alcoholic dementia.*

Figure 2 *Drinking too many alcoholic drinks can cause alcoholic dementia.*

How do scientific methods and theories develop over time?

For centuries, people have known that drinking too much alcohol causes memory loss, poor judgement, and personality changes. It is only recently that a new technology – magnetic resonance imaging (MRI) – has allowed scientists to observe directly the effect of alcoholism on the brain (Figure 1).

Robot spacecraft Cassini recently sent photographs of Saturn's moon, Enceladus, back to Earth (Figure 3). Its on-board instruments identified materials ejected by the moon's geysers. Scientists analysed this evidence and suggested a new explanation, that Enceladus has liquid water beneath its icy surface.

These examples show that developing technologies allow scientists to collect new evidence and develop new explanations.

> **A** Suggest how the invention of the telescope helped scientists to collect new evidence.

Figure 3 *An image of Enceladus from Cassini.*

Figure 4 *This is a model car, but you cannot see and hold all models in science.*

What are scientific models?

Models are central to science. They make scientific ideas easier to understand. They also help in making predictions and developing explanations. Scientists spend hours creating, testing, comparing, and improving models. Models are simplified versions of reality, so no model is perfect.

There are different types of models:

- Representational models use familiar objects to describe and explain observations. An example is using marbles to model water particles.

Synoptic link

You will find out how new technology led to improved models of atoms in C1.2.3 *Developing the atomic model.*

- Spatial models often represent things that are tiny, or enormous. An example is the metal model of DNA shown in Figure 5.
- Descriptive models use words and ideas to help you imagine something, or to describe something simply. An example is using chemical equations to represent reactions.
- Mathematical models use maths to describe systems and make predictions. Scientists have developed mathematical models to describe and predict the movements of planets and stars.
- Computational models are a type of mathematical model. At the Met Office in Exeter supercomputers process millions of pieces of data to predict the weather (Figure 6).

Figure 6 *A supercomputer.*

Figure 5 *Watson and Crick with the first model of DNA.*

B Give an example of a spatial model, other than the one shown here.

Why do scientists communicate?

Scientists describe their methods, share their results and explain their conclusions in scientific papers. Before a paper is published other expert scientists check it carefully and suggest improvements. This is called **peer review**. Scientists also tell other scientists about their work at conferences (Figure 7).

Scientists use internationally accepted names, symbols, definitions and units so that scientists everywhere understand their work. The international system of units (SI units) is a system built on seven base units, including the metre, kilogram, and second. The International Union of Pure and Applied Chemistry (IUPAC) publishes rules for naming substances. You will use SI units and IUPAC names when communicating your investigations.

Scientists may also tell doctors and journalists about their work. For example, when researchers found that paracetamol may reduce testosterone levels in male foetuses they asked journalists to warn pregnant women about this possible hazard.

Figure 7 *This scientist is studying scientific posters at a conference in York.*

C Suggest two reasons for peer review.

1 Explain why it is important to use internationally accepted units, symbols and definitions in science. *(1 mark)*

2 Suggest why, before the invention of the microscope, scientists did not know that cells make up living organisms. *(1 mark)*

3 A teacher uses marbles to model water particles. Use your knowledge of particles to evaluate this model. *(6 marks)*

Synoptic link

To find out more about SI units, look at *Maths for GCSE Chemistry: 11 Quantities and units.*

WS3 Asking scientific questions

Learning outcomes

After studying this lesson you should be able to:

- Describe how to develop an idea into a question to investigate
- Explain what a hypothesis is.

Figure 1 shows an early microwave oven. In the 1950s, engineer Percy Spencer accidentally discovered that microwaves cook food when he was experimenting with microwaves. He was standing in front of the microwave source and noticed the bar of chocolate in his pocket had melted.

Figure 1 *A 1961 microwave oven.*

Figure 2 *A modern microwave oven.*

What are scientific questions?

Scientists ask questions. They ask about unexpected observations, for example, *What made my chocolate melt?* They also ask questions to solve problems, such as: *Which malaria treatment is most effective?* Some questions arise from simple curiosity, for example, *What makes rocks move on Mars?* (Figure 3).

All these questions are **scientific questions**. You can answer them by collecting and considering evidence. Of course, scientists and other people also ask questions that science cannot answer, such as: *Who should pay for vaccines?* or *Why did the Big Bang happen?*

Figure 3 *The surface of Mars, taken by a camera on the Curiosity rover.*

A Write down a scientific question that you have investigated at school.

What is a hypothesis?

Imagine you have warm water in one container, and the same volume of cold water in another container. You put both containers of water in the freezer (Figure 4). Which freezes first? Surprisingly, the warm water freezes more quickly. This is called the Mpemba effect, after the Tanzanian school student who described and investigated this unexpected observation.

How can you explain the Mpemba effect? Is it because more of the warm water evaporates, so reducing the mass of water to be frozen? Is it because cooler water freezes from the top, forming a layer of ice that insulates

Figure 4 *Warm water freezes more quickly than cold water.*

the water below? Or is it to do with faster-moving convection currents in warmer water transferring energy to the surroundings more quickly?

Each of these suggested explanations is a **hypothesis**. A hypothesis is based on observations and is backed up by scientific knowledge and creative thinking. A hypothesis must be testable.

There is still no accepted explanation for the Mpemba effect. In 2012 the Royal Society of Chemistry ran a competition to find the best explanation. More than twenty thousand people entered.

B Use the paragraphs above to help you write down two possible hypotheses to explain the Mpemba effect.

How do scientists answer questions?

Answering a scientific question involves making observations to collect data, and using creative thought to explain the data. Science moves forward through cycles or stages like those shown below (Figure 5), building on what is already known. Of course scientists do not always follow these stages exactly.

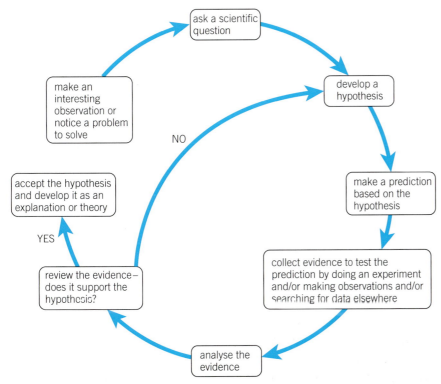

Figure 5 *Scientists may answer a scientific question by following these stages. Of course scientific advances do not always follow this route.*

C Write down these stages in the order shown by the cycle above, starting with *ask a scientific question*: make a prediction, ask a scientific question, collect evidence, develop a hypothesis.

1 State which of the following are scientific questions, giving reasons.
   a Which material is harder, glass or diamond?   *(1 mark)*
   b Which type of car travels fastest?   *(1 mark)*
   c Which type of car is best?   *(1 mark)*
   d How many years can you expect a robin to live?   *(1 mark)*

2 A student asks a scientific question, *Which has a higher boiling point, water or ethanol?* He knows that the forces holding the particles together in water in the liquid state are stronger than those in ethanol.
   a Use your scientific knowledge to suggest a hypothesis.   *(1 mark)*
   b Outline how the student could investigate his question, referring to the steps in Figure 5.   *(4 marks)*

3 Suggest how you could test the hypothesis that the Mpemba effect happens because cooler water freezes from the top, forming a layer of ice that insulates the water below.   *(4 marks)*

Learning outcomes

After studying this lesson you should be able to:

- identify different types of variable
- describe how to plan an investigation.

Figure 1 *Young people using a mobile phone.*

Figure 2 *In this investigation the dependent variable is the battery life.*

Do you ever wish your phone battery lasted longer?

How can you start an investigation?

Some students noticed that using social media seemed to make their phone battery run down quickly (Figure 1). They made up a scientific question to investigate:

How does the time spent on social media affect mobile phone battery life?

They used their observations, and scientific knowledge, to make a hypothesis:

Using social media involves downloading significant amounts of data. This means that extra energy is transferred as electricity from its chemical store in the battery to the thermal store of the surroundings. This shortens the battery life.

A Suggest a prediction based on this hypothesis.

How do you choose variables?

The students considered factors that might affect the outcome of their investigation. These are **variables**. There are three types of variable:

- The **independent variable** is the one you deliberately change.
- The **dependent variable** is the one you measure for each change of the independent variable.
- **Control variables** are ones that may affect the outcome, as well as the independent variable. Keep these variables the same for a **fair test**.

In the mobile phone investigation, the independent variable is the time spent on social media. The dependent variable is the battery life (Figure 2). Control variables include the model of the phone, as well as the number of texts sent.

You can collect data as words or numbers:

- A **continuous variable** can have any value, and can be measured. In this investigation, the time spent on social media is a continuous variable.
- A **discrete variable** has whole number values. The number of texts is a discrete variable.
- Values for a **categoric variable** are described by labels. The make and model of phone are categoric variables.

B Explain whether battery life is a continuous, discrete or categoric variable.

How do you plan an investigation?

The students used their hypothesis to say what they thought would happen. This is their **prediction**:

The more time spent on social media, the shorter the battery life.

The students planned how to test their prediction. They thought about the apparatus, and what to do with it. This is their plan:

1   People with the same model of phone fully charge their batteries and record the time.

2   People write down when, and for how long, they use social media.

3   Apart from social media, people can use the phone for texting only.

4   People record when the battery has run down.

> **C**   Write down two pieces of equipment needed for this investigation.

The students also considered hazards. They discovered that using a phone exerts high forces on the neck (Figure 3). They took precautions to reduce the risks from this hazard, including taking breaks.

What other types of investigation are there?

There are many other sorts of investigation. Scientists work out how to make new substances, and check that substances are what they should be. A scientist might check that data are correct, or use different types of trials to investigate the effects of a medicinal drug.

Figure 3 *Bending to use a phone exerts high forces on the neck.*

Figure 4 *This scientist is using gas chromatography to check food samples for contamination.*

Figure 5 *This scientist is inspecting sorghum plants being grown as part of a crop trial. Sorghum is used for grain and as an animal feed.*

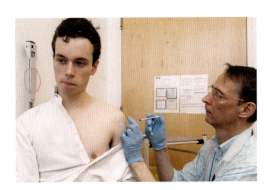

Figure 6 *This scientist is injecting an experimental Ebola vaccine into the arm of a volunteer who is testing the vaccine.*

> **1**   Compare continuous, discrete and categoric variables.   *(3 marks)*
>
> **2**   A student investigates the effect of the number of hours of sunlight on the number of units (kWh) of electricity generated by the solar panels on the school roof.
>   **a**   Write down the independent and dependent variables.   *(2 marks)*
>   **b**   Suggest a control variable and explain whether it is possible to keep this variable the same.   *(2 marks)*
>
> **3**   A student investigates whether shoe size affects speed of swimming. Identify the variables in the investigation, and use two labels such as *independent* and *discrete* to describe each one.   *(6 marks)*

WS5 Obtaining high-quality data

Learning outcomes

After studying this lesson you should be able to:

- describe how to obtain data that is accurate and precise
- compare the meanings of the terms repeatable and reproducible.

Figure 1 *Ethiopian coffee.*

Go further

Find out why water boils at a lower temperature in Addis Ababa than in London.

Figure 2 *The resolution of this thermometer is 0.5 °C because this is the smallest change in reading that you can see.*

A cup of coffee in Addis Ababa, Ethiopia (Figure 1), might taste delicious. But it is not as hot as a cup of coffee made in London.

How can you obtain accurate data?

In Addis Ababa or London, you can use a thermometer to measure the boiling point of water. In a scientific investigation, you need measurements that are close to the true value. These data are **accurate**.

You can collect accurate data by:

- using your thermometer carefully
- repeating measurements and calculating the mean
- repeating measurements with a different instrument, for example a temperature probe, and checking that the readings are the same.

> **A** A student measures the boiling point of water three times, and obtains these values: 101.0 °C, 99.0 °C, and 100.0 °C. Calculate the mean.

How can you obtain precise data?

Precise measurements give similar results if you repeat the measurement. If repeated measurements are precise, the **spread** of the data set is small. You can calculate the spread by subtracting the smallest measurement from the largest measurement for a set of repeats.

To get precise data you need a measuring instrument with a high **resolution**. The resolution of a measuring instrument is the smallest change in the quantity that gives a change in the reading that you can see (Figure 2).

When you are using a measuring instrument, it is important to record all the readings you make with this instrument in a particular experiment to the same number of decimal places. For example, two students working together on the same experiment record the following series of masses from a balance:

Student A:	11 g	9.8 g	8.65 g
Student B:	11.00 g	9.80 g	8.65 g

Student B has recorded each reading to two decimal places, so his set of readings is recorded correctly.

Precision

Students in Addis Ababa and London measured the boiling point of water. Which data set is more precise?

Table 1 *Temperature at which water boils in Addis Ababa and London.*

Place	Temperature at which water boils (°C)		
	First reading	Second reading	Third reading
Addis Ababa	94.0	93.0	94.0
London	97.0	100.0	99.0

Step 1: Calculate the spread of each data set by subtracting the smallest measurement from the largest measurement.

Addis Ababa   94.0 °C – 93.0 °C = 1.0 °C

London   100.0 °C – 97.0 °C = 3.0 °C

Step 2: Compare the spread of each data set to see which is smaller.

The spread for Addis Ababa is smaller, so their data set is more precise.

You can have data that are precise but not accurate. Precise data might not be close to the true value.

Two students used a burning fuel in a spirit burner (Figure 3) to heat water for two minutes. They measured the highest temperature the water reached. They repeated the experiment four times each. Their results are marked on the temperature scales in Figure 4.

The set of measurements on the left is precise but not accurate.

B Explain why the set of measurements on the right of Figure 4 is accurate but not precise.

What are repeatable and reproducible data?

If you repeat an investigation several times using the same method and equipment, and if you get similar results each time, your results are **repeatable**.

If someone else repeats your investigation, or if you do the same investigation with different equipment, and the results are similar, the investigation is **reproducible**.

C You do an investigation twice, with different equipment each time. Explain whether you expect your results to be repeatable or reproducible.

1 Compare the meanings of the terms *repeatable* and *reproducible*. (2 marks)

2 🔧 A student is investigating how the mass of salt added to water affects boiling temperature. Describe in detail how she can make accurate measurements of mass and temperature. (4 marks)

3 Two students make measurements of the time for an egg attached to a parachute to fall from a window. Their data are in the table.
a Explain which data set is more precise. (3 marks)
b Explain why you cannot tell which data set is more accurate. (2 marks)

Table 2 *Time taken for an egg to fall from a window to the ground.*

Student	Time for egg and parachute to reach the ground (s)			
	First measurement	Second measurement	Third measurement	Mean
A	3.0	2.8	3.2	3.0
B	3.2	3.4	3.3	3.3

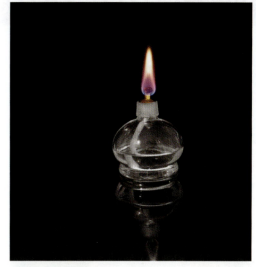

Figure 3 *A spirit burner.*

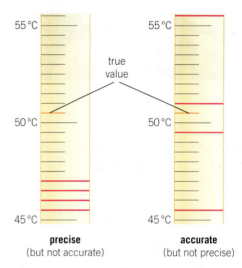

Figure 4 *Two sets of temperature data obtained when heating water with a burning fuel. Each piece of data is represented by a pink line.*

WS6 Presenting data

Figure 1 *A pet tortoise can live for 60 years.*

Table 1 *The lifespans of different pets.*

Pet species	Mean age at death (years)
cat	12
dog	10
guinea pig	5
mouse	2
tortoise	55

Maths link

Maths for GCSE Chemistry:
7 Mean values shows how to calculate the mean value from a set of data.

Tortoises are not easy to look after, but with proper care they can live for 60 years (Figure 1). A pet mouse has a much shorter lifespan.

How do you design a table to display data?

A group of students asked, *What is the lifespan of different pets?* They collected data by asking pet owners to complete a survey about the age of death of their pets.

The students calculated the mean lifespan for each pet. Table 1 summarises their results. In any table:

- Write the independent variable in the left column
- Write the dependent variable in the right column
- Write units in the column headings, not next to each piece of data.

A State which variable in the table is categoric.

When do you draw a bar chart?

If either variable is categoric, draw a **bar chart**. In the example the independent variable (pet species) is categoric, so the students draw a bar chart (Figure 2). In any bar chart:

- Write the independent variable on the *x*-axis and the dependent variable on the *y*-axis.
- Label the axes with the variable name and units (if there are any.)
- Choose a scale for the *y*-axis so that the chart is as big as possible, and make sure the scale is even.

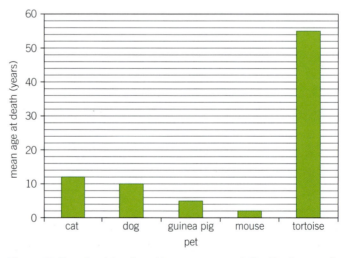

Figure 2 *Bar chart to show the mean age at death of a number of animals.*

B State whether the mean age at death is a continuous, discrete or categoric variable. Explain your answer.

Alternatively, when one variable changes the other variable might remain the same.

You can now write the first part of your conclusion. In the urine investigation, the first part of the conclusion from the second graph is:

The greater the mass of a mammal, the longer its urethra.

The scientists made a hypothesis to explain this relationship. They suggested that a longer urethra increases the gravitational force acting on the urine, so increasing the rate that urine leaves the body. The scientists used mathematical models to show that their hypothesis was correct.

B Complete the conclusion by adding a scientific explanation.

In this example, the longer urethra *causes* faster urination. However, the fact that there is a relationship does not necessarily mean that a change in one variable *caused* the change in the other. There could be some other reason for the change.

How do you evaluate an investigation?

When you evaluate an investigation, think about these two questions:

● How could you improve the method?

● What is the quality of the data?

In this investigation the scientists collected data by filming animals urinating. They also viewed videos online.

You can evaluate the quality of data by considering its accuracy, precision, repeatability, and reproducibility.

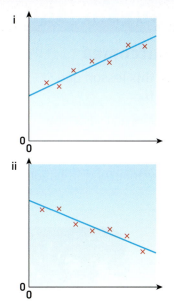

Figure 4 *These graphs show linear relationships. The lines of best fit are linear. The relationship in (i) is positive – as one variable increases, so does the other. The relationship in (ii) is negative – as one variable increases the other variable decreases.*

News report

Write the text for a news report to tell young people about the urination investigation.

1 Sketch graphs showing **a** a positive relationship and **b** a negative relationship. *(2 marks)*

2 Suggest how to collect accurate data for the duration of urination. *(3 marks)*

3 Table 1 shows data about urine flow rate and animal mass.
  a 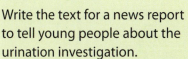 Plot the data on a graph, and draw a line of best fit. *(6 marks)*
  b Describe what the graph shows. *(2 marks)*

Table 1 *Data about the mass of an animal against urine flow rate.*

Animal	Mass of animal (kg)	Urine flow rate (cm³/s)
rat	0.3	2.0
small dog	3.5	1.0
goat	71	6.0
human	70	20
cow	640	450

Learning outcomes

After studying this lesson you should be able to:

- compare random and systematic errors
- explain the meaning of uncertainty
- explain the meaning of distribution.

Figure 1 *A herd of cows in a field.*

How long does it take for a cow (Figure 1) to empty its bladder?

What are random errors?

A student watched videos of three animals urinating. He timed the duration of urination for each animal. Table 1 shows his measurements.

Table 1 *The duration of urination for different animals.*

Animal	Duration of urination (s)			
	First measurement	Second measurement	Third measurement	Mean
cow	21.4	21.6	21.5	21.5
horse	22.4	22.0	22.8	22.4
sheep	20.8	21.0	20.9	20.9

Look at the times for the cow; the measurements are not the same. You cannot predict whether a fourth measurement would be higher or lower than the third measurement. The measurements are showing a **random error**.

Random errors are caused by known and unpredictable changes in the investigation, including changes to the environmental conditions. They are also caused by changes that occur in measuring instruments, or difficulties in being sure what values show. You cannot control the cause of random errors. However, you can reduce their effect by repeating measurements and calculating a mean.

A Suggest a cause of random error in this investigation.

What are systematic errors?

An error might also be a **systematic error**. This means that your measurements are spread about some value other than the true value. Each of your measurements differs from the true value by a similar amount; so all your values are too high, or too low. For example, a systematic error might be caused by an ammeter that does not read zero when there is no current.

Figure 2 *You can use different pieces of equipment to measure time.*

If you think that you have a systematic error, repeat the measurements with a different piece of equipment (Figure 2). Then compare the two sets of measurements.

B Suggest a cause of systematic error when measuring the boiling point of water.

What is uncertainty?

Your readings are only as good as your measuring instrument. On a thermometer it might be hard to tell if a reading is 65.5 °C, or 65.0 °C, or 66.0 °C. There is **uncertainty** in your measurement because of the thermometer you are using.

If the smallest scale division is 1.0 °C then you can estimate the uncertainty as ± 0.5 °C, which is half the smallest scale division.

A better definition of uncertainty is that it is the interval within which the true value can be expected to lie, with a given level of confidence. For example you could say that the temperature is 65.5 °C ± 0.5 °C with a confidence of 95%.

C A stopwatch has scale divisions of 0.2 seconds. Estimate the uncertainty of readings with this stopwatch.

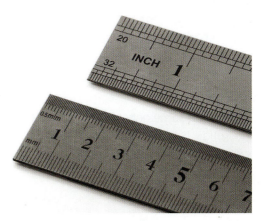

Figure 3 *There would be a greater uncertainty in a measurement made using a ruler measured in inches than one made with a metric ruler.*

Go further

What is distribution?

As you know, the spread of a set of repeated measurements is the difference between the highest and lowest values. The way the measurements are distributed between the highest and lowest values can take different forms. Often, the values are more likely to fall near the mean than further away. This may mean that the measurements have a **normal distribution** (Figures 4 and 5).

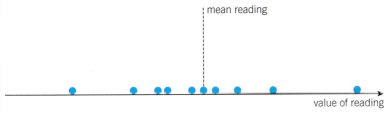

Figure 4 *Each blob represents one measurement. There are more measurements close to the mean.*

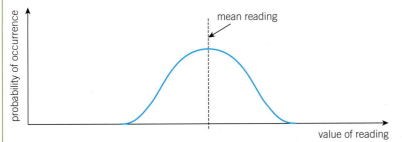

Figure 5 *This normal distribution shows that you are more likely to make measurements close to the mean.*

1 Explain the meaning of *uncertainty*.
(1 mark)

2 Compare the meaning of *random error* and *systematic error*. *(2 marks)*

3 Plot a blob diagram for these measurements of the duration of urination of a cow:
21.5, 21.4, 21.6, 21.3, 21.7, 21.4, 21.8, 21.2, 22.0, 20.5
Show the mean reading on your blob diagram, and explain what your plot shows about the distribution of the measurements. *(4 marks)*

Learning outcomes

After studying this lesson you should be able to:

- describe the main features of the particle model
- explain some general properties of solids, liquids, and gases.

Specification reference: C1.1a

Figure 1 *Orange squash and water mix together completely.*

Have you noticed how the orange colour in concentrated squash spreads out when you add water to it (Figure 1)? This happens because the two substances are made from tiny particles that can mix together.

What are particles?

You, and all the things you see around you, are made from **matter**. A **particle** is a tiny bit of matter. Particles can be large enough to see, like pieces of dust, or too small to see, even with a microscope.

Chemists are the scientists who study chemistry. When chemists talk about particles, they usually mean atoms, ions, and molecules. You will find out more about atoms, ions, and molecules later. For now, just imagine particles as hard, tiny spheres that are too small to see.

A Name three examples of particles in chemistry.

What is the particle model?

The **particle model** describes how particles are arranged and how they move in solids, liquids, and gases, as shown in Table 1.

Table 1 *Arrangement and movement of particles in the three states of matter.*

State	Diagram of particles	Arrangement of particles	Relative distance between particles	Main movement of particles
solid		regular	very close	vibrate about fixed positions
liquid		random	close	move around each other
gas		random	far apart	move quickly in all directions

Study tip

Remember that matter is not continuous and instead is made from particles. When you draw diagrams to represent particles, do not shade in between the particles.

The particle model explains why some of the properties of a substance depend on its **state**. For example:

- You cannot compress (squash) a substance in its solid state or its liquid state. This is because there is no space for the particles to move into.
- A substance in its solid state has a fixed shape and cannot flow. This is because its particles vibrate around fixed positions and cannot move from place to place.

B Describe the main difference between the arrangement of particles in the liquid state and particles in the gas state.

C Explain why you can compress a substance in its gas state.

Drawing 3D objects

A solid is a three-dimensional (3D) object. It has height, width, and depth. A piece of paper is so thin that it is almost a two-dimensional (2D) object. You have to make compromises when you represent a 3D object on paper. For example, Figure 2 shows an isometric diagram in which you can only see parts of the particles facing you, and some particles are hidden.

Models in science

Scientists use **models** to explain the things they see happening. When you think of a model, you might think about a smaller version of a real object, or something that you build from pieces. Some scientific models are like that. Other types of scientific model are chemical equations, diagrams, or analogies. Models are used to solve problems, to make predictions, and to develop scientific understanding.

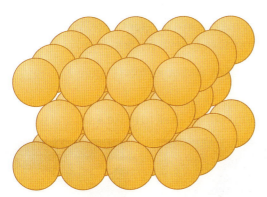

Figure 2 *An isometric diagram of particles in a solid.*

D Compare the isometric diagram (Figure 2) with the 2D diagram of particles in a solid in Table 1. Suggest the advantages and disadvantages of each type of diagram.

1 Use the particle model to explain why a substance flows in its gas state and completely fills its container. *(2 marks)*

2 Water can be poured into a glass, and takes the shape of the bottom of the glass. Use the particle model to explain these observations. *(2 marks)*

3 When a solid is heated, its particles vibrate more vigorously. Explain why a solid expands when it is heated. *(2 marks)*

C1.1.2 Chemical and physical changes

After studying this lesson you should be able to:

- describe differences between chemical and physical changes
- explain chemical and physical changes in terms of particles.

Specification reference: C1.1b

Figure 1 *Eggs frying.*

Synoptic link

You can learn more about physical changes in C2.3.2 *Changing state.*

Do you like fried eggs? You can cook them by melting butter in a pan, cracking open the eggs and adding them to the hot liquid butter (Figure 1). A minute or so later, your fried eggs are cooked and ready to eat. Which of the changes in this process are physical changes? Which are chemical changes?

What are chemical and physical changes?

Melting butter and cracking eggs are **physical changes**. A physical change happens when a substance changes state or shape, or breaks into pieces. No new substances are made. Many physical changes can be reversed, including:

- freezing juice to make an ice lolly
- mixing sand with water
- dissolving sugar in water.

> **A** Suggest how you could reverse each of the physical changes in the list above.

Many **chemical changes** happen in an egg as it cooks. A chemical change is a change that produces one or more new substances. The properties of the new substances are often very different from the properties of the original substances. For example, they may look or taste different, and they may be in a different state. Many chemical changes are difficult or impossible to reverse, including:

- cooking eggs, cakes, and other food (Figure 2)
- steel rusting
- an acid reacting with an alkali to make a salt and water.

Figure 2 *Chefs rely on a chemical change, natural gas burning, to cook food.*

> **B** Name two more chemical changes that are not described above.

Chemical – noun or adjective?

Some people use the word 'chemical' as a noun. They talk about 'chemicals' when they mean substances such as artificial fertilisers and disinfectants. However, all substances are chemicals, so the use of the word as a noun is not helpful. It is best to use the word 'chemical' as an adjective to describe something. So a change that makes new substances is described as a chemical change or a **chemical reaction**.

C The label on a bottle of shampoo states that it is 'chemical-free'. Explain why this cannot be true.

What happens to particles during changes?

When a substance changes state, its particles stay the same but their arrangement and movement change. For example, when an ice cube melts to make liquid water, its particles begin to move around each other instead of being in fixed positions. The particle arrangement becomes random instead of regular.

However, in a chemical change, particles break up and then join together in different ways. This is why new substances are made (see Figure 3).

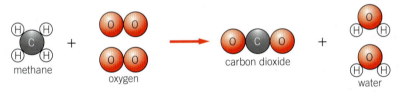

Figure 3 *During this chemical change, the particles in methane and oxygen join together in a different way to make carbon dioxide and water.*

D Suggest what must happen to the methane and oxygen particles so that they can make particles of carbon dioxide and water.

Go further

A baker is making cupcakes (Figure 4). He starts by mixing together flour, sugar, butter, eggs, and baking powder. Then he cooks the mixture in a hot oven. He makes icing by mixing together butter and icing sugar, and then ices the cakes.

Decide which of these changes are physical, and which are chemical. Then find out how baking powder works.

Figure 4 *Making cupcakes involves both physical and chemical changes.*

1 Name three examples of chemical changes, and three examples of physical changes. (*6 marks*)
2 Describe two differences between a chemical change and a physical change. (*2 marks*)
3 Use the particle model to explain in detail the differences between physical changes and chemical changes. Include examples to illustrate your answer. (*6 marks*)

Learning outcomes

After studying this lesson you should be able to:

- compare the sizes of particles to the distances between them
- describe the forces between particles
- explain the limitations of the particle model.

Specification reference: C1.1c

Figure 1 *Magnetic resonance imaging (MRI) scanners provide detailed images of the inside of the body.*

Have you ever had a helium-filled party balloon? Helium is not just useful for giving balloons lift; liquid helium also cools superconducting magnets in MRI scanners (used in hospitals) to a chilly −269 °C (Figure 1). Compared to other substances, helium particles are particularly small.

How big are particles?

The smallest particles that make up a substance are called **atoms**. Helium atoms are the smallest of all atoms, at just 62 pm (62×10^{-12} m) in diameter. This is so tiny that the diameter of a helium atom is about a billion (1×10^{9}) times smaller than the diameter of a tennis ball. The distance between two helium atoms in the gas state is about 55 times larger than the diameter of a helium atom. This makes it difficult to draw gas particles to scale.

A   The Earth is 1.27×10^{7} m in diameter. Calculate its diameter if this were a billion times smaller.

B   In a scale drawing, helium atoms are shown as 6 mm diameter circles. Calculate how far apart two atoms will be. Explain why this might be a problem.

The space between atoms

The mean distance between two helium atoms in the gas state is 3.4×10^{-9} m. Calculate the ratio of distance to diameter for helium atoms in the gas state.

Step 1: Write down the data you have been given.

mean distance between atoms = 3.4×10^{-9} m

diameter of atom = 6.2×10^{-11} m

Step 2: Put the numbers in the equation, then calculate the answer and its unit (if it has one).

$$\text{ratio of distance to diameter} = \frac{\text{distance between atoms}}{\text{diameter of atom}}$$

$$= \frac{3.4 \times 10^{-9}\,\text{m}}{6.2 \times 10^{-11}\,\text{m}}$$

$$= 0.548 \times 10^{2}$$

$$= 55$$

C   The mean distance between caesium atoms in the gas state is 5.1×10^{-9} m. The diameter of a caesium atom is 6.0×10^{-10} m. Calculate the ratio of distance to diameter for caesium atoms.

What are the forces between particles?

Chemists have identified attractive forces between the particles they study. Figure 3 shows how chemists represent these forces. These are **electrostatic forces** of attraction between positive and negative charges. The forces become weaker the further apart the particles are. They are strongest in solids, and weakest in gases.

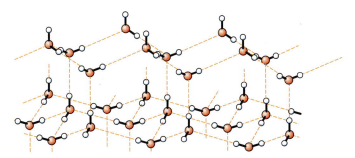

Figure 3 *The forces between particles are drawn as solid or dashed lines.*

Figure 2 *Electrostatic forces attract hair to a balloon.*

> **D** Describe what keeps particles relatively close together in liquids.

What are the limitations of the particle model?

The particle model is helpful, but is not perfect. It does not take into account:

- the forces between particles
- the size of particles
- the space between particles.

For example, the volume of a substance usually increases a little when it melts, as some of the forces of attraction are overcome and the particles can move around each other. The volume increases a great deal when a liquid becomes a gas, but some forces of attraction do remain.

> **Synoptic link**
>
> In the particle model, the collisions between particles are 'inelastic'. This means that energy is transferred from the particles to the surroundings when particles collide. In reality, particle collisions are 'elastic' because energy is not transferred to the surroundings during a collision. You can learn more about elastic and inelastic collisions in P2.2.6 *Momentum*.

> **Study tip**
>
> Particles *do not* change size during state changes or when the temperature changes.

1 Describe three limitations of the particle diagrams shown below.
(3 marks)

solid

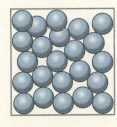
liquid

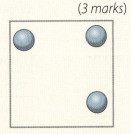
gas

2 When iron melts, its volume increases by 1.1%. Suggest what this indicates about the distances and attractive forces between iron atoms in the solid and liquid states. *(2 marks)*

3 Figure 3 shows the arrangement of water particles in ice. Suggest why liquid water expands when it freezes, unlike most other substances. *(3 marks)*

Particles

C1.1 The particle model

Summary questions

1 Draw diagrams to show how particles are arranged in substances in:
  a the solid state
  b the liquid state
  c the gas state.

2 Copy and complete Table 1.

Table 1 *The three states of matter.*

State	Arrangement of particles	Relative distance between particles	Main movement of particles
solid			
	random		move around each other
		far apart	

3 Write down simple definitions for the following terms:
  a chemical change
  b physical change.

4 Use the particle model to explain the following observations:
  a A substance can be compressed when it is in the gas state but not when it is in the solid state.
  b A substance has a fixed shape when it is in the solid state but not when it is in the liquid state.

5 a Describe what happens to the particles of a substance in the liquid state as it:
  i cools down but remains in the liquid state
  ii freezes.
  b Explain whether the changes described in **a** are chemical changes or physical changes.

6 For each of the following situations, explain whether the change is a chemical change or a physical change:
  a a glass beaker breaking
  b table salt dissolving in water to make a solution
  c iron railings becoming rusty
  d charcoal burning in a barbecue.

7 Figure 1 is a typical particle diagram.

Figure 1 *A particle diagram.*

  a Write down the state it represents.
  b Describe *one* way in which the diagram does not accurately model:
   i the arrangement of the particles
   ii the movement of the particles.
  c Apart from your answers to **b**, describe one other limitation of this model.

8 Argon is in the gas state at room temperature. The mean distance between argon atoms is 3.27 nm and the diameter of an argon atom is 144 pm.
  a Convert the mean distance between argon atoms to metres, m, giving your answer in standard form (1 nm = 10^{-9} m).
  b Convert the diameter of an argon atom to metres, m, giving your answer in standard form (1 pm = 10^{-12} m).
  c Calculate the ratio of distance to diameter for argon atoms, giving your answer to three significant figures.

9 The density of a substance is a measure of the mass contained in a given volume. Table 2 shows the density of gallium metal in its solid state and in its liquid state.

Table 2 *Properties of gallium.*

State	Density (g/cm³)
solid	5.91
liquid	6.09

  a Explain in which state, solid or liquid, gallium atoms are closer together.
  b Explain in which state, solid or liquid, the energy stored in the gallium atoms will be lower.

Revision questions

1 Which change of state (A, B, C, or D) happens when a substance condenses?
  A solid to liquid
  B liquid to gas
  C gas to liquid
  D gas to solid *(1 mark)*

2 What is the best description (A, B, C, or D) of the particles in a gas?

	Distance between particles	Movement of particles
A	close together	vibrating about a fixed point
B	close together	in continuous random motion
C	far apart	vibrating about a fixed point
D	far apart	in continuous random motion

(1 mark)

3 Which of the following (A, B, C, or D) describes changes of state?
  A They involve chemical changes and are reversible.
  B They involve physical changes and are reversible.
  C They involve chemical changes and are not reversible.
  D They involve physical changes and are not reversible. *(1 mark)*

4 Name the state of matter modelled in the diagram.

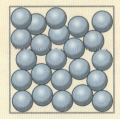

(1 mark)

5 Ice is water in the solid state.
  a Describe two features of ice that show that it is solid. *(2 marks)*

b Ice melts when it is heated.
  i Describe what happens to the amount of stored energy in the water particles during melting. *(1 mark)*
  ii Describe the changes to the arrangement and movement of water particles in ice during melting. *(2 marks)*

6 Methane is found in natural gas. The following diagram models three particles of methane.

methane   methane   methane

  a Methane can be stored under pressure as a liquid, but is used as a fuel when it is in the gas state.
  i Name the change of state from liquid to gas. *(1 mark)*
  ii Explain what happens to the methane particles during this change of state. *(2 marks)*
  b Methane takes part in a chemical change when it burns. In terms of particles, describe the differences between chemical changes and physical changes. *(2 marks)*

7* Dry ice is carbon dioxide in the solid state. It stops packets of frozen food thawing during transport. The carbon dioxide sublimes instead of melting. It changes directly from the solid to the gas state. Suggest an advantage of using dry ice instead of frozen water. Explain, in terms of particles and the forces between them, what happens when dry ice sublimes. *(6 marks)*

8 In the particle model, particles are represented as hard spheres. Explain the limitations of the particle model in terms of forces of attraction, particle size and the space between particles. *(3 marks)* **H**

Learning outcomes

After studying this lesson you should be able to:

- recall the typical sizes of atoms and molecules
- describe the structure of the atom
- recall the properties of subatomic particles.

Specification reference: C1.2b, C1.2c, C1.2d

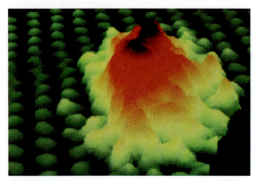

Figure 1 *Lines of carbon atoms with a pile of gold atoms on top.*

Study tip

Remember that molecules have different chemical and physical properties from the atoms they contain.

If you could go on a fantastic journey into a piece of gold, what would you find? If you could get small enough you would find that gold is made from gold atoms. You cannot see atoms, but a special microscope called a scanning tunnelling microscope can 'feel' them to make a picture like Figure 1.

What are atoms and molecules?

Gold and carbon are two different **elements**. An **atom** is the smallest particle of an element that still has its chemical properties. Different elements are made from different atoms.

A **molecule** is made from two or more atoms joined together. They are joined by attractive forces called chemical **bonds**.

The atoms in a molecule can be all the same or they can be different (Figure 2):

- an oxygen molecule is made from two oxygen atoms
- a carbon dioxide molecule is made from a carbon atom and two oxygen atoms.

Figure 2 *Models of an oxygen molecule and a carbon dioxide molecule.*

A Explain why O_2 is the formula for an oxygen molecule, but CO_2 is the formula for carbon dioxide.

How big are atoms, molecules, and bonds?

The size of an atom is given by its **atomic radius**, which is half its diameter. A **bond length** is the distance between the centres of two joined atoms.

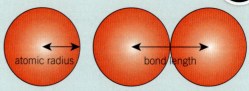

Figure 3 *Atomic radius and bond length.*

Atomic radii and bond lengths are typically around 10^{-10} m. For example, the atomic radius for oxygen is 0.73×10^{-10} m, and the oxygen–oxygen bond length is 1.21×10^{-10} m. One nanometre (1 nm) is 1×10^{-9} m, so the atomic radius for oxygen is 0.073 nm.

B Calculate the diameter of an oxygen atom, and the oxygen–oxygen bond length, in nanometres. Compare the two numbers; what do you notice? Suggest an explanation for this.

What is inside an atom?

While the inside of an atom is mostly empty space, it does contain three even smaller **subatomic particles**:

- **protons** and **neutrons**, joined together as the **nucleus** at the centre
- **electrons**, surrounding the nucleus in **shells**.

The radius of a nucleus is about 100 000 times less than the radius of an atom.

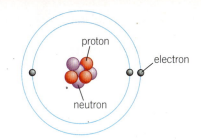

Figure 4 *A diagram of a lithium atom (not to scale).*

C   Suggest why the lithium atom in Figure 4 is not drawn to scale.

Table 1   *The subatomic particles*

Subatomic particle	Relative mass	Relative charge
proton	1	+1
neutron	1	0
electron	0.0005	−1

C   Explain why most of the mass of an atom is in its nucleus.

What does 'relative' mean?

The mass of a proton is 1.673×10^{-27} kg and the mass of a neutron is 1.675×10^{-27} kg. These small numbers can be difficult to use. If you just call the mass of a proton 1, the mass of a neutron is also 1 (1.001 to 4 significant figures). When you compare two quantities in this way, they are *relative* to each other (Figure 5).

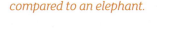

Figure 5 *The relative mass of a mouse is about 0.000 006 compared to an elephant.*

Figure 6 *If an atom were the size of the London Eye, its nucleus would be just 1 mm across!*

1   Describe the structure of an atom.   *(6 marks)*

2   The charge on an electron is -1.60×10^{-19} C. Calculate the charge on a proton.   *(1 mark)*

3   The mass of an electron is 9.109×10^{-31} kg. Calculate how many times greater the mass of a proton is compared to an electron. Compare your answer to the relative masses in Table 1, and suggest a reason for any difference.   *(4 marks)*

Learning outcomes

After studying this lesson you should be able to:

● calculate the numbers of subatomic particles in atoms and ions
● explain what isotopes are.

Specification reference: C1.2e

Figure 1 *This scientist is collecting samples of Antarctic ice.*

Go further

Find out how scientists work out past temperatures using the ratios of oxygen isotopes in ice cores.

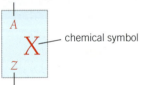

mass number =
number of protons + number of neutrons

$$_{Z}^{A}X$$ ——— chemical symbol

atomic number = number of protons

Figure 2 *The atomic number gives you the number of protons (and electrons) in an atom.*

Climate change is often in the news. Scientists ask questions about what the past was like to help them to predict what the future climate might be like. Antarctic ice (Figure 1) has built up over thousands of years. Different types of oxygen atoms in the water molecules of the ice provide evidence of the past climate.

How many subatomic particles do atoms have?

Every element has its own name and **chemical symbol**. A full chemical symbol for an atom gives its chemical symbol and two numbers (Figure 2). The bottom number is the **atomic number** for that atom. This is the number of protons in its nucleus. Every atom has equal numbers of protons and electrons, so the atomic number also tells you how many electrons it has. The full chemical symbol for a carbon-12 atom is $_{6}^{12}C$.

> **A** 🔲 Calculate the numbers of protons and electrons in a carbon atom.

The top number is the element's **mass number**. This is the total number of protons and neutrons:

mass number = number of protons (atomic number) + number of neutrons

You can use the mass number and atomic number to calculate the number of neutrons.

Calculating the number of neutrons

Calculate the number of neutrons in a $_{6}^{12}C$ atom.

Step 1: Write down what you know.

> mass number = 12
> atomic number = 6

Step 2: Rearrange the equation. Then put numbers in and calculate the answer.

> number of neutrons = mass number − atomic number
> = 12 − 6
> = 6

> **B** 🔲 Calculate the number of neutrons in a $_{11}^{23}Na$ atom.

What are isotopes?

Isotopes of an element are atoms with the same number of protons and electrons, but different numbers of neutrons. This means that isotopes have the same atomic number, but their mass number is different. You should identify isotopes using their full chemical symbols, but you may see isotopes named after their mass number. For example, $_1^2H$ is called hydrogen-2 (Figure 3).

C Calculate the number of protons, neutrons, and electrons in each isotope of hydrogen shown in Figure 3.

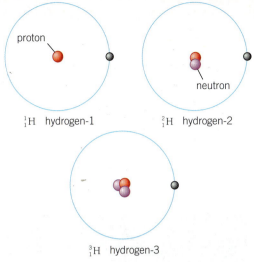

Figure 3 *Three isotopes of hydrogen.*

What are isotopes like?

The chemical properties of an element depend on the number of electrons in its atoms. All the isotopes of an element have the same number of electrons, so they have identical chemical properties. However, the different numbers of neutrons may affect their physical properties. For example, water vapour in air condenses and falls as rain more easily if its molecules contain $_8^{18}O$ atoms, rather than $_8^{16}O$ atoms.

What are ions?

Ions are charged particles. They are formed when atoms, or groups of atoms, lose or gain electrons. This can happen during chemical reactions, like the one between sodium and oxygen (Figure 4):

- a sodium atom loses one electron to become a sodium ion, Na^+
- an oxygen atom gains two electrons to become an oxide ion, O^{2-}.

For example, a $_{13}^{27}Al$ atom has 13 electrons, but a $_{13}^{27}Al^{3+}$ ion has 10 electrons (13 − 3 = 10).

Figure 4 Na^+ ions and O^{2-} ions form when sodium burns in air.

1 Define the terms:
 a atomic number   **b** mass number
 c isotope   **d** ion.   (*6 marks*)

2 Calculate the numbers of each type of subatomic particle in:
 a $_8^{16}O$ **b** $_8^{18}O$ **c** $_{26}^{56}Fe$ **d** $_{26}^{56}Fe^{2+}$. (*12 marks*)

3 Explain why $_{18}^{40}X$ and $_{20}^{40}Y$ are not isotopes of the same element. (*2 marks*)

Study tip

Only the number of electrons changes when an atom becomes an ion. The numbers of protons and neutrons stay the same.

Learning outcomes

After studying this lesson you should be able to:

- describe how and why the atomic model has changed over time.

Specification reference: C1.2a

Figure 1 *In blindfold tag, the blindfolded player must find their hidden friends.*

Figure 2 *Dalton's solid-atom model.*

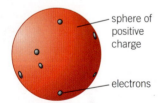

sphere of positive charge

electrons

Figure 3 *Thomson's plum-pudding model.*

How do you find objects that you cannot see? This is the problem scientists faced when developing models about atoms. A major advance happened because of an experiment that resembled subatomic blindfold tag.

What did Dalton do?

In 1803, John Dalton suggested that all matter is made from atoms. Other scientists had suggested this before, but Dalton's model was more detailed. It also explained evidence from his experiments on gases:

- all atoms of an element are identical
- different elements contain different types of atom.

Dalton did not know what atoms actually were, or even if they existed. He imagined them as tiny solid balls (Figure 2) – the technology of his time did not let him investigate atoms in detail.

> **A** In terms of subatomic particles, explain how accurate Dalton's model was.

What did Thomson do?

Joseph John (J. J.) Thomson discovered the first subatomic particle, the electron, in 1897. In his experiments, he found that beams of 'cathode rays' changed direction in electric and magnetic fields. He concluded that cathode rays were actually tiny particles that were negatively charged and much smaller than atoms. Thomson's model of the atom had to make sense of two observations:

- atoms contain electrons
- atoms are neutral overall.

In his **plum-pudding model** (Figure 3), Thomson suggested that atoms are spheres of positive charge with electrons dotted around inside, like pieces of fruit in a cake.

> **B** Suggest why protons and neutrons were not included in Thomson's model.

What did Rutherford do?

Ernest Rutherford worked with Hans Geiger and Ernest Marsden to test the plum-pudding model. In 1909 they pointed beams of positively charged particles, called alpha particles, at thin gold foil.

The scientists expected the particles to go straight through the foil. Most did, but many of them changed direction slightly. Some even came straight back. It was as if the alpha particles had been pushed away by something.

Rutherford explained the results by suggesting that an atom has a positively charged nucleus containing most of its mass. He also suggested that outside the nucleus, electrons orbit like planets in a solar system (Figure 4).

Figure 4 *Rutherford's planetary model of the atom.*

C Explain why the results that Rutherford, Geiger, and Marsden expected in their experiment are like playing blindfold tag in an empty field (Figure 1), and why the results they got are more like playing it in a field with a picnic basket in the middle.

Synoptic link

You can learn more about the atomic model in C2.3.2 *Changing state.*

Developing theories

The gold-foil experiment was devised to test a hypothesis, the plum-pudding model. Describe the scientific advances that were made because of its unexpected results.

What did Bohr do?

Niels Bohr realised that orbiting electrons would be attracted to the oppositely charged nucleus and would rapidly spiral inwards. In 1913 he used mathematical models to improve Rutherford's model. Bohr showed that electrons occupy fixed energy levels, or shells, around the nucleus (Figure 5).

Figure 5 *Bohr's model of the atom.*

D Explain the difference between the type of work Bohr did to form his model and the type of work that Thomson, Geiger, and Marsden did to form theirs.

1 Suggest why many scientists in the 19th century rejected Dalton's atomic theory. *(1 mark)*

2 An alpha particle is identical to a helium nucleus. Use the Periodic Table to predict what an alpha particle contains. *(2 marks)*

3 Explain how results from the Geiger–Marsden experiment suggest than an atom has a positive nucleus. *(4 marks)*

Go further

Make a timeline to show the development of theories about the atom. Find out who discovered protons and neutrons, and when they did this. Add this information to your timeline.

Particles

C1.2 Atomic structure

Summary questions

1 Write down definitions for the following terms:
   a atom
   b molecule
   c chemical bond
   d Periodic Table
   e atomic number
   f mass number
   g isotopes
   h ion.

2 Figure 1 represents a beryllium atom, Be. Name each of the parts labelled a–e.

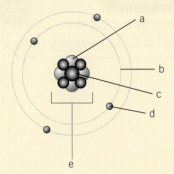

Figure 1 *A beryllium atom.*

3 Copy and complete Table 1.

Table 1 *Subatomic particles.*

Subatomic particle	Location in the atom	Relative mass	Relative charge
proton	nucleus		
		1	
			−1

4 Copy and complete Table 2.

Table 2 *Numbers of subatomic particles in isotopes.*

Isotope	Atomic number	Mass number	Number of: protons	neutrons	electrons
$_1^1H$					
$_1^2H$					
$_6^{12}C$					
$_6^{14}C$					
$_{17}^{35}Cl$					
$_{92}^{238}U$					

5 Copy and complete these sentences.
   a Atoms are neutral overall because they contain equal numbers of _____ and _____.
   b Ions are charged particles formed when atoms lose or gain _____.
   c Isotopes are atoms of an element that have the same number of _____ but different numbers of _____ in the nucleus.

6 Outline how each of the following scientists have contributed to the development of the atomic model:
   a John Dalton   **b** J. J. Thompson
   c Ernest Rutherford   **d** Niels Bohr.

7 Two scientists, Geiger and Marsden, carried out experiments to test the plum-pudding model. They directed beams of positively charged alpha particles at thin gold foil. Their results (Figure 2) showed that:
   ● most of the particles went straight through the foil
   ● some of them were deflected a little from their path
   ● a very small number were deflected a great deal from their path.

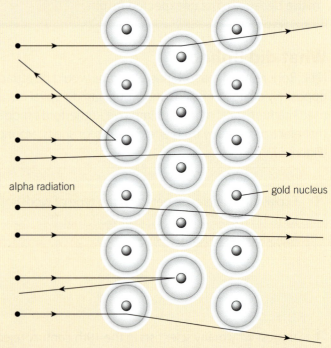

Figure 2 *The results of Geiger and Marsden's gold-foil experiment.*

   a Describe how these observations are evidence that a gold atom is mostly empty space.
   b Explain how these observations are evidence that a gold atom contains a small, positively charged nucleus.

Revision questions

1 Whose model of the atom was developed first?
   A John Dalton
   B JJ Thompson
   C Ernest Rutherford
   D Niels Bohr   (1 mark)

2 What is the approximate radius of an atom?
   A 1.3×10^{-10} m
   B 1.3×10^{-7} m
   C 1.3×10^{-6} m
   D 1.3×10^{-3} m   (1 mark)

3 Which description best describes the structure
   of an atom?
   A A nucleus containing protons and electrons,
      surrounded by neutrons.
   B A nucleus containing neutrons and electrons,
      surrounded by protons.
   C A nucleus containing protons and neutrons,
      surrounded by electrons.
   D A nucleus containing neutrons, surrounded by
      electrons and protons.   (1 mark)

4 Write down the meaning of the following terms.
   a Atomic number, Z   (1 mark)
   b Mass number, A   (1 mark)

5 An atom of phosphorus has the chemical symbol $^{31}_{15}P$.
   Write down how many of the following subatomic
   particles it contains.
   a protons   (1 mark)
   b neutrons   (1 mark)
   c electrons   (1 mark)

6 Explain, in terms of subatomic particles, why
   $^{35}_{17}Cl$ and $^{37}_{17}Cl$ are isotopes of an element.   (3 marks)

7 Protons, neutrons and electrons are subatomic
   particles found in atoms.
   Copy and complete the table to show the missing
   information.

Subatomic particle	Relative charge	Relative mass
	0	
		0.0005
		1

(3 marks)

8 A magnesium ion has the symbol $^{24}_{12}Mg^{2+}$.
   a Describe how this ion forms from a magnesium
      atom, $^{24}_{12}Mg$.   (2 marks)
   b Explain, in terms of the numbers of subatomic
      particles it contains, why:
      i   a magnesium atom has no charge overall
                                                      (2 marks)
      ii a magnesium ion has two positive charges.
                                                      (3 marks)

9* Describe how and why the atomic model has   H
   changed over time, from the simple Dalton model
   to the more complex Bohr model. In your answer,
   include your knowledge and understanding of the
   models of Dalton, Rutherford, and Bohr.   (6 marks)

Particles: Topic summary

C1.1 The particle model

- Describe the main features of the particle model.
- Explain some general properties of solids, liquids, and gases.
- Describe differences between chemical and physical changes.
- Explain chemical and physical changes in terms of particles.

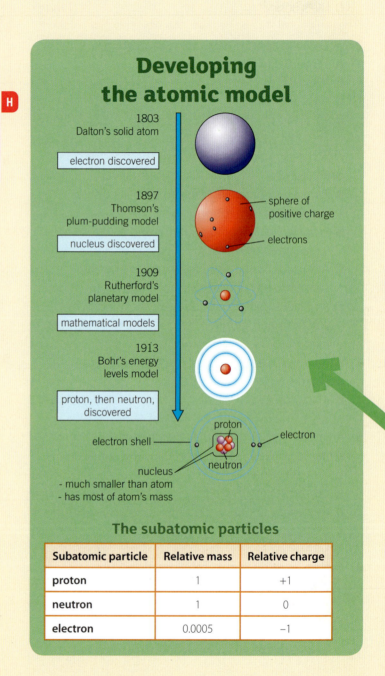

- Compare the sizes of particles to the distances between them. **H**
- Describe the forces between particles.
- Explain the limitations of the particle model.

C1.2 Atomic structure

- State the typical sizes of atoms and molecules.
- Describe the structure of the atom.
- State the properties of protons, neutrons, and electrons.
- Calculate the numbers of protons, neutrons, and electrons in atoms and ions.
- Explain what isotopes are.
- Describe how and why the atomic model has changed over time.

Developing the atomic model

1803
Dalton's solid atom

electron discovered

1897
Thomson's plum-pudding model
- sphere of positive charge
- electrons

nucleus discovered

1909
Rutherford's planetary model

mathematical models

1913
Bohr's energy levels model

proton, then neutron, discovered

electron shell
proton
electron
nucleus
neutron
- much smaller than atom
- has most of atom's mass

The subatomic particles

Subatomic particle	Relative mass	Relative charge
proton	1	+1
neutron	1	0
electron	0.0005	−1

The particle model

- particles represented as hard spheres
- properties of a substance depend on its state

solid
- very close together
- regular arrangement
- vibrate around fixed positions

liquid
- close together
- random arrangement
- move around each other

gas
- far apart
- random arrangement
- move quickly in all directions

Forces between particles [H]

- particles held together by electrostatic forces of attraction (bonds)

Limitations of the particle model [H]

- does not take into account:
  - forces between particles
  - size of particles
  - space between particles

Atomic structure

- two oxygen atoms joined together = oxygen molecule
- bond length and atomic radius in the order of 10^{-10} m (about 0.1 nm)

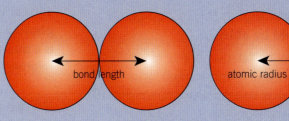

bond length   atomic radius

Chemical symbol

mass number = number of protons + number of neutrons

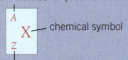

A
X ← chemical symbol
Z

atomic number = number of protons

Isotopes

- atoms of an element with:
  - the same number of protons and electrons
  - different numbers of neutrons
- examples: $^{35}_{17}Cl$ $^{37}_{17}Cl$

Ions

- atoms or groups of atoms that have lost or gained electrons
- examples: $^{23}_{11}Na^+$, $^{16}_{8}O^{2-}$

C1 The particle model

Chemical and physical changes

Physical

- reversible
- no new substances made

Chemical

- often difficult to reverse
- new substances made

methane + oxygen → carbon dioxide + water

C2 Elements, compounds, and mixtures
C2.1 Purity and separating mixtures
C2.1.1 Relative formula mass

Learning outcomes

After studying this lesson you should be able to:

- define 'relative atomic mass' and 'relative formula mass'
- calculate relative formula masses from formulae.

Specification reference: C2.1c

Figure 1 *The mass of the stone on the left is equal to the mass of the four stones on the right.*

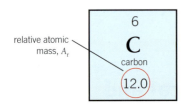

Figure 3 *This is the box that represents carbon in the Periodic Table.*

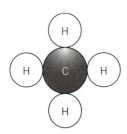

Figure 4 *A model of a methane molecule.*

What are the masses of the five balanced stones in Figure 1? You cannot determine their masses in kilograms just from the photograph, but you can say that the mean mass of a little stone is a quarter of the mass of the large stone. Scientists describe the relative masses of atoms and molecules in a similar way.

What is relative atomic mass?

The masses of subatomic particles can be described by relative masses – their masses compared to the mass of a proton. In a similar way, the masses of atoms can be described using relative *atomic* masses. Atoms are compared to a standard atom, rather than to a proton:

- **Relative atomic mass**, A_r is the mean mass of an atom of an element compared to $\frac{1}{12}$ the mass of a $^{12}_{6}C$ (carbon-12) atom.

The relative atomic mass of $^{12}_{6}C$ is defined as 12.0 exactly. This means that an atom has:

- less mass than a carbon-12 atom if its A_r is below 12.0
- more mass than a carbon-12 atom if its A_r is above 12.0.

A Use Figure 2 to write down the relative atomic masses of helium and titanium. Explain your answers.

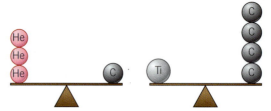

Figure 2 *Three helium atoms have the same mass as one carbon-12 atom, and one titanium atom has the same mass as four carbon atoms.*

The **Periodic Table** shows the relative atomic masses of the elements. The bottom number in each box is the A_r for that element's atoms (Figure 3).

What does a chemical formula tell you?

A **chemical formula** tells you how many atoms of each element there are in a unit of a substance. For example, the chemical formula of water is H_2O. This tells you that each molecule of water contains two hydrogen atoms and one oxygen atom joined together.

B Explain how Figure 4 shows that the chemical formula of methane is CH_4.

Brackets in chemical formulae

Some chemical formulae contain brackets. For example, the chemical formula of aluminium sulfate is $Al_2(SO_4)_3$. The 3 tells you that there are three times the number of each atom inside the brackets. A unit of aluminium sulfate contains three elements and 17 atoms:

- 2 aluminium atoms
- $(3 \times 1) = 3$ sulfur atoms
- $(3 \times 4) = 12$ oxygen atoms.

C Calculate the number of atoms of each element in a unit of ammonium sulfate, $(NH_4)_2SO_4$.

What is a relative formula mass?

The masses of substances made from two or more atoms can also be compared to the mass of a $^{12}_{6}C$ atom:

- **Relative formula mass**, M_r, is the mean mass of a unit of a substance compared to $\frac{1}{12}$ the mass of a $^{12}_{6}C$ atom.

You cannot find M_r values in the Periodic Table, so you must work them out. You add together the A_r values for all the atoms in the formula for the substance. A_r and M_r values have no units.

Molecules and relative molecular mass

The term 'relative formula mass' refers to all substances that consist of two or more atoms, but you can use the term **relative molecular mass** instead for substances that exist as molecules.

M_r is the symbol for both relative formula mass and relative molecular mass.

Calculating relative formula mass

Calculate the relative formula mass, M_r, of magnesium hydroxide, $Mg(OH)_2$.

Step 1: Write down the A_r values of the elements in the compound.

$Mg = 24.3, O = 16.0, H = 1.0$

Step 2: Work out the number of atoms of each element in $Mg(OH)_2$:

$Mg = 1, O = (2 \times 1) = 2, H = (2 \times 1) = 2$

Step 3: Multiply these values to calculate the relative formula mass.

M_r of $Mg(OH)_2 = (1 \times 24.3) + (2 \times 16.0) + (2 \times 1.0)$

$= 24.3 + 32.0 + 2.0$

$= 58.3$

D Calculate the relative formula masses of water, methane, and aluminium sulfate, $Al_2(SO_4)_3$. A_r values: $Al = 27.0, S = 32.1, O = 16.0$.

1 Write down the relative atomic masses of nitrogen, chlorine, and sodium. *(3 marks)*

2 Calculate the relative formula masses of oxygen, O_2, carbon dioxide, CO_2, ammonia, NH_3, and sodium chloride, NaCl. *(4 marks)*

3 Calculate the relative formula masses of aluminium hydroxide, $Al(OH)_3$ and ammonium carbonate, $(NH_4)_2CO_3$. *(2 marks)*

Learning outcomes

After studying this lesson you should be able to:

- calculate the empirical formula of a compound
- calculate relative formula masses in balanced equations.

Specification reference: C2.1c, C2.1d

Figure 1 *Butane is normally in the gas state at room temperature but is liquefied in pressurised gas canisters.*

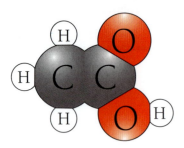

Figure 2 *A space-filling model of ethanoic acid.*

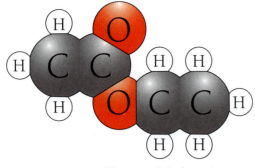

Figure 3 *A space-filling model of ethyl ethanoate.*

Hikers use butane as a cooking fuel (Figure 1). Each butane molecule contains four carbon atoms and ten hydrogen atoms, so its chemical formula is C_4H_{10}. Butane also has a simpler formula, the empirical formula.

What is an empirical formula?

An **empirical formula** shows the simplest whole-number ratio of the atoms of each element in a compound. The numbers in some chemical formulae are already the simplest whole numbers they can be, such as H_2O and CH_4. The numbers in other chemical formulae can be simplified using their highest common factor.

Calculating an empirical formula

The chemical formula for butane is C_4H_{10}. Calculate its empirical formula.

Step 1: Find the highest common factor.

The highest common factor of 4 and 10 is 2.

Step 2: Divide the chemical formula by the highest common factor:

$$C = \frac{4}{2} = 2 \qquad H = \frac{10}{2} = 5$$

Step 3: Write down the empirical formula:

$$C_2H_5$$

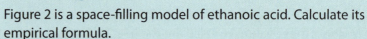

A   Calculate the empirical formula of Pb_2O_4.

Calculating empirical formulae from diagrams

Figure 2 is a space-filling model of ethanoic acid. Calculate its empirical formula.

Step 1: Work out the number of atoms of each element in the molecule by counting how many times each chemical symbol is shown:

$$C = 2 \qquad H = 4 \qquad O = 2$$

Step 2: Divide by the highest common factor:

$$C = \frac{2}{2} = 1 \qquad H = \frac{4}{2} = 2 \qquad O = \frac{2}{2} = 1$$

Step 3: Write down the empirical formula:

$$CH_2O$$

B   Figure 3 is a space-filling model of ethyl ethanoate. Calculate its empirical formula.

You have already used empirical formulae for compounds consisting of metals and non-metals. For example, a unit of sodium chloride contains many sodium and chlorine particles (Figure 4). It would make little sense to write a formula with huge numbers in it, so chemists use its empirical formula, NaCl.

How to work out relative formula mass from equations

A **balanced chemical equation** shows the formulae and number of units for all the substances in a reaction. For example, carbon reacts with oxygen to produce carbon dioxide. If you write down the A_r and M_r values for each substance, you find that the total on the left of the arrow equals the total on the right:

$$
\begin{array}{ccccc}
\text{C} & + & \text{O}_2 & \rightarrow & \text{CO}_2 \\
12.0 & + & 32.0 & = & 44.0 \\
(1 \times 12.0) & & (2 \times 16.0) & & (12.0 + 16.0 + 16.0)
\end{array}
$$

This gives you a way to check that you have correctly calculated the M_r values for all the substances in a balanced equation. It also gives you another way to work out the M_r for a substance, if you know the A_r and M_r values for the other substances in an equation.

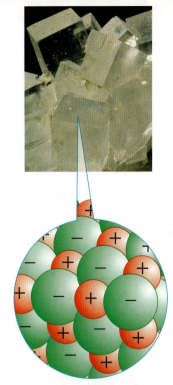

Figure 4 *Table salt (sodium chloride) contains sodium ions and chloride ions in a 1 : 1 ratio.*

> **C** Hydrochloric acid reacts with sodium hydroxide:
> $$\text{HCl} + \text{NaOH} \rightarrow \text{NaCl} + \text{H}_2\text{O}$$
> Calculate the M_r values of all the substances in this equation. Give your answers to 1 decimal place.
>
> **D** Suggest the meaning of M (as used in M_r), and explain why this could be misleading.

Use the Periodic Table to find the A_r values needed to answer these questions.

1 Calculate the empirical formulae for S_8, C_3H_6, and $C_6H_{12}O_6$. *(3 marks)*

2 Calculate the relative formula masses of CuO, HNO_3, and H_2O. *(3 marks)*

3 Copper oxide reacts with nitric acid:
$$\text{CuO} + 2\text{HNO}_3 \rightarrow \text{Cu(NO}_3)_2 + \text{H}_2\text{O}$$
Use your answers to question **2** to calculate the relative formula mass of copper nitrate, $\text{Cu(NO}_3)_2$. *(2 marks)*

Learning outcomes

After studying this lesson you should be able to:

- explain what purity means
- explain that many useful materials are mixtures
- use melting point data to distinguish pure substances from impure substances.

Specification reference: C2.1a, C2.1b, C2.1e

Water companies add chlorine to tap water to kill harmful microorganisms. Some people do not like its taste, so they use a filter jug (Figure 1) to try to remove chlorine and other impurities. But to a scientist the water is still not 'pure'.

Figure 1 *Water filter jugs remove chlorine from tap water.*

What does 'pure' mean?

In everyday life, 'pure' describes natural substances that have not been processed or changed. A tin of 'pure brilliant white' paint is not pure in the scientific sense – it just looks very white. The paint is a **mixture** of many substances, including water and white pigment. In science, a **pure substance** consists of just one element or compound.

This means that mixtures are **impure substances**, since they contain more than one element or compound.

> **A** The juice carton in Figure 2 states that it contains '100% pure squeezed fruit'. Explain why a scientist would say that it is not pure.

Figure 2 *The orange juice in this carton is not 'pure' in the scientific sense.*

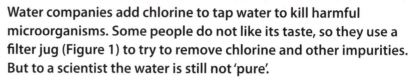

Figure 3 *The atoms in an alloy are usually different sizes.*

Can mixtures be useful?

It is difficult to obtain pure substances. Just one atom or molecule of something else makes a substance impure. Even the high-**purity** 'ultrapure' water used in the manufacture of computer chips is easily contaminated by carbon dioxide from the air. Air itself is a mixture of nitrogen, oxygen, and other substances.

Many useful materials are mixtures of different substances, often deliberately chosen to produce the desired properties. An **alloy** is a mixture of a metal with one or more other elements (Figure 3). Most of the metals you use are alloys. For example, pure gold is very soft, so a harder mixture of gold and copper is often used for jewellery.

> **B** Cough mixture contains active ingredients such as paracetamol, as well as water, colouring, and flavouring. Suggest why cough mixture is more useful than just its active ingredients alone.

How do you use melting points to determine purity?

The **melting point** of a substance is the temperature at which it changes from the solid state to the liquid state. The melting point of a pure substance is a single temperature. If a substance is impure:

● its melting point is less than that of the pure substance

● it often melts over a range of temperatures, not just one temperature.

The greater the difference between the measured melting point for a substance and its accepted melting point, the lower its purity is likely to be.

> **C** The melting point of pure gold is 1064.18 °C, but a sample from a gold ring melts at 915 °C to 963 °C. Explain why the ring cannot be made from pure gold, and suggest why the melting point for pure gold is given to two decimal places.

How do you determine melting point?

You can determine the melting point of a substance by heating it. You then either measure the temperature at which it melts, or measure its temperature at regular time intervals and plot a graph like that in Figure 4. It is important to:

● heat the substance slowly

● stir the substance as it melts.

Heating a substance slowly allows the temperature of the whole sample to increase. Mixing ensures that the entire sample is at the same temperature. These two actions improve the accuracy of a measurement of the melting point of a sample.

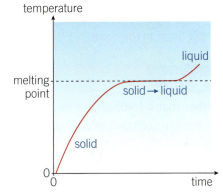

Figure 4 *The temperature of a pure substance remains constant while it melts.*

1 Explain why water poured from a filter jug is not pure in the scientific sense. *(2 marks)*

2 Explain why seawater freezes at a different temperature to fresh water. *(2 marks)*

3 ⓘ The metals you know as 'steels' are alloys of iron and carbon. Describe what the data in Table 1 shows about these steels, and justify your answer. *(6 marks)*

Table 1 *Melting points of iron and different steels*

Metal	Melting point (°C)
iron	1538
low carbon steel (0.2% C)	1466
medium carbon steel (0.4% C)	1424
high carbon steel (1.6% C)	1352

Learning outcome

After studying this lesson you should be able to:

- describe and explain how filtration and crystallisation work.

Specification reference: C2.1f

Figure 1 *These workers are collecting sea salt at a salt farm on the coast.*

Some people flavour their food with sea salt. Seawater is a solution of several salts. You can separate the salts from water by evaporation (Figure 1). This is just one way to separate individual components of mixtures.

What happens when a substance dissolves?

A **solution** forms when one substance dissolves in another. The **solute** is the substance that dissolves, and the **solvent** is the substance it dissolves in. When a substance **dissolves**, its particles separate and become completely mixed with the particles of the solvent.

If a substance can dissolve in a particular solvent, it is **soluble** in that solvent. If it cannot dissolve, it is **insoluble** in that solvent. Substances can be soluble in one solvent but insoluble in another solvent. For example, nail varnish dissolves in nail-varnish remover but not in water.

> **A** Copper sulfate crystals are dissolved in water. Name the solvent and the solute here.

How does filtration work?

Filtration separates an insoluble substance in the solid state from substances in the liquid state. It works because filter paper has tiny, microscopic holes. When you filter a mixture of sand and water, water molecules are small enough to pass through the filter paper, but the larger grains of insoluble sand cannot. The sand stays behind in the filter paper as the **residue** while the water passes through as the **filtrate** (Figure 2). Particles of a dissolved solute are also small enough to pass through filter paper.

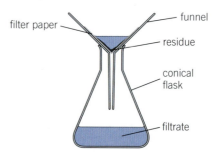

Figure 2 *Filtration using filter paper and a filter funnel.*

> **B** Copper oxide is black and copper sulfate solution is blue. Explain why you obtain a blue filtrate and a black residue if you filter a mixture of copper sulfate solution and copper oxide.

Fluting filter paper

You may have folded filter paper using two folds, but chemists 'flute' their filter paper (Figure 3). This produces a larger surface area for the filtrate to pass through.

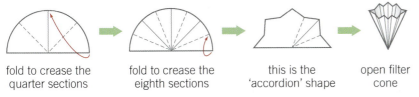

fold to crease the quarter sections   fold to crease the eighth sections   this is the 'accordion' shape   open filter cone

Figure 3 *Practise fluting filter paper using these instructions.*

How does crystallisation work?

If you heat a solution, the solvent **evaporates** leaving the solute behind. If you heat the solution too strongly, you get a powder, but if you allow the solvent to evaporate slowly, you get regularly shaped crystals (Figure 4).

Figure 4 *Copper sulfate crystals.*

evaporating basin

copper sulfate solution

boiling water

Bunsen burner

Figure 5 *A water bath heats solutions slowly, enabling regular crystals to form.*

Crystallisation takes patience because you need to heat the solution gently until it becomes a **saturated solution**. A solution is saturated when no more solute can be dissolved at that temperature: Crystals will start forming at this point, so you then let the solution cool slowly. As the solution cools, the **solubility** of the solute decreases, so more crystals form (Figure 5). You can separate them from the remaining solution by filtration, and dry them in a warm oven or by patting them with filter paper.

> **C**   Explain why crystals form on a glass rod that is dipped in a hot saturated solution, then taken out and cooled.

> **1**   Sea-salt manufacturers may filter seawater before evaporation. Suggest why they do this and describe what happens to the salts during filtration.   *(3 marks)*

> **2**   Copper oxide powder is insoluble in water. It reacts with dilute sulfuric acid to produce copper sulfate solution. Outline how to produce dry copper sulfate crystals using this reaction.   *(6 marks)*

> **3**   Sugar is soluble in ethanol and water. Sodium chloride is insoluble in ethanol. Suggest how you could separate a mixture of sugar and sodium chloride.   *(4 marks)*

Learning outcome

After studying this lesson you should be able to:

- describe and explain how simple distillation and fractional distillation work.

Specification reference: C2.1f

Figure 1 *This factory produces bioethanol from plant material such as sugar cane and wheat.*

Petrol in the UK and many other countries is actually a mix of petrol and bioethanol. Bioethanol is produced from plant material by fermentation (Figure 1). Bioethanol is purified using fractional distillation.

How does simple distillation work?

Simple distillation separates a solvent from a solution. It relies on the solvent having a much lower **boiling point** than the solute. When the solution is heated, the solvent boils but the solute does not. The solvent escapes from the solution in its gas state. It is then cooled and **condensed** back to its liquid state by a **condenser**, a piece of apparatus that is kept cold using a flow of cold water.

Simple distillation is useful in the laboratory if you want to purify a solvent. For example, it can provide purified water for chemists. Tap water contains small amounts of sodium chloride and other dissolved salts. These interfere with some chemical tests. The boiling point of water is 100 °C, much lower than the boiling point of sodium chloride, 1413 °C. During simple distillation, water **vapour** (water in its gas state) escapes from the tap water, leaving the dissolved salts behind.

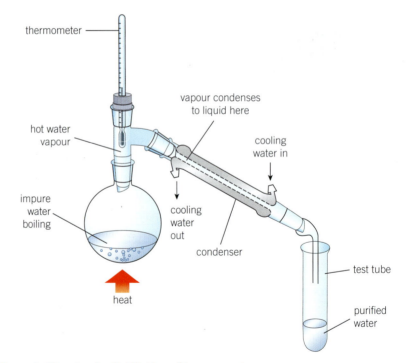

Figure 2 *The simple distillation of impure water.*

A Suggest the reading that would be shown on the thermometer in Figure 2, and justify your answer.

How does fractional distillation work?

Fractional distillation separates two or more substances from a mixture in the liquid state. It relies on each substance having a different boiling point and uses a special piece of equipment called a **fractionating column**, as shown in Figure 3. Fractional distillation is used industrially to separate bioethanol from water for use as a fuel. The boiling point of ethanol is 78.4°C, which is lower than the boiling point of water. During fractional distillation of an ethanol–water mixture, ethanol vapour and water vapour both leave the liquid mixture. In the fractionating column:

- the vapours condense on the inside surface, heating it up
- when the temperature inside reaches the boiling point of ethanol, ethanol vapour cannot condense any more, but water vapour can
- water droplets fall back into the flask, and ethanol vapour passes into the condenser.

In the condenser, the ethanol vapour is cooled and condensed back to its liquid state. It drips into the collecting container.

> **B** Suggest why it is safer to heat the ethanol–water mixture with an electric hot plate than with a Bunsen burner.

Fractions

Each substance separated by fractional distillation is called a **fraction**, because it is just a part of the original mixture.

You could carry out fractional distillation without the fractionating column. However, the fractionating column improves the separation of the mixture. It has a large surface area on which the vapours can continually condense. During fractional distillation, the column becomes hottest at the bottom and coolest at the top.

> **C** Explain why ethanol passes into the condenser during the fractional distillation of an ethanol–water mixture, while water drips back into the mixture.

1. Name the physical property on which simple distillation and fractional distillation rely. *(1 mark)*
2. Ink is a mixture of water and a soluble coloured pigment. Explain why ink gradually turns darker in colour during simple distillation. *(2 marks)*
3. Explain why, during fractional distillation of an ethanol–water mixture, water vapour eventually passes to the condenser. *(2 marks)*

Synoptic link

You can learn more about the fractional distillation of crude oil in C6.2.5 *Alkanes from crude oil*.

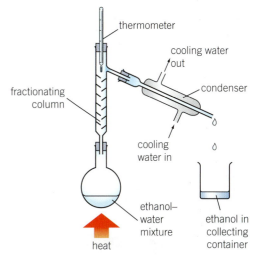

Figure 3 *The fractional distillation of an ethanol–water mixture.*

Go further

Steam distillation extracts fragrant plant oils, such as lavender oil (Figure 4). Find out how it works, and why ordinary simple distillation is not suitable for these oils.

Figure 4 *Lavender oil is used in aromatherapy and to make scented candles.*

Figure 1 *These scientists are preparing samples for gas chromatography.*

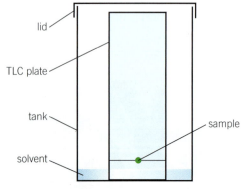

Figure 2 *The apparatus used for thin-layer chromatography (TLC).*

B Calculate the R_f value for substance B in Figure 3, giving your answer to two decimal places.

How can scientists tell whether food contains traces of a harmful substance, or whether a sample of paint found at a crime scene matches a suspect's car? They use chromatography to analyse mixtures (Figure 1).

How does chromatography work?

Chromatography relies on two different chemical **phases**:

- a **stationary phase** that *does not* move
- a **mobile phase** that *does* move.

A phase is a substance in the solid, liquid, or gas state. You may have already used **paper chromatography** to separate the coloured substances in ink or sweets. In paper chromatography:

- the stationary phase is absorbent paper
- the mobile phase is a solvent in the liquid state, such as water or propanone.

Thin-layer chromatography works in the same way, but the stationary phase is a thin layer of silica or alumina powder spread over a plate of glass or plastic.

Thin-layer chromatography

You can use thin-layer chromatography to separate a sample into its components for identification or analysis.

1. Put the solvent into a chromatography tank to a depth of about 1 cm (Figure 2). If the solvent is flammable, make sure that there are no naked flames, and that the room is well ventilated.
2. Add a small amount of the sample to the baseline, taking care not to damage the powder on the plate.
3. Let the solvent travel through the powder, and take the plate out before it reaches the top.
4. Analyse the pattern of coloured spots, which is called a **chromatogram**.

The pattern produced depends on how each component is distributed between the two phases. A component travels further up the plate if it forms stronger bonds with the mobile phase than with the stationary phase. A component will not travel very far if it forms stronger bonds with the stationary phase than with the mobile phase.

A For thin-layer chromatography, suggest why the baseline should be above the solvent level.

Scientists use R_f **values** to compare the different spots on a chromatogram. If two spots have the same R_f value and are the same colour, they are likely to be identical.

How does gas chromatography work?

Gas chromatography is different to paper chromatography and thin-layer chromatography:

- the stationary phase is silica or alumina powder packed into a metal column
- the mobile phase is an unreactive **carrier gas** such as nitrogen, which does not react with the sample.

Gas chromatography separates the components of a mixture and also measures their amounts, as shown in Figure 4.

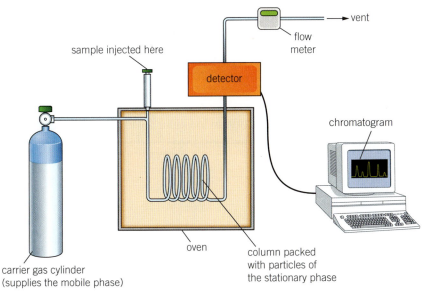

Figure 4 *The metal column in gas chromatography is coiled so that it can fit into a very hot oven.*

The sample is turned into the gas state when it is injected into the column. The carrier gas pushes the sample through the column. The different components take different times to travel through the column, depending on how strongly they bond to the stationary phase. A detector sends a signal to a computer as each component leaves the column. The computer produces a chromatogram in which each component is a peak plotted against the travel time.

Calculating an R_f value

Figure 3 shows the thin-layer chromatogram for two samples. Calculate the R_f value for substance A, giving your answer to two decimal places.

Step 1: Measure the distance travelled by the substance, and the distance travelled by the solvent.

From the diagram, this is 8 cm and 12 cm.

Step 2: Calculate the R_f value:

$$R_f = \frac{\text{distance travelled by substance}}{\text{distance travelled by solvent}}$$

$$R_f = \frac{8\,\text{cm}}{12\,\text{cm}} = 0.67$$

R_f values vary from 0 to 1, and have no units.

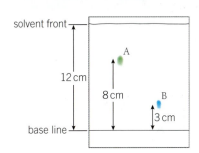

Figure 3 *A thin-layer chromatogram for substances A and B.*

1 Figure 5 shows a chromatogram produced when scientists analysed a coloured drink.
   a Explain which permitted colour is in the drink. *(2 marks)*
   b Suggest why further analysis of the drink may be needed. *(2 marks)*
   c Calculate the R_f values for C1 and C3 in Figure 5, giving your answer to two decimal places. *(2 marks)*
2 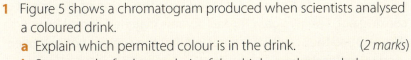 Compare thin-layer chromatography with gas chromatography. *(6 marks)*
3 Suggest why helium can be used as a carrier gas but oxygen cannot. *(2 marks)*

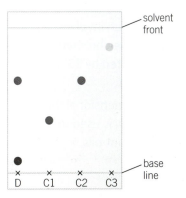

Figure 5 *D is a coloured drink, C1, C2, and C3 are permitted artificial colours.*

C2.1.7 Purification and checking purity

Figure 1 *Food leaves residues on pots and pans that can be analysed centuries later.*

What did our ancestors eat? One way to find out involves analysing the insides of ancient cooking pots to identify the food substances left behind (Figure 1).

How can you tell if a substance is pure?

You can use paper chromatography or thin-layer chromatography to tell if a substance is pure, but thin-layer chromatography has some advantages:

- it is quicker
- it is more sensitive, so a smaller sample can be used
- there is a larger range of stationary phases and solvents to choose from.

You can also scrape an individual spot from a thin-layer chromatogram for further analysis, for example by gas chromatography.

> **A** Explain how the chromatogram in Figure 2 shows that the brown food colouring is not a pure substance.

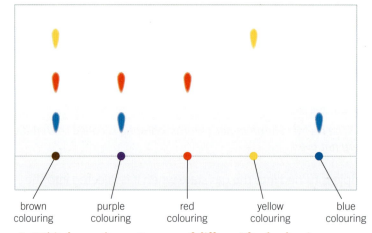

Figure 2 *A thin-layer chromatogram of different food colourings.*

Go further

The separation of a mixture can be improved using two-dimensional chromatography, which involves two different solvents. The chromatography paper or plate is rotated by 90° after using the first solvent.

Find out how two-dimensional thin-layer chromatography is used to separate different plant oils.

Information from a gas chromatogram

Scientists used gas chromatography to analyse wax from cabbage leaves and material scraped from the inside of an ancient cooking pot. The two gas chromatograms matched almost exactly (Figure 3), so the scientists suggested that the ancient pot had been used to cook cabbage.

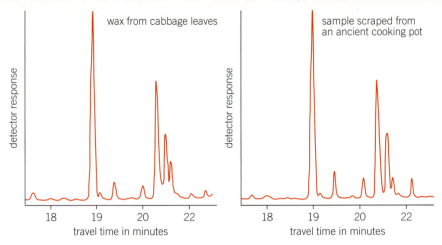

Figure 3 *Chromatograms from cabbage leaves (left) and an ancient cooking plot (right).*

B Explain how you can tell, from the chromatograms in Figure 3, that the wax from cabbage leaves is not a pure substance.

How to choose a separation method

You need to know some of the properties of the individual components in a mixture to work out how to separate them. Gas chromatography separates many different kinds of substance, but is not a method you are likely to use in school. Table 1 summarises some of the separation methods you are likely to use.

Table 1 *Different separation methods.*

Mixture contains	Separation method
insoluble and soluble substances	dissolving followed by filtration
a solute dissolved in a solvent (a solution)	crystallisation to obtain the solute, simple distillation to obtain the solvent
two or more substances in the liquid state	fractional distillation
coloured soluble substances	paper chromatography or thin-layer chromatography

You may need to use different combinations of separation methods. For example, a school technician drops a glass container of copper sulfate powder on the floor and it smashes. He brushes up the mixture of broken glass and powder, and adds it to hot water. He then filters the mixture to remove the glass, and obtains copper sulfate by crystallisation.

C Explain why the technician chose this combination of separation methods.

D Suggest why the copper sulfate obtained by the technician may not be pure.

1 Methanol and ethanol mix completely with one another. Their boiling points are 65.0 °C and 78.4 °C, respectively. Write down a suitable method you could use to separate these two substances, and justify your answer. *(2 marks)*

2 Suggest why the red, yellow, and blue colourings in the thin-layer chromatogram (Figure 2) might not be pure substances. Describe how a food scientist could carry out a further analysis to check whether they are pure or impure. *(4 marks)*

3 The coloured pigments in grass leaves are soluble in propanone. Outline how these pigments could be extracted and separated in a school laboratory. *(4 marks)*

Elements, compounds, and mixtures

C2.1 Purity and separating mixtures

Summary questions

1 For each of the following substances, write down whether they are pure substances or mixtures.
   - **a** milk
   - **b** calcium carbonate
   - **c** seawater
   - **d** silver
   - **e** crude oil
   - **f** distilled water.

2 The apparatus shown in Figure 1 can be used to separate sand from water. The labels A, B, C, and D represent the following parts: filtrate, funnel, filter paper, and residue. Write down the part that each label represents.

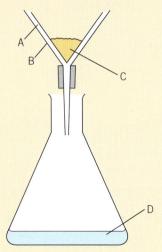

Figure 1 *Apparatus used to separate sand from water.*

3 Write down the most appropriate method (filtration, distillation, crystallisation, or chromatography) to separate the following mixtures.
   - **a** different coloured substances in a sample of black ink
   - **b** chalk from a mixture of chalk and water
   - **c** pure water from seawater
   - **d** salt from seawater.

4 Write down simple definitions for the following terms:
   - **a** relative atomic mass, A_r
   - **b** relative formula mass, M_r.

5 Ammonia reacts with nitric acid to form ammonium nitrate, which is used as a fertiliser:
$$NH_3 + HNO_3 \rightarrow NH_4NO_3$$
   - **a** Calculate the relative formula masses of NH_3 and HNO_3.
     Use these A_r values: H = 1, N = 14, O = 16.
   - **b** Use your answers to **a** to calculate the relative formula mass of ammonium nitrate, NH_4NO_3.

   - **c** Write down the number of different elements present in a unit of ammonium nitrate.
   - **d** Write down the total number of atoms present in a unit of ammonium nitrate.

6 Calculate the relative formula masses of the following compounds.
   - **a** lithium chloride, LiCl
   - **b** calcium bromide, $CaBr_2$
   - **c** calcium hydroxide, $Ca(OH)_2$
   - **d** iron(III) sulfate, $Fe_2(SO_4)_3$.

7 **a** Explain what the term **empirical formula** means.
   b Write down the empirical formula of:
      - **i** decene, $C_{10}H_{20}$
      - **ii** benzene, C_6H_6
      - **iii** dinitrogen tetroxide, N_2O_4.

8 A food scientist analysed a sample of curry sauce using thin-layer chromatography. The results are shown in Figure 2. The scientist used four known food colours (E102, E110, E122, and E124). The sample of curry sauce is labelled S.
   - **a** Calculate the R_f values for:
      - **i** E102
      - **ii** the second spot in the sample, S.
   - **b** Use your answers to **a** to explain whether the curry sauce contains E102.
   - **c** Without carrying out any calculations, explain which food colour(s) the curry sauce contains, and which one(s) it does not contain.
   - **d** Suggest one other observation you would need to make to be sure of your conclusions in part **c**.

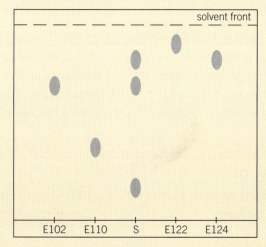

Figure 2 *A chromatogram of different E numbers.*

Revision questions

1 The molecular formula of butane is C_4H_{10}. What is the empirical formula of butane?

   A C_8H_{20}

   B C_4H_{10}

   C C_2H_5

   D $CH_{2.5}$ (*1 mark*)

2 A student is using paper chromatography to separate the dyes in some coloured sweets. He dissolves the sweets in water, and puts a spot of the coloured mixture onto a piece of filter paper. He then dips the filter paper in a boiling tube of water.

What name is given to the filter paper in this experiment?

   A liquid phase

   B stationary phase

   C mobile phase

   D absorption phase   (*1 mark*)

3 The formula of chloroethene is C_2H_3Cl.
What is the relative formula mass, M_r, of chloroethene?
(The relative atomic mass, A_r, of C is 12.0, of H is 1.0 and of Cl is 35.5.)

   A 48.5

   B 50.5

   C 60.5

   D 62.5   (*1 mark*)

4 Different mixtures can be separated using different methods, including filtration, crystallisation, simple distillation, and fractional distillation.

For each of the following mixtures, write down the best method to separate the underlined substance in the mixture.

   a <u>sodium chloride</u> dissolved in water   (*1 mark*)

   b <u>water</u> from copper(II) sulfate solution   (*1 mark*)

   c <u>sand</u> from a mixture of sugar solution and sand   (*1 mark*)

   d <u>ethanol</u> from a mixture of water and ethanol   (*1 mark*)

5 A biologist needs to separate the oils found in sweat using thin layer chromatography. He dissolves a sample of the oils in a liquid solvent. He then adds a small spot of the mixture to a 'base line' drawn lightly across the bottom of the chromatography plate.

a Describe what the biologist needs to do to the plate so that the oils in the spot separate from one another.   (*3 marks*)

b The oils are not coloured, but they turn brown when exposed to iodine vapour. The diagram shows the results obtained by the biologist.

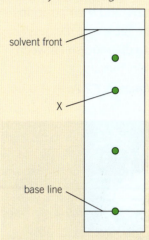

i Suggest why one of the oils in the mixture stayed on the base line.   (*1 mark*)

ii Calculate the R_f value for the spot labelled **X** on the chromatogram.   (*3 marks*)

iii Explain whether the sweat was a pure or an impure substance.   (*1 mark*)

6* A student needs to separate a mixture of water and **H** two other substances, **X** and **Y**.
Substance **X** and substance **Y** are both soluble in water. The table below gives some information about these substances.

Substance	Melting point (°C)	Boiling point (°C)
water	0	100
X	−114	78.5
Y	110	decomposes (breaks down)

Explain how the student can separate the mixture to get pure samples of each substance (water, substance X and substance Y).   (*6 marks*)

C2.2.1 Metals and non-metals

Learning outcomes

After studying this lesson you should be able to:

- describe the properties of metals and non-metals, and their position in the Periodic Table
- explain how to distinguish between metals and non-metals.

Specification reference: C2.2a

Figure 1 *Magnesium in this sparkler burns in air with a bright white flame, producing magnesium oxide in its solid state.*

Group numbers

There are two ways to number the columns (groups) in the Periodic Table:

- an older system, from 1 to 0 (in brackets in Figure 2)
- a newer IUPAC system, from 1 to 18.

Sparklers are hand-held fireworks that burn at over 1000 °C. Sparklers contain metals such as magnesium and aluminium. When ignited they react rapidly with a non-metal in the air, oxygen, emitting showers of sparks (Figure 1). What are metals and non-metals like, and how can you tell them apart?

What are metals and non-metals like?

Metal elements and **non-metal** elements have different characteristic **physical properties** (Table 1). A physical property is a characteristic that can be observed or measured.

Table 1 *Typical properties of metal and non-metal elements.*

Physical property	Metal elements	Non-metal elements
appearance	shiny	dull
melting point and boiling point	usually high	usually low
state at room temperature	solid	about half are solid about half are gas
malleable or brittle when solid	**malleable** (they bend without shattering)	**brittle** (they shatter when hammered)
ductile or non-ductile when solid	**ductile** (they can be pulled into wires)	non-ductile (they snap when pulled)
thermal and electrical conductivity	good **conductors**	poor conductors (they are **insulators**)

A Find out what 'IUPAC' stands for.

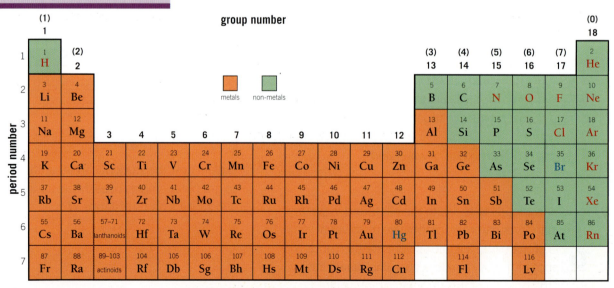

Figure 2 *At room temperature, the elements in the solid state are in black text, the two elements in the liquid state are in blue text, and the elements in the gas state are in red text.*

The **Periodic Table** is a table in which all the elements are arranged in rows and columns, in order of increasing atomic number (Figure 2). Metal elements are placed on the left-hand side of the Periodic Table, and non-metal elements are placed on the right-hand side. They can be separated by a zigzag boundary. Some of the elements on this boundary have properties of both metals and non-metals.

B Use the Periodic Table in Figure 2 to describe what is unusual about mercury, Hg, and bromine, Br.

C Graphite is in the solid state at room temperature. It is a brittle and shiny black form of carbon, C, and it is an electrical conductor. Explain what is unusual about these properties.

What are the chemical properties of metals and non-metals?

Metal elements and non-metal elements have different characteristic **chemical properties**. A chemical property is a characteristic of a substance that can only be determined by studying its chemical reactions. For example:

- Metals lose electrons to form positive ions, but non-metals gain electrons to form negative ions.
- Metals do not react with each other (they mix to form alloys), but non-metals react with each other to produce compounds that consist of molecules.

Metals and non-metals may react with oxygen to produce **oxides** (Figure 3), but these have different properties. If they dissolve in water:

- metal oxides produce **alkaline** solutions
- non-metal oxides produce **acidic** solutions.

D Describe two differences in the properties of magnesium oxide and sulfur dioxide.

Figure 3 *Blue fire from a volcano – sulfur is burning in air here, producing sulfur dioxide in its gas state.*

Synoptic link

You should have used universal indicator in your Key Stage 3 studies to distinguish between acidic and alkaline solutions.

1 Make a table outlining the differences in the chemical properties of metals and non-metals. *(6 marks)*

2 Outline how you would use a battery, lamp, wires, and crocodile clips to distinguish between metals and non-metals. *(4 marks)*

3 Air surrounds an open electric switch in a circuit but no current flows. Explain what this tells you about air, and the elements and compounds it contains. *(2 marks)*

4 An element, solid at room temperature, burns in oxygen to produce white clouds of its oxide. Describe how you could test the oxide to see if the element was a metal or a non-metal. *(4 marks)*

Learning outcomes

After studying this lesson you should be able to:

- describe the electronic structure of the first 20 elements
- explain how the position of an element in the Periodic Table relates to its electronic structure.

Specification reference: C2.2b, C2.2c

Figure 1 *Books are arranged on shelves at different heights from the floor. Similarly, electrons are arranged around the nucleus in shells with different energy levels.*

Table 1 *Maximum numbers of electrons in each shell.*

Shell	Maximum number of electrons
first	2
second	8
third	8
fourth	18

Livermorium, Lv, is a metallic element discovered at the start of this century. Only a few atoms of it have been made, and each has 116 electrons. How are all the electrons arranged in atoms?

How are elements arranged in the Periodic Table?

The rows and columns within the Periodic Table have special names:

- a **period** is a horizontal row
- a **group** is a vertical column.

The groups and periods are numbered to help identify them. If you look from left to right across a period, you will see that the atomic number increases by 1 going from one element to the next. This means that the number of electrons in each atom also increases by 1 each time.

The elements in a group have similar chemical properties. This is due to the arrangement of their electrons.

A Write down the number of elements in Periods 1, 2, 3, and 4.

How are electrons arranged in atoms?

The electrons in atoms are arranged around the nucleus in shells.

The **electronic structure** of an element shows how the electrons are arranged in its atoms. The outermost occupied shell in an atom is called the **outer shell**. The electrons in atoms are arranged around the nucleus in shells. Different shells can hold different numbers of electrons (Table 1).

Finding the electronic structure of an atom

Lithium has an atomic number of 3, so has three electrons in total. Find its electronic structure.

Step 1: Fill each of the shells in turn, starting with the first shell.

Lithium will have two electrons in its first shell and one electron its second shell.

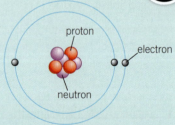

Figure 2 *The structure of a lithium atom.*

Step 2: Write down the electronic structure.

The electronic structure of lithium is 2.1 (the dot shows that you are going from one shell to the next).

You can work out electronic structures by counting from hydrogen, period by period, until you reach the element you want, as shown in Figure 3.

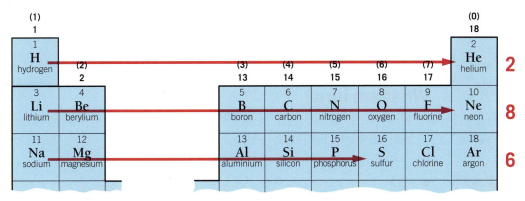

Figure 3 *Finding the electronic structure of sulfur, S.*

> **B** Write down the electronic structures for oxygen and magnesium.

How is electronic structure related to the Periodic Table?

There are links between the numbers in the electronic structure of an element and its position in the Periodic Table:

- The last number equals the non-IUPAC group number. However, atoms of Group 0 elements (**IUPAC Group** 18) all have full outer shells.

- The number of numbers equals the period number.

- The sum of the numbers equals the atomic number.

For example, the electronic structure of oxygen is 2.6 so you can say that oxygen is in Group 6 (IUPAC Group 16) and Period 2, and its atomic number is 8 (2 + 6).

This gives you another way to work out electronic structures. For example, calcium is in Group 2 and Period 4, and its atomic number is 20. Its electronic structure is 2.8.8.2 using the ideas above.

> **C** The electronic structure of potassium is 2.8.8.1. Write down what this tells you about potassium.

> **1** Explain what periods and groups are in the Periodic Table. *(2 marks)*
>
> **2** Phosphorus is in Group 5 (IUPAC Group 15). Write down the number of electrons in its outer shell. *(1 mark)*
>
> **3 a** The electronic structure of element X is 2.8.8. Write down the period and group to which X belongs, and its atomic number. Explain your answer. *(6 marks)*
>
> **b** Use the Periodic Table to name element X. *(1 mark)*
>
> **4** Write down the electronic structures of nitrogen, aluminium, and chlorine. *(3 marks)*

Learning outcomes

After studying this lesson you should be able to:

- explain how ions form
- draw electron diagrams for atoms and ions.

Specification reference: C2.2d

Figure 1 *Lithium-ion batteries power smartphones, tablet computers, and cameras.*

Study tip

For most simple ions, the number of charges on a positive ion is equal to the (non-IUPAC) group number of the original metal atom.

The number of charges on a simple negative ion can be calculated as:

$$\text{charge on negative ion} = 8 - \text{group number of the original non-metal atom}$$

Do you have a mobile phone? If you do, one of the reasons it can be so small is its lithium-ion battery (Figure 1). The chemical reactions in these rechargeable batteries involve ions, electrically charged particles.

What are ions?

An **ion** is an electrically charged particle formed when an atom, or group of atoms, loses or gains electrons:

- metal atoms lose electrons to form positive ions
- non-metal atoms gain electrons to form negative ions.

The numbers of protons and neutrons do not change when an atom forms an ion. Just like atoms, you cannot see individual ions. Table 1 describes how four different ions are formed from their atoms.

Table 1 *The formulae for ions are easy to spot. They have a + or − sign at the top right, showing whether the ion is positively charged or negatively charged.*

Atom	Ion	How the ion forms
Na	Na^+	atom loses its outer electron
Mg	Mg^{2+}	atom loses its two outer electrons
Cl	Cl^-	atom gains one electron to complete its outer shell
O	O^{2-}	atom gains two electrons to complete its outer shell

A Explain how Al^{3+} and N^{3-} ions form.

What are the electronic structures of ions?

You can work out an ion's electronic structure if you know the electronic structure of the original atom, and the number of electrons lost or gained.

For example, the electronic structure of a sodium atom is 2.8.1. It loses its outer electron when it forms a sodium ion. The full second shell is now the outer shell and the ion's electronic structure is 2.8. In a similar way, the electronic structure of a chloride ion is 2.8.8 because it forms when a chlorine atom (2.8.7) gains one electron and completes its outer shell. Atoms and ions with full outer shells are stable.

B Write down the electronic structures of Mg^{2+} ions and O^{2-} ions.

What are electron diagrams?

An **electron diagram** represents the electronic structure of an atom or ion. You draw a circle to represent each shell, and dots or crosses to represent its electrons (Figure 2). Ions go inside brackets with the charge written at the top right, and the element's symbol may be written at the centre instead of showing a nucleus (Figure 3).

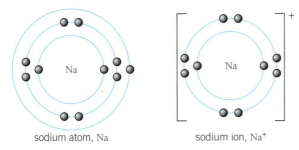

sodium atom, Na   sodium ion, Na^+

Figure 2 *Electron diagrams for sodium atoms (2.8.1) and sodium ions (2.8).*

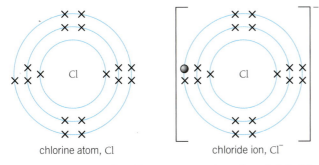

chlorine atom, Cl   chloride ion, Cl^-

Figure 3 *Electron diagrams for chlorine atoms (2.8.7) and chloride ions (2.8.8).*

Note that electrons are all the same in these diagrams. The use of dots or crosses lets you model which atom provided a particular electron.

C   Draw electron diagrams for Mg atoms and Mg^{2+} ions.

---

1   Draw electron diagrams for O atoms and O^{2-} ions.   *(4 marks)*

2   Lithium ions, Li^+, and helium atoms, He, have the same electronic structure. Explain why lithium does not become helium when it forms ions.   *(2 marks)*

3   Hydrogen usually forms hydrogen ions, H^+, but sometimes it forms hydride ions, H^-.

   a   Write down the electronic structures for H atoms and H^- ions.   *(2 marks)*

   b   Describe what is unusual about the H^+ ion.   *(2 marks)*

   c   Suggest why hydrogen can be difficult to classify as a metal or a non-metal.   *(2 marks)*

C2.2.4 Ionic compounds

Learning outcomes

After studying this lesson you should be able to:

- explain how ionic compounds form
- draw dot-and-cross diagrams to represent ionic compounds
- describe the structure and bonding in ionic compounds
- describe the limitations of different models of ionic compounds.

Specification reference: C2.2d, C2.2e, C2.2f, C2.2g

Figure 1 *Sodium reacts vigorously with oxygen in air to produce sodium oxide.*

Sodium fires are difficult to put out. Sodium reacts vigorously with oxygen, igniting and burning with an orange flame (Figure 1). Electrons transfer from sodium atoms to oxygen atoms in the reaction, forming an ionic compound called sodium oxide.

How do ionic compounds form?

When a metal reacts with a non-metal, electrons are transferred from the metal atoms to the non-metal atoms so both achieve more stable electronic structures. The metal atoms become positive ions and the non-metal atoms become negative ions. For example, when sodium reacts with chlorine, the outer electron of each sodium atom transfers to the outer shell of a chlorine atom, forming Na^+ ions and Cl^- ions (Figure 2).

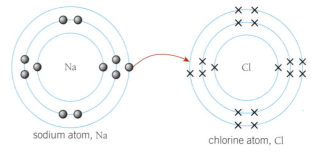

sodium atom, Na   chlorine atom, Cl

Figure 2 *The outer electron from a sodium atom transfers to the outer shell of a chlorine atom.*

You can model the ions in the **ionic compound** that is formed using a **dot-and-cross diagram**. You show the electrons from one atom as dots, and the electrons from the other atom as crosses (Figure 3).

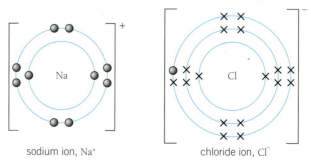

sodium ion, Na^+ chloride ion, Cl^-

Figure 3 *A dot-and-cross diagram for sodium chloride. You can see that the chloride ion contains an electron gained from a sodium atom.*

> **A** Sodium oxide, Na_2O, contains Na^+ ions and O^{2-} ions. Explain how this ionic compound forms when sodium and oxygen react together.

What is the structure and bonding in ionic compounds?

Ionic compounds in their solid state contain positive and negative ions arranged in a regular way. This arrangement is called a **giant ionic lattice**. The ions are held in place by **ionic bonds**, which act in all directions. Ionic bonds are strong electrostatic forces of attraction between oppositely charged ions. **Space-filling models** such as Figure 4 are one way of representing ionic compounds.

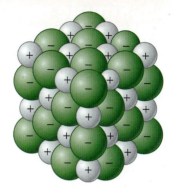
Figure 4 *A space-filling model of sodium chloride.*

What does 'giant ionic lattice' mean?

The word 'giant' tells you that the arrangement is repeated many times, not that it is huge. The word 'ionic' reminds you that the structure contains ions, and 'lattice' means that the arrangement is regular and not random.

B Explain the meaning of the term 'ionic bond'.

C Draw a diagram to represent one face of the space-filling model of sodium chloride. Show each ion as a circle with a + or – inside it.

Models of giant ionic lattices

A giant ionic lattice exists in three dimensions, but you can only draw it in two dimensions. You could make a **ball-and-stick model** like the one in Figure 5. Each plastic ball represents an ion and each plastic link represents an ionic bond. These models give you a clearer idea of the structure and shape of the lattice, but they do have limitations. Remember that ions are close together, and that bonds are forces rather than physical objects made from matter.

D Compare drawn space-filling models with ball-and-stick models.

Figure 5 *A ball-and-stick model of the giant ionic lattice of sodium chloride.*

1 Draw a dot-and-cross diagram to show the ions present in magnesium oxide, MgO. *(4 marks)*

2 Magnesium chloride, $MgCl_2$, contains Mg^{2+} ions and Cl^- ions. Explain how this ionic compound forms when magnesium and chlorine react together. *(4 marks)*

3 Potassium bromide, KBr, has a similar structure to sodium chloride. Describe the structure and bonding in potassium bromide. *(3 marks)*

Go further

Sodium chloride, NaCl, is a binary ionic compound. Its positively charged ions come from a single metal element (sodium), and its negatively charged ions come from a single non-metal element (chlorine).

Find some other examples of binary ionic compounds.

Learning outcomes

After studying this lesson you should be able to:

● explain how covalent bonds form
● draw dot-and-cross diagrams for simple molecules
● describe the structure and bonding in simple molecules
● describe the limitations of dot-and-cross diagrams, and of ball-and-stick models.

Specification reference: C2.2d, C2.2e, C2.2f, C2.2g

Figure 1 *This engineer monitors the emissions from these chimneys to make sure they contain acceptable amounts of air pollutants.*

Study tip

Hydrogen only forms one covalent bond. For other non-metals, the number of bonds can be calculated as:

number of bonds = 8 – their non-IUPAC group number

Chlorine in Group 7 (IUPAC Group 17) forms 1 bond, and oxygen in Group 6 (IUPAC Group 16) forms 2 bonds. Group 0 (IUPAC Group 18) elements are **inert** (unreactive), so they do not form bonds.

How clean is the air you breathe (Figure 1)? Air is a mixture of elements and compounds, most of which exist as simple molecules. For example, carbon dioxide, a greenhouse gas, exists as simple molecules. What are its molecules like?

What are covalent bonds?

A **covalent bond** is a shared pair of electrons. Covalent bonds form between two non-metal atoms when the atoms get close enough to share electrons in their outer shells. By sharing electrons, the atoms complete their outer shells.

Covalent bonds can be modelled using dot-and-cross diagrams (Figure 2). The electrons from one of the bonded atoms are shown as dots, and the electrons from the other bonded atom are shown as crosses. Each pair of electrons in the shared area between the overlapping circles represents a covalent bond. Only the outer shells are usually shown.

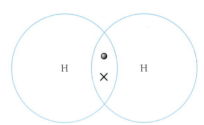

Figure 2 *A dot-and-cross diagram for hydrogen, H$_2$.*

A Explain what the dot-and-cross diagram in Figure 2 tells you about hydrogen.

What are simple molecules?

Hydrogen, oxygen, water, and carbon dioxide all exist as simple molecules. A molecule is a particle in which non-metal atoms are joined to each other by covalent bonds. A **simple molecule** is a molecule that only contains a few atoms. These substances can all be modelled using dot-and-cross diagrams (Figure 3).

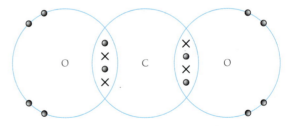

Figure 3 *A dot-and-cross diagram for carbon dioxide, CO$_2$. The non-bonding oxygen electrons are also shown.*

B Explain how the dot-and-cross diagram for carbon dioxide shows that the carbon–oxygen bond is a double covalent bond.

What is the structure and bonding in simple molecules like?

Covalent bonds involve electrostatic forces of attraction, just like ionic bonds do. However, for a covalent bond, the forces are between the nucleus of each bonded atom and the shared electrons. The covalent bonds between atoms in a simple molecule are strong, but the **intermolecular forces** between the molecules are weak.

C Explain what a covalent bond is in terms of forces.

Models of simple molecules

You can make **ball-and-stick models** for simple molecules, just as you can for ionic compounds (Figure 5). This model has limitations: both the sizes of the atoms and the length of the bonds are exaggerated, and it suggests that the electrons that make the bonds do not move.

You can also draw a **displayed formula** for simple molecules (Figure 6). In these formulae, each atom is represented by its chemical symbol, and each covalent bond by a straight line. Notice that simple molecules have shapes. However, this model is limited as it does not show the three-dimensional shape of the molecule.

D Compare the advantages and disadvantages of the three models of a water molecule (Figures 4–6), and write down the piece of information missing from each of them.

Synoptic link

You can learn more about how covalent bonds and intermolecular forces determine physical properties of molecular substances in C2.3.2 *Changing state*.

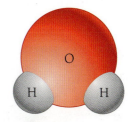

Figure 4 *A space-filling model of a water molecule.*

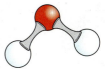

Figure 5 *A ball-and-stick model of a water molecule.*

Figure 6 *A displayed formula for water.*

1 Explain what a covalent bond is in terms of electrons. (*1 mark*)
2 **a** Draw a dot-and-cross diagram of a water molecule. (*2 marks*)
  b Describe advantages and disadvantages of this model with the other models of water shown in Figures 4–6. (*3 marks*)
3 Draw dot-and-cross diagrams of a chlorine molecule, Cl_2, and an oxygen molecule, O_2. (*4 marks*)
4 Describe the bonds present in a sample of carbon dioxide, and compare their strengths. (*3 marks*)

Learning outcome

After studying this lesson you should be able to:

- describe the structure and bonding in giant covalent structures.

Specification reference: C2.2d, C2.2g

Figure 1 *Sand grains are made from giant molecules.*

The empirical formula for silica, the main compound found in sand, is SiO_2 (Figure 1). You might think that silica exists as simple molecules, but it actually exists as giant covalent structures.

What are giant covalent structures?

A **giant covalent structure** consists of very many non-metal atoms joined by covalent bonds and arranged in a repeating regular pattern called a giant lattice. These structures are also called giant covalent lattices.

Diamond is a form of carbon. It exists as a giant covalent structure in which each carbon atom is joined to four other carbon atoms by covalent bonds (Figure 2).

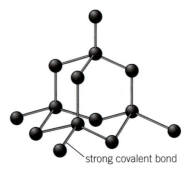
strong covalent bond

Figure 2 *A ball-and-stick model of part of the structure of diamond.*

A Explain why diamond can be described as having a *giant lattice* structure.

What are the chemical formulae of giant molecules?

Just like in ionic compounds, where very many ions are involved, giant covalent structures have very many atoms. It would make little sense for you to try to write a chemical formula with these huge numbers in it. Instead, you use the empirical formula for the substance. You will remember that this formula shows the simplest whole number ratio of atoms of each element. This is why the formula for diamond is given in chemical equations as C.

B Graphite is a form of carbon that exists as a giant covalent structure. Explain why its chemical formula is C.

C Suggest why it would be difficult to represent a giant covalent structure using dot-and-cross diagrams.

Empirical formulae of giant covalent structures

Figure 3 shows part of the giant covalent structure of silica. Each silicon atom is covalently bonded to four oxygen atoms, so you might think that the empirical formula is SiO_4. However, each oxygen atom is also bonded to two silicon atoms, which would give the empirical formula Si_2O. Taken over the whole structure each silicon atom is, on average, bonded to two oxygen atoms.

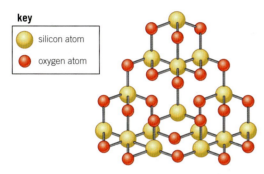

Figure 3 *The atoms in silica are joined in a giant covalent structure.*

D Explain why the formula of silica is given as SiO_2. Suggest what information is missing from the diagram and justify your answer.

Go further

As the number of atoms in a molecule increases, when does a large simple molecule become a small giant covalent structure? Carbon does not just exist as diamond or graphite. It also forms a vast range of molecules called fullerenes. The first of these, buckminsterfullerene, was discovered in 1985.

Study the structure of buckminsterfullerene (Figure 4) carefully, and explain why its formula is C_{60} rather than C.

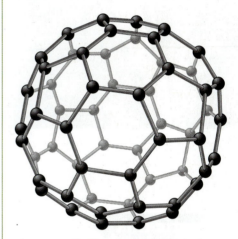

Figure 4 *A ball-and-stick model of buckminsterfullerene, a football-shaped hollow molecule.*

1 Describe the structure and bonding in a giant covalent structure. *(3 marks)*

2 Nanotubes are a form of carbon. Part of a nanotube is shown in Figure 5. Explain why a nanotube is not a simple molecule but has a giant covalent structure. *(2 marks)*

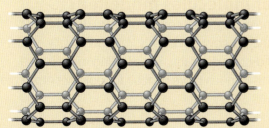

Figure 5 *A ball-and-stick model of a section of part of a hollow nanotube.*

3 Compare the structures of diamond and silica. *(3 marks)*

C2.2.7 Polymer molecules

Figure 1 *An electron microscope image of human hair with split ends.*

Synoptic link

You can learn more about the two different types of reaction used to make polymers in C6.2 *Organic Chemistry*. Poly(ethene) and poly(propene) are *addition polymers* made by addition reactions. Complex carbohydrates, DNA, proteins, and nylon are *condensation polymers* made by condensation reactions.

Plastics such as nylon are artificial polymers, but your body is full of natural polymers. Complex carbohydrates, DNA, and proteins in your skin are all polymers. Take good care of your hair (Figure 1) – it is made from polymers too!

What are polymers?

All **polymers**, whether they are artificial or natural, are made from many smaller molecules called **monomers**. These monomers are able to join end to end in chemical reactions, producing longer polymer molecules (Figure 2). For example:

- many ethene molecules can join end to end to make poly(ethene), often called polythene
- many amino acid molecules can join end to end to make a protein, such as the keratin in hair.

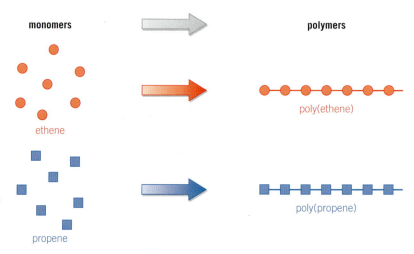

Figure 2 *A simple model showing how polymers form.*

A Explain how a molecule of poly(propene) forms (Figure 2).

B Suggest what the straight lines in the diagrams of poly(ethene) and poly(propene) represent in Figure 2.

How are polymer molecules modelled?

Monomers are simple molecules. They consist of a few non-metal atoms joined to each other by covalent bonds. You can usually model monomers using dot-and-cross diagrams, space-filling models, and ball-and-stick models. This is much more difficult for polymer molecules (Figure 3).

Figure 3 *A space-filling model for a section of starch, a complex carbohydrate.*

C Suggest why it is difficult to represent polymers with the models used for simple molecules.

You will often see diagrams like Figure 4, in which each polymer molecule is drawn as a wavy line. These sometimes have straight lines between them to represent covalent bonds between individual polymer molecules. There are weak intermolecular forces between polymer molecules but these are not shown.

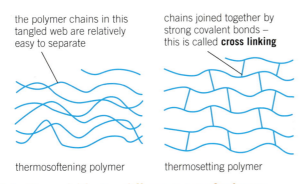

the polymer chains in this tangled web are relatively easy to separate

chains joined together by strong covalent bonds – this is called **cross linking**

thermosoftening polymer

thermosetting polymer

Figure 4 *A simple model for two different types of polymer.*

Some polymers are made from several different types of monomer, for example proteins and DNA. However, other polymers are made from one type of monomer, for example poly(ethene) is only made from ethene, C_2H_4. Poly(ethene) can be modelled using the idea of a **repeating unit** (Figure 5). This is a section of the polymer molecule that is repeated over and over again, rather like links on a chain.

Go further

Thermosoftening polymers can be heated to soften them, then moulded into a new shape that sets when they cool down. *Thermosetting* polymers cannot be remoulded. Instead, they char or burn when heated. Find out about these two types of polymers and what they are used for.

Figure 5 *Ethene and the repeating unit for poly(ethene).*

1 Name two natural polymers, and two artificial polymers. *(4 marks)*

2 Describe the bonding between atoms in a polymer molecule, and the bonding between polymer molecules. *(3 marks)*

3 Discuss the advantages and limitations of the model shown in Figure 4 for the thermosoftening and thermosetting polymers. *(6 marks)*

Learning outcomes

After studying this lesson you should be able to:

● describe the structure and bonding in metals

● describe the limitations of different models of metals.

Specification reference: C2.2d

Figure 1 *The Infinity Bridge and its reflection form the infinity symbol, ∞, used in mathematics.*

The Infinity Bridge at Stockton-on-Tees in the north-east of England (Figure 1) is made from steel and steel-reinforced concrete. The structure and bonding in steel makes it strong enough to be useful for constructing buildings and vehicles, as well as bridges.

What is the structure of metals like?

All metals, apart from mercury, are in the solid state at room temperature. Their atoms are packed together in a regular way, forming a **giant metallic lattice**. This is modelled by drawing circles or spheres arranged in a regular pattern, touching each other (Figure 2).

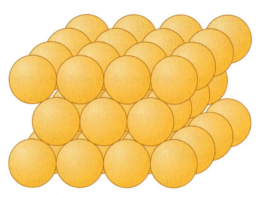

Figure 2 *Metal atoms packed in a giant metallic lattice.*

A Describe the arrangement of atoms in mercury in the solid state, and justify your answer.

What are metallic bonds like?

You will remember that, in reactions with non-metals, metal atoms lose electrons from their outer shell to form positively charged ions. This also happens in metals themselves. Electrons leave the outer shells of the metal atoms, forming a 'sea' of electrons around positively charged metal ions (Figure 3). These electrons are free to move through the structure of the metal, so they are called **delocalised electrons**.

Metallic bonds are the strong electrostatic forces of attraction between the delocalised electrons and the closely packed, positively charged metal ions.

B Describe metallic bonds.

C Explain which properties of delocalised electrons are missing from Figure 3.

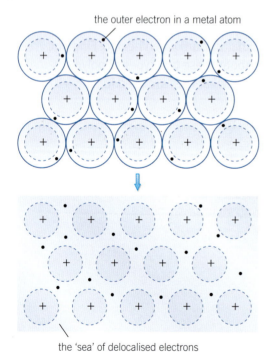

the outer electron in a metal atom

the 'sea' of delocalised electrons

Figure 3 *The outer electrons become delocalised in a metal.*

Localised and delocalised

Something that is 'localised' is restricted to a particular place. Something that is 'delocalised' is free to move from its usual place. For example, when you sit in your usual place in the laboratory you are localised, but you become delocalised when you move around the laboratory during a practical activity.

Models of metals

A metallic structure extends in three dimensions, and its metallic bonding extends in all directions. You lose some information when you represent the structure and bonding in metals in two dimensions, as you can see by comparing Figure 4 and Figure 5.

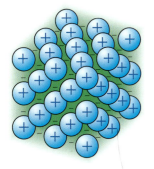

Figure 4 *An isometric diagram representing the structure of metals.*

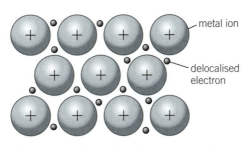

metal ion

delocalised electron

Figure 5 *An elevation view representing the structure of metals.*

D Compare the two models in Figures 4 and 5, and describe the advantages and limitations of each model.

1 Describe the structure and bonding present in metals. (*6 marks*)

2 Compare ionic bonds with metallic bonds. In what ways are they similar or different? (*3 marks*)

3 Some students who looked at Figures 4 and 5 did not realise that they applied to all metals. Instead, they thought that the figures only showed a metal from Group 1 of the Periodic Table. Suggest why they thought this, and describe changes the students might have made so that the diagrams showed a metal from Group 2. (*5 marks*)

Learning outcome

After studying this lesson you should be able to:

● explain how Mendeleev's arrangement was refined into the modern Periodic Table.

Specification reference: C2.2h

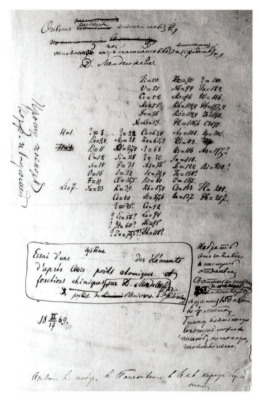

Figure 1 *Mendeleev's first Periodic Table from 17 February 1869. He also wrote chemistry notes on the back of a letter about a cheese factory inspection.*

Figure 2 *Gallium, discovered in 1875, is used to make LED lights. Its properties closely match the properties for 'eka-aluminium', one of Mendeleev's predicted elements.*

What would you rather do if you were busy writing a chemistry text book, worry about inspecting a cheese factory for the Russian Free Economic Society, or work out a way to arrange the elements for your book? The Russian chemist Dmitri Mendeleev faced this decision in February 1869, and thankfully decided to work on arranging the elements.

What did Mendeleev do?

Mendeleev's arrangement (Figure 1) was not the first attempt by chemists to organise the elements into a table. However, it was the most successful and it led to our modern Periodic Table.

Mendeleev had been considering his own and other scientists' evidence about the elements for several years before his first table. This included:

● the atomic weights of the known elements, similar to our modern relative atomic masses

● knowledge of the chemical reactions of different elements

● knowledge of physical properties, such as melting points and boiling points.

Mendeleev arranged all the elements known at the time in order of increasing atomic weight, and he grouped together the ones with similar chemical properties. However, he swapped the positions of tellurium and iodine because he felt that this matched their chemical properties better. Mendeleev left spaces for elements he thought would exist but were not yet discovered, and predicted their properties from those of nearby elements.

A Suggest why some other chemists criticised Mendeleev's table when it was published.

What did Mendeleev do next?

Mendeleev's first Periodic Table showed groups as rows, not columns. By 1871 he had rotated his table so that groups were in columns, as in the modern Periodic Table. Three of his predicted elements were discovered between 1875 and 1886. They were found to have similar properties to the ones he had predicted years before.

B Explain why the discovery of gallium (Figure 2) and the other predicted elements gave support to Mendeleev's ideas.

Why is the modern table in order of atomic number?

Mendeleev developed his table without knowing about atomic structure. During Mendeleev's lifetime, the atomic number of an element was just its position in the Periodic Table. Mendeleev died in 1907, before the proton was discovered. It was not until 1913 that Henry Moseley, an English physicist, discovered that an atom's atomic number was actually the number of protons in its nucleus.

Moseley's work showed that there were seven gaps left to fill in the 1913 Periodic Table (Figure 3). It also explained why Mendeleev had been right to swap tellurium and iodine around.

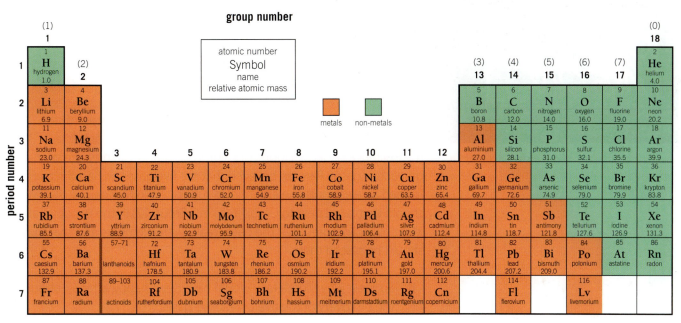

Figure 3 *In Mendeleev's table and the modern Periodic Table, tellurium is in Group 6 (IUPAC Group 16) and iodine is in Group 7 (IUPAC Group 17).*

C Explain why tellurium is placed before iodine in the modern Periodic Table.

1 Explain why Mendeleev should have placed iodine before tellurium in his Periodic Table, following the pattern of the rest of his table. *(2 marks)*

2 The chemical properties of an element are determined by its electronic structure, which in turn depends upon its atomic number. Explain why Mendeleev did not consider electronic structures when he developed his tables. *(2 marks)*

3 Other scientists developed similar Periodic Tables around the same time, but Mendeleev published his first. Suggest why you rarely hear about the work of the other scientists. *(2 marks)*

Learning outcome

After studying this lesson you should be able to:

- explain patterns of chemical properties in the Periodic Table in terms of atomic structure.

Specification reference: C2.2i

Figure 1 *Neon, used in advertising signs, is an element in Group 0 (the noble gases, IUPAC Group 18).*

(0)

18
2
He
helium
4.0
10
Ne
neon
20.2
18
Ar
argon
39.9
36
Kr
krypton
83.8
54
Xe
xenon
131.3
86
Rn
radon

Figure 2 *Group 0 (IUPAC Group 18) is on the right of the Periodic Table. Radon was discovered in 1900, and isolated by Ramsay in 1910.*

Imagine that you have just finished your life's work, the Periodic Table. Scientists then announce that they have discovered a group of elements that you completely failed to predict. How would you feel? This happened to Dmitri Mendeleev when the noble gases were discovered.

A new group – what happened?

Lord Rayleigh and William Ramsay discovered argon in 1894. No-one had predicted its existence, not even Mendeleev. Ramsay discovered helium the following year. Both gases are inert. Mendeleev thought that elements should be able to react with other elements, so he was reluctant to believe that helium and argon were elements.

> **A** Suggest why it would have helped Mendeleev if helium and argon were not elements.

The situation changed during 1898 when Ramsay discovered three more new elements. Just like argon, Mendeleev had not predicted the existence of neon (Figure 1), krypton, or xenon. All three are very unreactive gases. Ramsay believed that the five gases he had discovered formed a new group of elements. He suggested to Mendeleev that they should be placed next to Group 7 (IUPAC Group 17) in the Periodic Table. Mendeleev accepted that the gases really were elements and that they should be placed there.

> **B** Suggest why Mendeleev would have been happy, after all, about the discovery of the five unreactive gases.

What are the patterns of chemical properties?

To understand the patterns of chemical properties in the Periodic Table, you need to remember that:

- Elements are arranged in order of increasing atomic number.
- The atomic number is the number of protons in an atom.
- The number of electrons in an atom is equal to the number of protons.
- Electronic structure is determined by the number of electrons (Figure 3).
- The electronic structure of an element determines its chemical properties.

This means that there is a link between the position of an element in the Periodic Table and its chemical properties.

Table 1 shows properties of the elements in four groups. Elements in Groups 1 and 2 become *more* reactive as you go down the group, and Group 7 (IUPAC Group 17) elements become *less* reactive.

Table 1 *Properties of elements in Groups 1, 2, 7 (IUPAC Group 17), and 0 (IUPAC Group 18).*

Group	Type of element	Reactivity	Electronic structures	Ions formed in reactions
1	metal	very reactive	end in 1	+1
2	metal	reactive	end in 2	+2
7	non-metal	very reactive	end in 7	−1
0	non-metal	very unreactive	outer shells are full	do not react

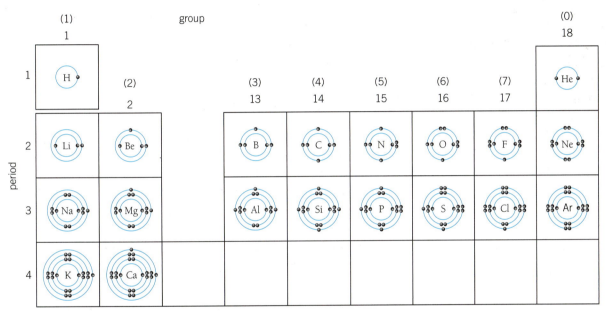

Figure 3 *The first 20 elements with their electron diagrams in a 'short form' of the Periodic Table.*

C Describe the feature of the atomic structure of Group 0 (IUPAC Group 18) elements that make them very unreactive.

1 Name the common feature of the atomic structures of elements in Group 6 (IUPAC Group 16). *(1 mark)*

2 Three elements have the atomic numbers 4, 12, and 20.
  a Write their electronic structures. *(3 marks)*
  b Use your answer to **a** to explain why these elements form ionic compounds with oxygen. *(4 marks)*

3 A historian has suggested that Mendeleev's Periodic Table succeeded because it could fit in a new group, rather than because Mendeleev made accurate predictions. Evaluate this statement. *(5 marks)*

Elements, compounds, and mixtures

C2.2 Bonding

Summary questions

1 Elements may be divided into metals and non-metals. Copy and complete the sentences below using words from the box. Each word may be used once, more than once, or not at all.

brittle	conductors	dull	gas
high	insulators	liquid	low
malleable	shiny		solid

   a Metals are _____ in appearance, but non-metals are usually _____.
   b Metals usually have _____ melting points and non-metals usually have _____ melting points.
   c Most metals are in the _____ state at room temperature. One non-metal is _____ at room temperature, about half of non-metals are in the _____ state, and the rest are in the _____ state.
   d Metals are _____ and ductile, but non-metals are _____ in the solid state.
   e Metals are usually good _____ of heat and electricity whereas non-metals are usually _____.

2 This question is about the Periodic Table.
   a Explain what groups are.
   b Explain what periods are.
   c Write down the name and chemical symbol for the element in Period 3, Group 4 (IUPAC Group 14).
   d Explain the order in which elements are arranged in the Periodic Table.

3 Figure 1 represents the Periodic Table. The letters A–G indicate the positions of eight different elements.

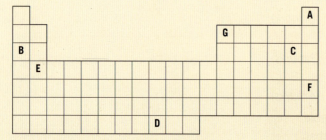

Figure 1 *An outline of the Periodic Table.*

Write down the letters (from Figure 1) for:
   a two metals
   b two non-metals
   c one element in Period 2
   d one element in Group 1.

4 Write down the electronic structures for:
   a fluorine, F
   b magnesium, Mg
   c potassium, K.

5 a Describe how a sodium ion, Na^+, forms from a sodium atom, Na.
   b Describe how a sulfide ion, S^{2-}, forms from a sulfur atom, S.

6 Draw dot-and-cross diagrams to model the ions that make up:
   a sodium chloride, NaCl
   b magnesium oxide, MgO.

7 Draw dot-and-cross diagrams to model the following simple molecules:
   a hydrogen, H_2
   b hydrogen chloride, HCl
   c methane, CH_4
   d oxygen, O_2.

8 Figure 2 models four different types of substance. The particles are not drawn to scale.

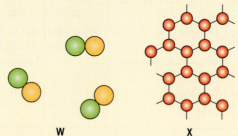

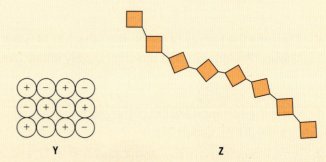

Figure 2 *Four different types of substance.*

Using the letters W, X, Y, or Z, write down which substance or substances:
   a contain ionic bonds
   b contain covalent bonds
   c have a giant structure
   d exist as simple molecules
   e represent a polymer molecule.

Revision questions

1 Sulfur is a non-metal. Which statement about sulfur is true because it is a non-metal?
   A Sulfur is in Group 6 of the Periodic Table.
   B Sulfur is in Period 3 of the Periodic Table.
   C Sulfur melts when it is heated.
   D Sulfur is brittle when it is in the solid state.   *(1 mark)*

2 In what order are the elements arranged in the modern Periodic Table?
   A increasing atomic weight
   B increasing relative atomic mass
   C increasing atomic number
   D increasing number of neutrons   *(1 mark)*

3 What type of bonds are present in diamond, a form of the element carbon?
   A ionic
   B covalent
   C metallic
   D intermolecular   *(1 mark)*

4 The table shows four models of a molecule of ammonia.

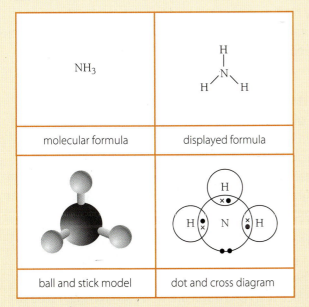

| molecular formula | displayed formula |
| ball and stick model | dot and cross diagram |

   a Describe the information shown by the molecular formula for ammonia.   *(1 mark)*
   b Name the type of bonding present in a molecule of ammonia, and explain your answer.   *(2 marks)*

c Describe one advantage and one disadvantage of the ball and stick model compared to the displayed formula.   *(2 marks)*
d Describe two pieces of information given by the dot and cross diagram but not by the other three models.   *(2 marks)*

5 In terms of electrostatic forces of attraction, describe the nature of the following types of chemical bond.
   a ionic bonds   *(2 marks)*
   b covalent bonds   *(2 marks)*
   c metallic bonds   *(2 marks)*

6 Carbon is a non-metal and calcium is a metal. Both elements react with oxygen to form oxides.
   These oxides dissolve in water to form solutions that are not neutral.
   a Describe one difference between the two solutions.   *(2 marks)*
   b The electronic structure of carbon is 2.4.
      Draw a dot and cross diagram to show the covalent bonding in a molecule of carbon dioxide, CO_2.   *(2 marks)*
   c The electronic structure of calcium is 2.8.8.2 and the electronic structure of oxygen is 2.6.
      Draw dot and cross diagrams, including the charges on the ions, to show the ionic bonding in calcium oxide.   *(3 marks)*

7* A potassium atom can be represented by this chemical symbol: $^{39}_{19}K$
   Explain, as fully as you can, what can be deduced about the structure of this potassium atom and the position of potassium in the Periodic Table.   *(6 marks)*

8* Describe and compare the nature and arrangement of the bonds in these three substances:
   ● sodium fluoride
   ● diamond
   ● poly(ethene)   *(6 marks)*   **H**

Learning outcomes

After studying this lesson you should be able to:

- explain how carbon is able to form different families of compounds
- explain the properties of the different forms of carbon.

Specification reference: C2.3a, C2.3b, C2.3c

Figure 1 *Chicken wire is made from interlocking hexagons.*

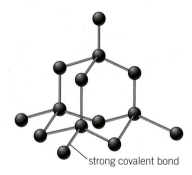
strong covalent bond

Figure 2 *This structure is repeated many times in diamond.*

Diamond, graphite, graphene, and fullerenes are different forms of carbon. Remarkably, apart from diamond, the arrangement of the atoms in the other forms resembles chicken wire (Figure 1).

Why can carbon form many different compounds?

Carbon is in Group 4 (IUPAC Group 14), so its atoms have four electrons in their outer shell and can form four covalent bonds. A carbon atom can join with other carbon atoms to form chains and rings. Carbon atoms in these structures can also form covalent bonds with other elements, such as hydrogen and oxygen, producing many different compounds.

What is diamond like?

Allotropes are different forms of an element in the same state but with different atomic arrangements. For example, **diamond** and **graphite** are two allotropes of carbon. Diamond is transparent and very hard, while graphite is grey-black and soft. The differences are because of their structure and bonding.

Diamond exists as a giant covalent structure in which each carbon atom is covalently bonded to four other carbon atoms (Figure 2). Covalent bonds are strong, and diamond has very many of them. This means that diamond has a very high melting point and is very hard. This hardness makes diamond suitable for the tips of dental drills. All the outer electrons in its atoms are shared, forming covalent bonds. Diamond has no delocalised electrons, so it does not conduct electricity.

What is graphite like?

Graphite exists as a giant covalent structure in which each carbon atom is covalently bonded to just three other carbon atoms. This means that one electron in the outer shell of each atom is not involved in bonding. It becomes delocalised instead. The delocalised electrons in graphite are free to move through the structure, so graphite conducts electricity even though it is a non-metal.

Graphite has a layered structure (Figure 3). The atoms in each layer form interlocking hexagons like the knots in chicken wire. The many strong covalent bonds in the layers give graphite a very high melting point. However, the forces between each layer are weak, so layers can slide over each other easily. This is why graphite is slippery, and why some of the graphite tip of a pencil rubs off on paper.

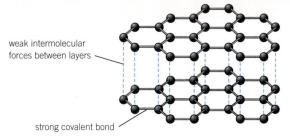

Figure 3 *This layered structure is repeated many times in graphite.*

A Explain why graphite and diamond have high melting points.

B Explain why graphite has delocalised electrons but diamond does not.

What are graphene and fullerenes like?

Graphene is a carbon allotrope that resembles a single layer of graphite. It is almost transparent, extremely strong, and conducts electricity. Scientists are excited by graphene's properties and are researching ways to use it (Figure 4).

Fullerenes form a large family of carbon allotropes in which the molecules are shaped like tubes or balls. A **nanotube** (Figure 5) resembles a sheet of graphene rolled into a tube. Nanotubes are strong and are used to reinforce some sports equipment. A **buckyball** (Figure 6) resembles a sheet of graphene closed to make a hollow ball, but the carbon atoms may be in pentagons as well as hexagons. Buckyballs have potential uses as lubricants, with the molecules acting like tiny ball bearings. Their small size allows buckyballs to pass through cell membranes, and they might one day deliver medical drugs directly to cells.

Figure 4 *Graphene could be used to make flexible touch screens.*

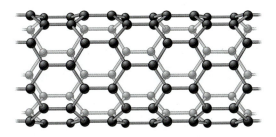

Figure 5 *A model of a section of a nanotube.*

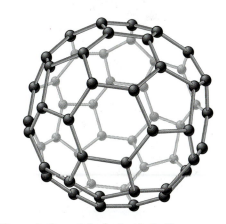

Figure 6 *A model of a buckyball.*

C Suggest why graphene and nanotubes conduct electricity.

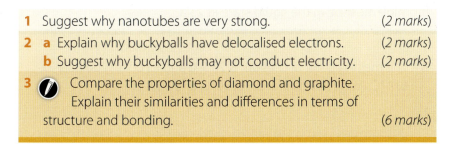

1 Suggest why nanotubes are very strong. *(2 marks)*

2 a Explain why buckyballs have delocalised electrons. *(2 marks)*
   b Suggest why buckyballs may not conduct electricity. *(2 marks)*

3 Compare the properties of diamond and graphite. Explain their similarities and differences in terms of structure and bonding. *(6 marks)*

Learning outcomes

After studying this lesson you should be able to:

- use data to predict the states of a substance
- explain how different substances change state.

Specification reference: C2.3d, C2.3e

Figure 1 *Ice cream melts when it is heated.*

Synoptic link

You can learn about changes of state in C1.1.2 *Chemical and physical changes*.

Ice cream stays in the solid state in the freezer. Once you take it out, the ice cream begins to melt as forces of attraction between its particles are overcome (Figure 1).

What happens when substances change state?

When a substance **melts** or **boils**, forces of attraction between its particles are overcome. In other words, some or all of the bonds between its particles break:

- some bonds break going from the solid to the liquid state
- all remaining bonds break going from the liquid to the gas state.

Depending on the type of substance, these will be metallic bonds, ionic bonds, covalent bonds, or intermolecular forces. The stronger the bonds, and the more of them there are, the more energy must be transferred from the surroundings to break them. Substances have high melting points if they have many strong bonds in the solid state, and high boiling points if they have many strong bonds in the liquid state.

Bonds form when a substance **condenses** or **freezes**:

- some bonds form going from the gas to the liquid state
- many bonds form going from the liquid to the solid state.

Stored chemical energy is transferred to the surroundings, usually by heating, when chemical bonds form.

A In terms of the number of attractive forces overcome between water molecules, explain why the boiling point of water is higher than its melting point.

Predicting states

The melting point of bromine is −7 °C and its boiling point is 59 °C. Predict its state at 25 °C.

Step 1: Is the temperature below the melting point? If it is, the substance will be in the solid state.

  25 °C is above −7 °C, so it is not in the solid state.

Step 2: Is the temperature above the boiling point? If it is, the substance will be in the gas state.

  25 °C is below 59 °C, so it is not in the gas state.

Bromine must be in the liquid state at 25 °C.

B The melting point of caesium is 28.5 °C and its boiling point is 671 °C. Predict its state at 25 °C, 30 °C, and 700 °C.

Which bonds are broken?

To decide if a substance is likely to have a low or high melting point or boiling point, you need to know the type of bonds involved in state changes (Figure 2).

Table 1 *Examples of bonds involved in state changes.*

Type of substance	Bonds involved in state changes	Relative strength	Examples of substances
metal	metallic bonds	strong	iron, mercury
ionic compound	ionic bonds	strong	sodium chloride
giant covalent structure	covalent bonds	strong	diamond, silica
simple molecule	intermolecular forces	weak	oxygen, water, wax

These substances are usually in the solid state at room temperature:

- metals
- ionic compounds
- giant covalent substances.

Simple molecular substances are in the liquid or gas state at room temperature, or in the solid state but easily melted.

> **C** Suggest what type of bonds break when ice cream melts. Explain your answer.

Sublimation

Some substances can **sublime** – change directly from the solid to the gas state below their melting and boiling points. Iodine is like this. It exists as simple molecules attracted to each other by weak intermolecular forces. The molecules pack together in a regular way in the solid state (Figure 3), forming shiny black crystals. Iodine molecules easily separate from one another at room temperature, forming a purple vapour. The opposite of sublimation is called *deposition*.

> **D** 'Dry ice' is carbon dioxide in the solid state. Explain why it sublimes.

1 The melting point of methane is −182 °C and its boiling point is −161.5 °C. Predict its state at −183 °C and at −162 °C. *(2 marks)*

2 Covalent bonds exist between oxygen and hydrogen atoms in a water molecule, and intermolecular forces exist between water molecules. Explain which type of bonds break when water boils. *(2 marks)*

3 Explain why most of the substances that form air must be simple molecules. *(2 marks)*

4 Iron, sodium chloride, and silica have high boiling points but oxygen has a low boiling point. Explain, in terms of the bonds broken, why this is. *(6 marks)*

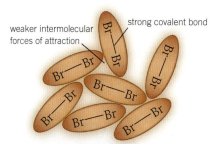

Figure 2 *Weak forces between molecules break when bromine boils, not the strong covalent bonds in its molecules.*

Synoptic link

You can learn about metallic bonds in C2.2.8 Structure of metals, ionic bonds in C2.2.4 Ionic compounds, and giant covalent structures in C2.2.6 Giant covalent structures.

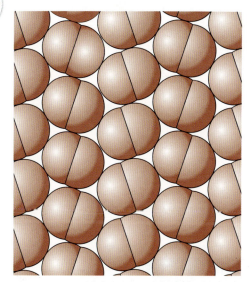

Figure 3 *A space-filling model of iodine molecules in the solid state.*

Go further

Find out how freeze-drying, which is used to produce instant coffee and dried food, works.

Learning outcome

After studying this lesson you should be able to:

- explain how the bulk properties of materials are related to the nature and arrangement of their bonds.

Specification reference: C2.3f

Figure 1 *Some materials are malleable but others, such as the glass on this phone, are brittle.*

Why do some materials shatter when you drop them, while others just bend or dent (Figure 1)? Properties like these are to do with the types of bond a substance contains, how strong these are, and how they are arranged.

Why are some substances brittle?

A brittle substance cracks or breaks when an external force is applied. On the other hand, a malleable substance can change shape without cracking or breaking. The difference depends on how easily the particles in the substance can change their positions in the **lattice** structure.

Metals are malleable even though metallic bonds are strong. Metal ions are held in a lattice by forces that attract them to a 'sea' of delocalised electrons. When a large enough external force is applied, layers of metal ions slide over one another, as shown in Figure 2. However, since the delocalised electrons are free to move, overall no bonds are broken.

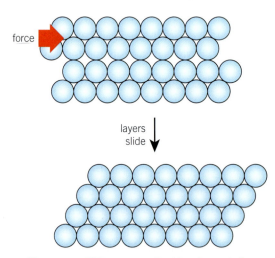

Figure 2 *Layers of ions can slide over each other in metals.*

Giant covalent structures contain very many atoms held together in a giant lattice by strong covalent bonds. If a large enough force is applied, many covalent bonds break at once and the substance breaks. A similar situation exists in ionic compounds, where oppositely charged ions are held together in a giant lattice by strong ionic bonds. Ionic compounds are usually brittle too.

A Explain why a diamond will shatter if hit with a hammer.

> ### Study tip
>
> Individual atoms do not have the physical properties of the materials they form. For example, a single copper atom does not conduct electricity and a single carbon atom is not brittle. Any property that results from the behaviour of many atoms acting together is called a bulk property.

Simple molecules, and polymer molecules, are attracted to each other by weak intermolecular forces. These forces are easily broken. If these substances are in the solid state and their molecules are arranged in a lattice, they may be brittle. Iodine crystals and some polymers are like this. If the molecules are not arranged in a lattice, the substance may be soft or flexible. Wax is like this, particularly when it is warm (Figure 3).

Figure 3 *The molecules in wax crayons have weak intermolecular forces and are not arranged in a lattice.*

B Poly(ethene) bags are soft and flexible. Suggest how their polymer molecules are arranged and the type of bonding between them.

Why do some substances conduct electricity?

A substance can conduct electricity if it has charged particles that are free to move. Metals conduct electricity in the solid and liquid states because their delocalised electrons are free to move through the lattice. Simple molecules, most polymers, and substances with a giant covalent structure do not conduct electricity because they have no delocalised electrons.

C Explain why graphite conducts electricity, even though it has a giant covalent structure.

Ionic compounds contain oppositely charged ions. These are free to move when an ionic compound is in the liquid state or dissolved in a solvent such as water, but not when the ionic compound is in the solid state. Ionic compounds conduct electricity when molten or in solution, but not when they are in the solid state.

1 Explain why sodium chloride crystals are brittle. *(2 marks)*
2 Metals in the gas state consist of individual atoms. Do metals in the gas state conduct electricity? Justify your answer. *(2 marks)*
3 Explain, in terms of properties and the reasons for them, the materials used for electricity cables (Figure 4). *(6 marks)*

Figure 4 *Electricity cables consist of copper wire surrounded by a polymer sheath.*

Go further

Metalloids have properties of both metals and non-metals. For example, they may conduct electricity better than most non-metals but not as well as metals.

Find out which elements are metalloids and where they are located in the Periodic Table.

Learning outcomes

After studying this lesson you should be able to:

● recall the relative size of nanoparticles

● describe and explain the properties of nanoparticles

● relate the properties of nanoparticles to their uses and possible risks.

Specification reference: C2.3g, C2.3h, C2.3i, C2.3j

Figure 1 *An electron microscope image of nanoparticles on cotton fibres.*

Maths link

You can learn more about units, standard form, and significant figures in *Maths for GCSE Chemistry: 2 Standard form* and *Maths for GCSE Chemistry: 5 Significant figures.*

Figure 2 *Scientists are researching ways for nanoparticles to deliver anti-cancer drugs directly to cancer cells.*

Would you like a T-shirt that never gets dirty or smelly? Thanks to nanoparticles, you might get one (Figure 1). Scientists have discovered how to treat fabric with these tiny particles so that it is cleaned by sunlight alone.

How big are nanoparticles?

A **nanoparticle** is a particle between 1 nm and 100 nm across, and consists of just a few hundred atoms. Table 1 shows the size of nanoparticles compared to the sizes of other small objects.

Table 1 *Relative size of nanoparticles.*

Object	Diameter (nm)
helium atom	0.062
methane molecule	0.41
nanoparticle	1–100
red blood cell	7000
human hair	100 000

A The diameter of a buckminsterfullerene molecule is 1.1 nm. Explain why it is a nanoparticle, but a methane molecule is not.

B A nanoparticle has a diameter of 31 nm. Calculate the number of these nanoparticles that would fit across a human hair, giving your answer to two significant figures.

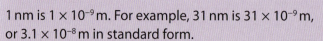

Nanometres

1 nm is 1×10^{-9} m. For example, 31 nm is 31×10^{-9} m, or 3.1×10^{-8} m in standard form.

C Show the diameter of a methane molecule in metres, giving your answer in standard form.

What gives nanoparticles their useful properties?

A material made from nanoparticles is described as **nanoparticulate**. Grains, lumps, and sheets are materials 'in bulk'. Nanoparticulate materials have different properties to the same substance in bulk. Some of these properties are due to the very small size of nanoparticles. For example, titanium dioxide is white in bulk but transparent when it is nanoparticulate. The small size of nanoparticles also makes them useful for:

● new paints

● new cosmetics, medicines (Figure 2), and sunscreens.

Many properties of nanoparticles are due to their very large surface area to volume ratio compared with the same substance in bulk (Figure 3). These properties are leading to exciting developments, including:

- new catalysts to speed up industrial chemical reactions
- self-cleaning windows, ovens, and clothes.

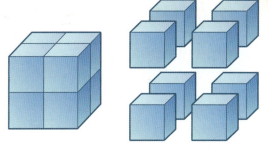

Figure 3 *The eight smaller cubes have the same total volume as the large cube, but twice the surface area.*

Calculating surface area to volume ratio

A cube-shaped nanoparticle has sides of length 100 nm. Calculate the surface area to volume ratio of this nanoparticle.

Step 1: Calculate its surface area (cubes have six equal faces).

$$area = 6 \times (length)^2$$

$$area = 6 \times (100\,nm)^2 = 60\,000\,nm^2 = 6 \times 10^4\,nm^2$$

Step 2: Calculate its volume.

$$volume = (length)^3$$

$$volume = (100\,nm)^3 = 1\,000\,000\,nm^3 = 1 \times 10^6\,nm^3$$

Step 3: Calculate its surface area to volume ratio.

$$ratio = \frac{surface\ area}{volume}$$

$$ratio = \frac{6 \times 10^4\,nm^2}{1 \times 10^6\,nm^3} = 6 \times 10^{-2}\,/nm$$

D Calculate the surface area to volume ratio for a cube-shaped nanoparticle with sides of 10 nm.

What are some possible risks of nanoparticles?

Nanoparticles are so tiny that they may be breathed in, absorbed by the skin, or pass into cells. Nanoparticles may take a long time to break down once released into the environment, and toxic substances may stick to their surfaces. Some scientists are concerned that nanoparticles may be harmful to health and the environment in ways that are difficult to predict, with risks that are difficult to determine.

Study tip

Surface area to volume ratio can also affect the properties of substances in bulk. For example, powders dissolve and react faster than lumps of the same material.

---

1 Explain why a nanoparticulate material has different properties to those of the same substance in bulk. *(2 marks)*

2 Silver has antibacterial properties. Silver nanoparticles are used in some sports socks. Suggest an advantage of these socks, and describe a possible hazard to the environment that they may cause. *(3 marks)*

3 Titanium dioxide absorbs harmful ultraviolet light from the Sun. Evaluate the use of nanoparticulate and bulk titanium dioxide in sunscreen. *(6 marks)*

Elements, compounds, and mixtures

C2.3 Properties of materials

Summary questions

1 Write down simple definitions for the following terms:
   a melting
   b boiling
   c condensing
   d freezing.

2 Table 1 shows the melting points and boiling points of six different substances.

Table 1 *Melting and boiling points.*

Substance	Melting point (°C)	Boiling point (°C)
A	−39	357
B	0	100
C	801	1413
D	−102	−34
E	29	2400
F	−130	35

   a Write down the letter of the substance that has:
   i the lowest melting point
   ii the highest boiling point.
   b Write down the letters of the substance or substances in the:
   i solid state at 25 °C
   ii liquid state at 25 °C
   iii gas state at 25 °C.

3 Diamond, graphite, and graphene are three forms of the element carbon. Copy and complete Table 2 to describe their structure and bonding.

Table 2 *Different forms of carbon.*

Substance	Number of covalent bonds between carbon atoms (1, 2, 3, or 4)	Layered structure (✓ or ✗)	Delocalised electrons (✓ or ✗)
diamond			
graphite			
graphene			

4 Explain why graphite conducts electricity but diamond does not, even though they are both forms of carbon.

5 Buckminsterfullerene is a form of carbon. It exists as football-shaped molecules with the formula C_{60}.

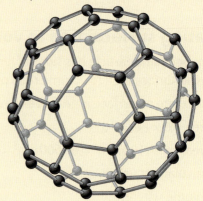

Figure 1 *Buckminsterfullerene.*

   a i Write down the type of bond present between each carbon atom in the molecule.
   ii Write down the number of bonds formed by each carbon atom in buckminsterfullerene.
   b Use your answers to a to explain whether or not a molecule of buckminsterfullerene contains delocalised electrons.
   c Buckminsterfullerene is soft, like graphite, and in the solid state at room temperature. Explain the type of bond or force that exists between buckminsterfullerene molecules.

6 Use the words in the box to name the type of bond broken or overcome when the following substances change from the liquid state to the gas state:

metallic   covalent   ionic   intermolecular

   a sodium chloride   b diamond
   c water   d sodium.

7 Nanodiamonds are nanoparticles 5 nm across. They are found naturally in meteorites and can also be manufactured. Nanodiamonds are absorbed by the skin, but the diamonds used in jewellery are not.
   a Write down the size of a nanodiamond in metres, m, in standard form (to two significant figures).
   b i Explain why nanodiamonds are absorbed by the skin but jewellery diamonds are not.
   ii Explain why large amounts of substances, such as those used in skin-moisturising products, can attach to nanodiamonds.
   iii Suggest a use of nanodiamonds, and explain your answer.

Revision questions

1 How many covalent bonds can a carbon atom form?

   A one

   B two

   C three

   D four   *(1 mark)*

2 Sodium chloride is an ionic compound.
Which row correctly describes its ability to conduct electricity?

	Sodium chloride when solid	Sodium chloride when liquid	Sodium chloride when dissolved in water
A	does not conduct	does not conduct	conducts
B	does not conduct	conducts	conducts
C	conducts	does not conduct	conducts
D	conducts	conducts	does not conduct

       (1 mark)

3 Nanoparticles are similar in size to simple molecules. Which of the following best describes the size of a nanoparticle? **S**

   A 9×10^{-3} m

   B 9×10^{-6} m

   C 9×10^{-9} m

   D 9×10^{-12} m   *(1 mark)*

4 Magnesium oxide is a compound formed by the reaction of magnesium with oxygen.
The following table shows the melting points of these three substances.

Substance	Melting point (°C)
magnesium oxide	2852
oxygen	−218
magnesium	650

In terms of the bonds it contains and the way in which these are arranged, explain why:

   a magnesium oxide, an ionic compound, has a very high melting point   *(3 marks)*

   b oxygen has a very low melting point   *(3 marks)*

   c magnesium has a high melting point.   *(3 marks)*

5 A 'self-cleaning window' keeps itself free of dirt. Its glass is coated with a thin layer of titanium dioxide nanoparticles. These act as a catalyst for reactions that break down dirt. **S**

   a Explain why titanium dioxide nanoparticles can act as a catalyst.   *(2 marks)*

   b Explain the possible risks associated with some nanoparticles.   *(2 marks)*

6* Diamond and graphite are two different forms of carbon. The table shows some of their properties.

Property	Diamond	Graphite
melting point	very high	very high
hardness	very hard	soft
electrical conductivity	does not conduct	good conductor

Explain these properties using your knowledge and understanding of the structure and bonding in diamond and graphite.   *(6 marks)*

7 This question is about three substances: copper, sodium chloride, and water. **H**

   a Copy and complete the table below to show some properties of the three substances.   *(2 marks)*

Substance	Melting point (°C)	Does it conduct electricity in the solid state?	Does it conduct electricity in the liquid state?	Is it brittle in the solid state?
copper	1085		Yes	No
sodium chloride	801	No		
water	0		No	Yes

   b* Use your knowledge and understanding of the structure and bonding of the three substances to explain the differences in properties shown in your table.   *(6 marks)*

C2.1 Purity and separating mixtures

- Define the terms *relative atomic mass, relative formula mass,* and *relative molecular mass.*
- Calculate relative formula masses from formulae and from balanced equations.
- Calculate the empirical formula of a compound.
- Explain what *purity* means, and that many useful materials are mixtures.
- Explain how to use melting point data to distinguish pure from impure substances.
- Describe and explain how different purification methods work, including filtration, crystallisation, simple distillation, and fractional distillation.
- Suggest suitable purification methods using information about the substances involved.
- Explain how different chromatography methods work.
- Calculate R_f values from chromatograms.
- Suggest suitable chromatography methods to distinguish pure from impure substances.

C2.2 Bonding

- Explain how to distinguish between metals and non-metals using their properties.
- Draw and write down the electronic structures of the first 20 elements.
- Explain how an element's position in the Periodic Table is linked to its electronic structure.
- Explain how ions and covalent bonds form.
- Draw dot-and-cross diagrams for ionic compounds and simple molecules.
- Describe the structure and bonding in ionic compounds, simple molecules, giant covalent structures, polymers, and metals.
- Describe the limitations of different models of ionic compounds, molecules, and metals.
- Explain how Mendeleev's arrangement of elements was refined into the modern Periodic Table.
- Explain patterns of chemical properties in the Periodic Table in terms of atomic structure.

C2.3 Properties of materials

- Explain why carbon forms different families of compounds.

- Explain the properties of the different forms of carbon.
- Use melting point data and boiling point data to predict the states of a substance at given temperatures.
- Explain how different substances change state.
- Explain how the bulk properties of materials are related to their structure and bonding.

- Recall the relative size of nanoparticles. **S**
- Describe and explain the properties of nanoparticles.
- Relate the properties of nanoparticles to their uses and possible hazards.

Formulae

- relative atomic mass (A_r)
- relative formula mass (M_r)
- relative molecular mass (M_r)
- empirical formula

Metals and non-metals

Metal elements	Non-metal elements
shiny	dull
solid at room temperature (mercury is liquid)	about half are solid, half are gas (bromine is liquid)
malleable	brittle
good conductors	insulators
form + ions	form – ions
metallic bonding	covalent bonding

Ionic bonding

- metal + non-metal
- strong electrostatic force of attraction between oppositely charged ions

Pure and impure

pure substance
- contains one element or compound
- single sharp melting point

impure substance
- melts over a range of temperatures

The Periodic Table

1869

elements arranged in order of ↑ relative atomic mass

proton discovered

elements arranged in order of ↑ atomic number

today

- elements with similar properties grouped
- some pairs reversed
- periods (rows of elements)
- groups (columns of elements)

Carbon allotropes

Diamond	Graphite
4 covalent bonds per C atom	3 covalent bonds per C atom
giant covalent lattice	layered structure with weak forces between layers
hard	soft and slippery
no delocalised electrons	delocalised electrons
does not conduct electricity	good conductor of electricity

Fullerenes

Nanoparticles

- 1 nm to 100 nm across (just a few hundred atoms)
- very large surface area to volume ratio

Separating mixtures

Filtration
- separates a solid from a liquid
- insoluble substance forms residue
- soluble substance goes through filter paper as filtrate

Crystallisation
- produces dry sample of soluble substance
- solution is heated
- solvent evaporates

Simple Distillation
- separates a solvent from a solution
- evaporation → cooling → condensing

Fractional Distillation
- separates a liquid from a mixture of liquids
- evaporation → cooling at different temperatures → condensing

Chromatography
- stationary phase
- mobile phase
- substances distributed between phases

$$R_f = \frac{\text{distance travelled by substance}}{\text{distance travelled by solvent}}$$

C2 Elements, compounds, and mixtures

Bonding, structure, and change of state

Metals	Ionic compounds	Giant covalent structures	Simple molecules
• strong	• strong	• strong	• weak
• high melting points and boiling points	• high melting points and boiling points	• high melting points and boiling points	• low melting points and boiling points
• conduct electricity when solid or liquid	• conduct electricity when molten or dissolved	• do not conduct electricity (except graphite)	• conduct electricity

Learning outcomes

After studying this lesson you should be able to:

- use names and chemical symbols to write the formulae of elements
- use names and chemical symbols to write the formulae of simple covalent compounds.

Specification reference: C3.1a, C3.1c

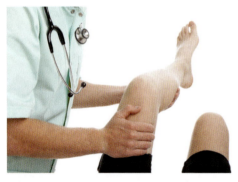

Figure 1 *Titin, $C_{169\,723}H_{270\,464}N_{45\,688}O_{52\,243}S_{912}$, is the third most abundant muscle protein. The human body contains about 0.5 kg of it.*

How long would it take for you to say the chemical name of titin, the largest known protein? In 2012, Dmitry Golubovskiy took over three hours to do this, since the name has 189 819 letters. He would have saved a lot of time if he had given its chemical formula instead (see Figure 1).

How do you write the formulae of metal elements?

The Periodic Table shows you the names of the elements, and the chemical symbols for their atoms. Each chemical symbol:

- has one, two, or three letters
- starts with a capital letter, with any other letters in lowercase.

For example, the chemical symbol for mercury is Hg. It is not HG, hg, or hG.

The formulae for metal elements are always written as empirical formulae because metals exist as giant metallic lattices. You do not need to include numbers in their formulae because these would be huge, and would vary depending on the amount of metal.

> **A** Write down the chemical formulae for sodium, magnesium, aluminium, iron, and copper.

How do you write the formulae of non-metal elements?

The non-metal elements in Group 0 (IUPAC Group 18) exist as individual atoms, attracted to each other by weak intermolecular forces. Their formulae are the same as their chemical symbols. For example, helium is He and not He_2.

The non-metal elements in Group 7 (IUPAC Group 17) exist as **diatomic molecules**, attracted to each other by weak intermolecular forces. A diatomic molecule contains two atoms covalently bonded together. This means that the formulae for Group 7 (IUPAC Group 17) elements all have a subscript 2 in them. For example, the formula for chlorine is Cl_2 (Figure 2). Hydrogen, nitrogen, and oxygen are not in Group 7 (IUPAC Group 17), but they also exist as diatomic molecules.

> **B** Write down the chemical formulae for oxygen, bromine, nitrogen, argon, and hydrogen.

Figure 2 *Chlorine gas exists as diatomic molecules, Cl_2.*

Other non-metal elements exist as giant covalent structures, such as carbon and silicon, or as simple molecules. For example, sulfur exists as S_8 molecules. In chemical equations, the formulae for all these elements are given as their empirical formulae. Note that phosphorus exists as P_4 molecules, shown as P_4 in chemical equations.

How do you write the formulae of simple covalent compounds?

The **molecular formula** for a simple covalent compound shows:

- the symbols for each element it contains
- the number of atoms of each element in one of its molecules.

For example, the molecular formula for carbon dioxide is CO_2. It shows that each carbon dioxide molecule contains one carbon atom and two oxygen atoms. Take care not to write CO^2 or Co_2 (Co is the chemical symbol for cobalt, a metal).

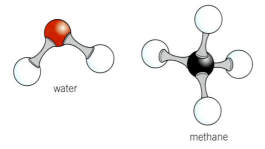

water

methane

Figure 3 *Ball and stick models of H_2O and CH_4.*

> **C** Use the displayed formula shown in Figure 4 to write down the molecular formula for pentane.

$$
\begin{array}{c}
\quad\ \ \text{H}\ \ \ \text{H}\ \ \ \text{H}\ \ \ \text{H}\ \ \ \text{H} \\
\quad\ \ |\ \ \ \ |\ \ \ \ |\ \ \ \ |\ \ \ \ | \\
\text{H}-\text{C}-\text{C}-\text{C}-\text{C}-\text{C}-\text{H} \\
\quad\ \ |\ \ \ \ |\ \ \ \ |\ \ \ \ |\ \ \ \ | \\
\quad\ \ \text{H}\ \ \ \text{H}\ \ \ \text{H}\ \ \ \text{H}\ \ \ \text{H}
\end{array}
$$

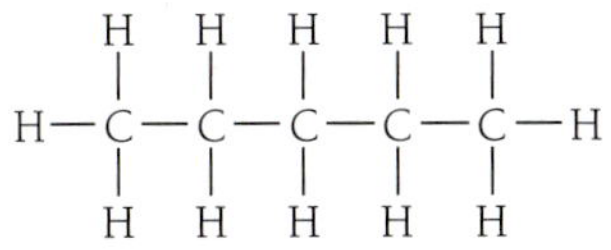

Figure 4 *The displayed formula of pentane.*

1 Explain why the formula for bromine in the liquid state is Br_2 and not Br. *(2 marks)*

2 Write down the formulae for iodine, neon, phosphorus, lead, and zinc. *(5 marks)*

3 Write down the molecular formulae for:
 a carbon monoxide, sulfur dioxide, and sulfur trioxide. *(3 marks)*
 b the simple molecules shown below. *(3 marks)*

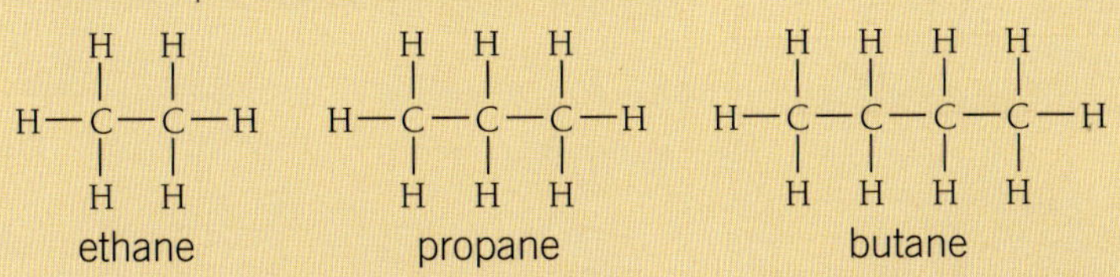

ethane   propane   butane

Learning outcome

After studying this lesson you should be able to:

- use the formulae of common ions to deduce the formulae of ionic compounds.

Specification reference: C3.1a, C3.1d

Figure 1 *Fireworks exploding in the night sky in London.*

Fireworks explode in a brilliant pyrotechnic display, lighting up the night sky (Figure 1). The gunpowder they use contains potassium nitrate. How can you work out that its chemical formula is KNO_3?

What are the formulae of common ions?

The Periodic Table does not show you the formulae of ions, but you can work many of them out:

1 Hydrogen ions have a +1 charge.

2 Metals in Groups 1, 2, and 3 (IUPAC Group 13) produce positive ions in which the number of charges is the same as their (non-IUPAC) group number.

3 Transition metals are in the block of elements between Groups 2 and 3 (they make up the IUPAC Groups 3 to 12). They produce positive ions that usually have a 2+ charge (silver and iron(III) are common exceptions).

4 Non-metals in Groups 5, 6, and 7 (IUPAC Groups 15, 16, and 17) produce negative ions in which the number of charges is equal to eight minus the (non-IUPAC) group number.

Compound ions are ions that contain more than one element.

Table 1 *The names and formulae of some common ions. It is useful to know the formulae of compound ions, which are shaded here.*

Positive ions	
ammonium	NH_4^+
hydrogen	H^+
lithium	Li^+
sodium	Na^+
potassium	K^+
silver	Ag^+
barium	Ba^{2+}
calcium	Ca^{2+}
copper(II)	Cu^{2+}
iron(II)	Fe^{2+}
lead(II)	Pb^{2+}
magnesium	Mg^{2+}
zinc	Zn^{2+}
aluminium	Al^{3+}
iron(III)	Fe^{3+}

Negative ions	
chloride	Cl^-
bromide	Br^-
iodide	I^-
hydroxide	OH^-
nitrate	NO_3^-
oxide	O^{2-}
carbonate	CO_3^{2-}
sulfate	SO_4^{2-}

A Suggest the meanings of the Roman numbers for 2 and 3 in the names, iron(II) and iron(III).

How do you write the formulae for ionic compounds?

In any ionic compound, the total number of positive charges must equal the total number of negative charges (Figure 2). This is easy to see when the ions involved have the same number of charges. For example, each unit of:

- sodium chloride, NaCl, contains one Na^+ ion and one Cl^- ion
- magnesium oxide, MgO, contains one Mg^{2+} ion and one O^{2-} ion.

It can be more difficult when the ions have different numbers of charges. For example, each unit of:

- magnesium chloride, $MgCl_2$, contains one Mg^{2+} ion and two Cl^- ions
- sodium oxide, Na_2O, contains two Na^+ ions and one O^{2-} ion
- aluminium oxide, Al_2O_3, contains two Al^{3+} ions and three O^{2-} ions.

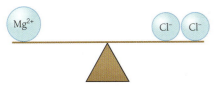

Figure 2 *You balance the number of charges, not the number of ions.*

Study tip

When you write the formula for an ionic compound, write the ions without their charges and put the positive ion first.

B Write down the formulae for iron(II) chloride and iron(III) oxide.

What do you do with compound ions?

You still need to balance the charges for ionic substances with compound ions. However, you must remember to write the compound ion in brackets if you need two or more in the formula. For example:

- sodium hydroxide is NaOH because each unit contains one Na^+ ion and one OH^- ion
- magnesium hydroxide is $Mg(OH)_2$ because each unit contains one Mg^{2+} ion and two OH^- ions.

1 a Write down the formulae for fluoride ions, caesium ions, and strontium ions. Use the Periodic Table to help you. *(3 marks)*
  b Write down the formulae for hydrogen fluoride, caesium bromide, and strontium oxide. *(3 marks)*

2 Write down the formulae for ammonium hydroxide, aluminium nitrate, and iron(III) sulfate. *(3 marks)*

3 The formula for dilute nitric acid is HNO_3 and the formula for dilute sulfuric acid is H_2SO_4. Name the ions they contain. *(3 marks)*

4 Explain, as fully as you can, what the formula $(NH_4)_2CO_3$ tells you. *(6 marks)*

C3.1.3 Conservation of mass

Learning outcomes

After studying this lesson you should be able to:

- recall and use the law of conservation of mass
- explain mass changes using the particle model.

Specification reference: C3.1i, C3.1j

Figure 1 *Firefighters working hard to put out a forest fire.*

Figure 3 *Measuring the mass before and after reacting silver nitrate solution with sodium chloride solution.*

A forest fire burns almost everything in its path (Figure 1). Trees disappear in flames and smoke, leaving just ashes behind. What has happened to their atoms – have they been destroyed by the fire?

What is the law of conservation of mass?

Atoms cannot be created or destroyed by chemical reactions. The same atoms are present at the start and end of a reaction (Figure 2). They are just joined in a different way, so the total mass stays the same during a chemical reaction. This is called the **law of conservation of mass**.

Figure 2 *No atoms are created or destroyed when carbon reacts with oxygen.*

You can investigate conservation of mass by carrying out a reaction in a **closed system**. This is a container in which no substances can enter or leave during the reaction. Examples include:

- a beaker of reactants in solution where the products are not gases
- a flask attached to a gas syringe to stop products in the gas state escaping.

Figure 3 shows the result of an investigation using silver nitrate solution and sodium chloride solution. They react to produce a white **precipitate** of silver chloride in silver nitrate solution.

> **A** Explain, in terms of particles, why the reading on the balance is the same in the two photographs in Figure 3.

Why does the mass seem to change during some reactions?

Substances can leave or enter the reaction mixture in a **non-enclosed system**. This usually happens in an open container when the reaction involves a substance in the gas state. For example, magnesium reacts with dilute hydrochloric acid to form magnesium chloride and hydrogen. The hydrogen escapes because it is in the gas state, so the mass appears to go down during the reaction (Figure 4).

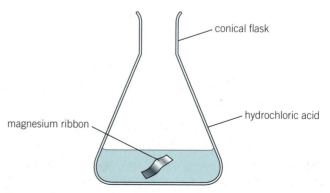

Figure 4 *A non-enclosed system.*

B Explain why a wooden log has a greater mass than the ashes formed after it has burnt.

The mass may appear to go up in some reactions, such as when a metal reacts with oxygen in the air. The metal atoms combine with oxygen atoms from the air to form a metal oxide in the solid state (Figure 5). If you could measure the mass of oxygen atoms used, you would find that the total mass before the reaction would still equal the total mass after.

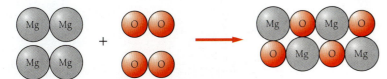

Figure 5 *No atoms are created or destroyed when magnesium reacts with oxygen. Here, four atoms of magnesium react with two molecules of oxygen to form some magnesium oxide.*

Calculating masses

2.40 g of magnesium is heated in air. It forms 4.00 g of magnesium oxide. Calculate the mass of oxygen gained.

Step 1: Write down what you know using the law of conservation of mass.

mass of magnesium + mass of oxygen = mass of magnesium oxide

Step 2: If necessary, rearrange the equation and substitute the masses given.

mass of oxygen = mass of magnesium oxide – mass of magnesium

mass of oxygen = 4.00 g – 2.40 g

= 1.60 g

C 4.40 g of carbon dioxide forms when 1.20 g of carbon reacts with oxygen. Calculate the mass of oxygen used.

1 Explain what is meant by the *law of conservation of mass*. (*2 marks*)

2 Copper carbonate decomposes when heated, forming copper oxide and carbon dioxide.

   a Explain why the mass appears to go down if this reaction is carried out in an open crucible. (*2 marks*)

   b 1.03 g of copper carbonate forms 0.660 g of copper oxide. Calculate the mass of carbon dioxide formed. (*2 marks*)

3 Explain, in terms of particles, why the mass of the solid increases when copper is heated in air. (*3 marks*)

Human joints should last a lifetime but some of us may eventually need replacements (Figure 1). Titanium is a strong, low-density metal that does not react with body tissues. It is extracted from its compounds using chemical reactions that can be modelled by equations.

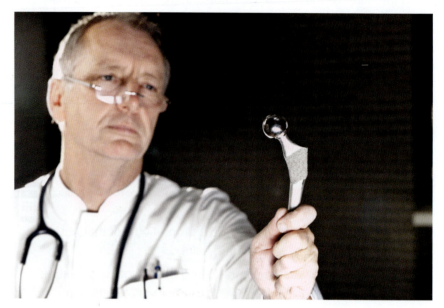

Figure 1 *Part of a titanium artificial hip joint.*

What are balanced equations?

A **word equation** is a simple model for a chemical reaction. It shows the names of the reactants and products involved. A **balanced equation** is a more detailed model. It shows:

- how the atoms are rearranged in a reaction
- the relative amounts of each substance involved.

For example, when carbon reacts with oxygen to produce carbon dioxide:

$$C + O_2 \rightarrow CO_2$$

The equation shows that one carbon atom reacts with one oxygen molecule to produce one carbon dioxide molecule. It models in symbols what happens to the atoms in the reaction (Figure 2).

Figure 2 *A particle model for the reaction: carbon + oxygen → carbon dioxide.*

A Sulfur reacts with oxygen to form sulfur dioxide: $S + O_2 \rightarrow SO_2$. Describe what the equation shows.

How do you write balanced equations?

An equation is **balanced** when there are equal numbers of atoms of each element on both sides of the arrow. For example:

- $Na + O_2 \rightarrow Na_2O$ is not balanced: the numbers of atoms on each side are not equal.
- $4Na + O_2 \rightarrow 2Na_2O$ is balanced: on each side as there are four atoms of Na and two atoms of O.

Balancing an equation

One stage in the manufacture of titanium involves reacting titanium(IV) chloride with magnesium, producing titanium and magnesium chloride. Write a balanced equation for the reaction.

Step 1: Write the word equation (if you are not given it) and write the formula for each substance underneath.

titanium(IV) chloride + magnesium $\rightarrow$ titanium + magnesium chloride

$$TiCl_4 + Mg \rightarrow Ti + MgCl_2$$

Step 2: Add numbers on the left of one or more formulae, if necessary, making equal numbers of atoms on each side.

$$TiCl_4 + 2Mg \rightarrow Ti + 2MgCl_2$$

B Explain fully why the answer to the example shown in the 'Balancing an equation' box is balanced.

State symbol	Meaning
(s)	solid
(l)	liquid
(g)	gas
(aq)	aqueous solution

What are state symbols?

State symbols show the physical state of each substance in a chemical reaction. For example, sodium metal reacts with water in the liquid state (Figure 3), producing sodium hydroxide dissolved in water, and hydrogen in the gas state:

$$2Na(s) + 2H_2O(l) \rightarrow 2NaOH(aq) + H_2(g)$$

C Write down the name of the solvent present when (aq) is used in an equation. Justify your answer.

1 Explain why balanced equations rely on the law of conservation of mass. *(4 marks)*

2 Balance these equations:
 a $Mg + O_2 \rightarrow MgO$ *(1 mark)*
 b $Fe + Cl_2 \rightarrow FeCl_3$ *(1 mark)*
 c $Ca + H_2O \rightarrow Ca(OH)_2 + H_2$ *(1 mark)*
 d $C_2H_5OH + O_2 \rightarrow CO_2 + H_2O$ *(1 mark)*

3 Compare the detail provided by these models: word equations, balanced equations without state symbols, and balanced equations with state symbols. *(6 marks)*

Figure 3 *Sodium reacts violently with water.*

C3.1.5 Half equations and ionic equations

Learning outcomes

After studying this lesson you should be able to:

- write half equations
- construct balanced ionic equations

Specification reference: C3.1b, C3.1e

Go further

Half equations are often used to model electrolysis reactions. Find out how copper is purified commercially by electrolysis, including the half equations for the changes involved.

Sodium reacts vigorously with chlorine to produce white clouds of sodium chloride (Figure 1). This reaction can be modelled by half equations, not just a balanced equation.

Figure 1 *The reaction between sodium and chlorine transfers energy to the surroundings as light and heat.*

What are half equations?

A **half equation** is a model for the change that happens to one reactant in a chemical reaction. The balanced symbol equation for the reaction between sodium and chlorine is:

$$2Na(s) + Cl_2(g) \rightarrow 2NaCl(s)$$

Sodium chloride, NaCl, is an ionic compound that contains Na^+ ions and Cl^- ions. Sodium ions form when sodium atoms lose electrons. This half equation models the change:

$$Na \rightarrow Na^+ + e^-$$

Chloride ions form when chlorine atoms gain electrons. However, chlorine atoms are joined together by covalent bonds to make diatomic chlorine molecules, so this is the half equation that models the change:

$$Cl_2 + 2e^- \rightarrow 2Cl^-$$

Notice that you must balance the charges in half equations, not just the numbers of atoms and the ions they form.

A Magnesium, Mg, reacts with oxygen, O_2, to produce magnesium oxide, MgO. Magnesium oxide contains Mg^{2+} ions and O^{2-} ions. Write two half equations to model the reaction.

What are ionic equations?

A complete **ionic equation** shows the ions present in a reaction mixture. It usually also includes the formulae of any molecular substances present, or substances in their solid state. For example, dilute hydrochloric acid reacts with sodium hydroxide solution to produce sodium chloride solution and water:

$$HCl(aq) + NaOH(aq) \rightarrow NaCl(aq) + H_2O(l)$$

This ionic equation also models the reaction:

$$H^+(aq) + Cl^-(aq) + Na^+(aq) + OH^-(aq) \rightarrow Na^+(aq) + Cl^-(aq) + H_2O(l)$$

Water is a molecular substance, so it is modelled as H_2O.

Notice that Cl^- ions and Na^+ ions appear in the same form on both sides of the equation. They are called **spectator ions** because they are in the reaction mixture but do not take part in the reaction. A net ionic equation leaves out the spectator ions. In this example, the ionic equation becomes:

$$H^+(aq) + OH^-(aq) \rightarrow H_2O(l)$$

This type of ionic equation is often used to model **precipitation** reactions. In these reactions, an insoluble product or **precipitate** forms when two solutions are mixed. For example, a red precipitate forms when silver nitrate solution and sodium chromate solution are mixed (Figure 2):

$$2AgNO_3(aq) + Na_2CrO_4(aq) \rightarrow Ag_2CrO_4(s) + 2NaNO_3(aq)$$

The red precipitate in Figure 2 is insoluble silver chromate, $Ag_2CrO_4(s)$, formed from silver ions and chromate ions. This ionic equation models the change:

$$2Ag^+(aq) + CrO_4^{2-}(aq) \rightarrow Ag_2CrO_4(s)$$

Notice that you need equal numbers of positive and negative charges to produce the neutral product.

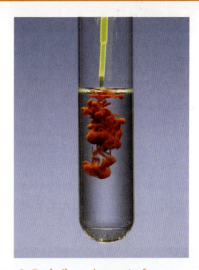

Figure 2 *Red silver chromate forms immediately when sodium chromate solution is added to silver nitrate solution.*

> **Study tip**
>
> A precipitate is always in the solid state and shown by (s) in equations.

B Barium chloride solution reacts with sodium sulfate solution to form a white precipitate of barium sulfate:

$$BaCl_2(aq) + Na_2SO_4(aq) \rightarrow BaSO_4(s) + 2NaCl(aq)$$

Write an ionic equation to show how barium sulfate forms from Ba^{2+} ions and SO_4^{2-} ions.

C Write down the formulae for the spectator ions in the reaction described in **B**.

---

1 Describe what half equations and ionic equations show. *(3 marks)*

2 Calcium reacts with bromine, Br_2, to form calcium bromide, $CaBr_2$. Write two half equations to model this reaction. *(2 marks)*

3 Lead nitrate solution, $Pb(NO_3)_2(aq)$, reacts with sodium iodide solution to form a precipitate of lead iodide, $PbI_2(s)$, and sodium nitrate solution.
  a Write a balanced equation for this reaction. Include state symbols. *(3 marks)*
  b Write an ionic equation to model the formation of lead iodide. Include state symbols. *(3 marks)*

Learning outcomes

After studying this lesson you should be able to:

- explain the meanings of the terms 'mole' and 'Avogadro constant'
- calculate the mass of a mole of a substance.

Specification reference: C3.1g, C3.1h

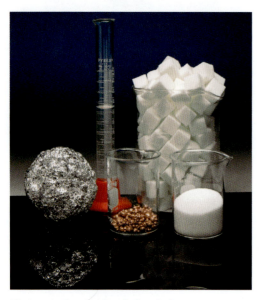

Figure 1 *One mole of aluminium, water, copper, table sugar (sucrose), and sodium chloride.*

Go further

The mole is an SI (Système International d'Unités) base unit. Six of the seven base units are described in *Maths for Chemistry GCSE: 11.1 International system of units*.

Find out what the missing seventh base unit is.

What does 18 g of water molecules have in common with 342 g of table sugar (sucrose)? The answer is that they both contain one mole of molecules (Figure 1).

What is a mole?

The **mole** is the unit for amount of substance. One mole of anything contains the same number of things. For example, a mole of sheets of paper is the same number of sheets as the number of carbon atoms in a mole of carbon atoms. This number is huge. If you could stack a mole of sheets of paper on top of each other, they would reach a quarter of the way to the centre of our galaxy (Figure 2). It makes little sense to use the mole in everyday life, but it is useful for tiny particles like atoms, molecules, ions, or electrons.

Figure 2 *The centre of our galaxy is about 26 000 light years away.*

The mole is defined as the amount of any substance that contains the same number of entities as there are atoms in 12.0 g of carbon-12, $^{12}_{6}C$, atoms. The word 'entities' means the particles that make up a substance, such as its atoms, ions, or molecules.

The word 'mole' is abbreviated to 'mol' in calculations. For example, 1 mol of water contains 1 mol of water molecules. It also contains 3 mol of atoms, because each H_2O molecule contains three atoms (two hydrogen atoms and one oxygen atom).

> **A** Write down the amount, in mol, of oxygen atoms in 2 mol of O_2 molecules.

How many entities are there in one mole?

The number of entities in 1 mol is called the **Avogadro constant**. It is 6.02×10^{23}/mol (to three significant figures). No one could sit down and successfully count out 1 mol of something. Even if you could count a million atoms per second, it would take you longer than the age of the Universe to complete your task. Luckily, there is a simpler way to measure 1 mol of a substance, as you are about to find out.

> **B** Calculate the number of aluminium atoms in 2.54 mol of aluminium. Express your answer in standard form to three significant figures.

How can you measure a mole of a substance?

The mass of 1 mol of a substance is its relative atomic mass, or relative formula mass, in grams. This is also its **molar mass**, measured in grams per mole (g/mol). For example:

- the A_r of carbon is 12.0, so its molar mass is 12.0 g/mol
- the M_r of carbon dioxide is 44.0, so its molar mass is 44.0 g/mol.

You do not need to count atoms to obtain a mole of a substance – the A_r or M_r in grams gives you 1 mol of the substance.

> **C** Explain why Figure 3 shows 1 mol of carbon atoms.

Figure 3 *One mole of carbon atoms.*

> **1** ✏ Define the terms 'mole' and 'Avogadro constant'. *(4 marks)*

> **2** Calculate the number of atoms in the following, giving your answer in standard form to three significant figures:
> **a** 12.0 mol of copper, Cu *(2 marks)*
> **b** 1.00 mol of sulfur dioxide molecules, SO_2 *(2 marks)*
> **c** 0.500 mol of table sugar molecules, $C_{12}H_{22}O_{11}$. *(2 marks)*

> **3** 🖩 Calculate the molar mass of each substance in Figure 1. *(5 marks)*

Learning outcomes

After studying this lesson you should be able to:

- use a balanced equation to calculate masses of reactants or products
- explain the effect of a limiting amount of a reactant
- calculate the stoichiometry of an equation.

Specification reference: C3.1l, C3.1k

Figure 1 *Cement is produced in factories like this one.*

Enough cement is made every year to give everyone in the world over half a tonne on their birthday (Figure 1). Cement factories use mole calculations to work out the mass of raw materials needed.

How do you calculate masses of reactants and products?

The mass of a substance is related to its molar mass and amount by this expression:

$$\text{mass (g)} = \text{molar mass (g/mol)} \times \text{amount (mol)}$$

A   Write down an expression to calculate the molar mass of a substance from its mass and amount.

Using moles in calculations

Nitrogen reacts with hydrogen to produce ammonia:

$$N_2(g) + 3H_2(g) \rightarrow 2NH_3(g)$$

Calculate the mass of ammonia made from 84.0 g of nitrogen.

Step 1: Calculate the molar masses using A_r values.

molar mass of N_2 = 2 × 14.0 = 28.0 g/mol

molar mass of NH_3 = 14.0 + (3 × 1.0) = 17.0 g/mol

Step 2: Calculate the amount of N_2 in 84 g by rearranging:

mass = molar mass × amount

$$\text{amount} = \frac{\text{mass}}{\text{molar mass}} = \frac{84.0\,\text{g}}{28.0\,\text{g/mol}}$$

amount = 3.0 mol

Step 3: Calculate the amount of NH_3 made from this amount of N_2.

From the balanced equation:

1 mol of N_2 makes 2 mol of NH_3

3.0 mol of N_2 makes (3.0 × 2 mol) = 6.0 mol of NH_3

Step 4: Calculate the mass of NH_3 in this amount.

mass = molar mass × amount

mass = 17.0 g/mol × 6.0 mol = 102 g

What is a limiting reactant?

One reactant is usually in **excess** in a reaction mixture. More of it is present than is needed to react with the other reactant, so some is left at the end. The other reactant is in a **limiting** amount. The amount of product formed is determined by the amount of the limiting reactant, not by the one in excess.

B   Calculate the mass of hydrogen needed, in reaction with excess nitrogen, to produce 6.8 g of ammonia.

How do you calculate the stoichiometry of an equation?

You balance a symbol equation by putting numbers to the left of the formulae as needed. The **stoichiometry** describes the relative amounts of each substance involved, and is to do with these balancing numbers. You can use mole calculations to work out the stoichiometry of an equation, in effect balancing it.

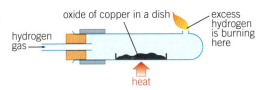

Figure 2 *Making copper from an oxide of copper.*

Calculating stoichiometry

An oxide of copper is heated with excess hydrogen, H_2, forming 6.40 g of copper and 0.900 g of water (Figure 2). Calculate the balanced equation. Molar masses: $Cu = 63.5$ g/mol, $H_2O = 18.0$ g/mol.

Step 1: Calculate the amount of each measured substance.

$$\text{amount of } Cu = \frac{\text{mass}}{\text{molar mass}} = \frac{6.40\,g}{63.5\,g/mol} = 0.10\,mol$$

$$\text{amount of } H_2O = \frac{0.90\,g}{18.0\,g/mol} = 0.05\,mol$$

Step 2: Simplify the ratio of these substances.

$$Cu : H_2O = 0.10 : 0.05$$
$$= 2 : 1$$

The right-hand side of the equation must be: $2Cu + H_2O$

Hydrogen is H_2, so the equation must be:

$$Cu_2O + H_2 \rightarrow 2Cu + H_2O$$

C 4.8 g of magnesium is heated with excess oxygen, O_2, forming 8.0 g of magnesium oxide, MgO. Calculate the equation for the reaction.

1 A cement factory wants to produce 28 tonnes of calcium oxide from calcium carbonate:

$$CaCO_3(s) \rightarrow CaO(s) + CO_2(g)$$

Calculate the mass of calcium carbonate needed. *(4 marks)*

2 1.8 g of carbon reacts with excess steam to produce 6.6 g of carbon dioxide and 0.60 g of hydrogen. Determine the equation for the reaction. *(6 marks)*

3 Hydrogen reacts with oxygen to form water:

$$2H_2 + O_2 \rightarrow 2H_2O$$

A reaction mixture containing 1.0 g of hydrogen and 4.0 g of oxygen produces 4.5 g of water.

  a Calculate the amount, in mol, of each substance. *(3 marks)*

  b Use your answer to part **a** to explain which reactant is in excess, and which is limiting. *(3 marks)*

Chemical reactions

C3.1 Introducing chemical reactions

Summary questions

1 Copy and complete Table 1.

Table 1 *Number of atoms in a molecule.*

Name of substance	Chemical formula	Number of atoms of each element in a molecule
fluorine	F_2	2 fluorine atoms
ammonia		1 nitrogen atom and 3 hydrogen atoms
		1 carbon atom and 2 oxygen atoms
hexane	C_6H_{14}	
water		
	SO_2	

2 Write down the chemical formula of each of the ionic compounds below. Use Table 1 from C3.1.2 *Formulae of ionic compounds* to help you.
   a hydrogen bromide
   b magnesium chloride
   c copper(II) carbonate
   d zinc hydroxide
   e ammonium sulfide
   f iron(III) nitrate.

3 a Explain the law of conservation of mass in terms of particles.
   b Nitrogen monoxide, NO, reacts with oxygen to form a toxic brown gas, nitrogen dioxide, NO_2:
   $$2NO + O_2 \rightarrow 2NO_2$$
   9.6 g of nitrogen dioxide forms from 3.2 g of oxygen. Calculate the mass of nitrogen monoxide that must have reacted.

4 Copy these equations and balance them:
   a $N_2 + O_2 \rightarrow NO$
   b $Na + Cl_2 \rightarrow NaCl$
   c $C_2H_2 + H_2 \rightarrow C_2H_6$
   d $Fe + Cl_2 \rightarrow FeCl_3$
   e $CO + NO \rightarrow CO_2 + N_2$

5 Aluminium powder burns in oxygen gas to form solid aluminium oxide:
   a Write down the word equation to model this reaction.
   b Write down the formulae of oxygen gas and aluminium oxide.
   c Write down a balanced equation for the reaction. Include state symbols in your answer.

   d In the reaction, aluminium atoms become aluminium ions, and oxygen gas becomes oxide ions. Write balanced half equations to model these changes. **H**

6 Calculate the relative formula mass, M_r, of:
   a oxygen, O_2
   b calcium oxide, CaO
   c hydrochloric acid, HCl
   d zinc chloride, $ZnCl_2$.

7 a Calculate, in mol: **H**
   i the amount of atoms in 2 mol of sulfur trioxide molecules, SO_3
   ii the amount of molecules in 2.5 mol of benzene molecules, C_6H_6.
   b The Avogadro constant has a value of 6.02×10^{23} /mol. Calculate:
   i the number of molecules in 0.500 mol of water molecules, H_2O
   ii the number of atoms in 1.00 mol of methane, CH_4.

8 Calcium burns in oxygen to form calcium oxide:
   $$2Ca(s) + O_2(g) \rightarrow 2CaO(s)$$
   a Write down the molar masses, in g/mol, of calcium, oxygen, and calcium oxide. Your answers to **6** will help you.
   b i Calculate the amount, in mol, of calcium in 160 g of calcium.
   ii Use the balanced equation to help you calculate the amount, in mol, of calcium oxide that could be made from this amount of calcium.
   c Use your answers to parts **a** and **bii** to calculate the mass of calcium oxide that could be made from 160 g of calcium.
   d Calculate the mass of oxygen needed to react completely with 160 g of calcium.

Revision questions

1 The formula of an aluminium ion is Al^{3+} and the formula of a carbonate ion is CO_3^{2-}.
What is the formula of aluminium carbonate?
A Al_2CO_3
B $Al_3(CO_3)_2$
C $Al_2(CO_3)_3$
D $(Al)_2(CO_3)_3$ (1 mark)

2 Calcium reacts with chlorine to make calcium chloride. Which of the following is the balanced symbol equation for this reaction?
A $Ca + Cl_2 \rightarrow CaCl_2$
B $2Ca + Cl_2 \rightarrow 2CaCl$
C $Ca + Cl \rightarrow CaCl$
D $Ca_2 + Cl_2 \rightarrow 2CaCl_2$ (1 mark)

3 Lithium metal reacts with water. Lithium hydroxide solution and hydrogen gas are made.
a Balance this equation, and complete it by adding state symbols.
$$Li + H_2O \rightarrow LiOH + H_2$$ (3 marks)

b A teacher measured the volume of hydrogen produced by different masses of lithium added to 1 dm^3 of water.
The following table shows her results.

Mass of lithium (g)	Volume of hydrogen (cm^3)
0.05	85
0.10	170
0.20	340

Explain these results. (2 marks)

4 Iron reacts with copper(II) sulfate solution. Iron(II) sulfate and copper are made.
a In the reaction, copper(II) ions are changed into copper atoms.
Balance this half equation for the reaction.
$$Cu^{2+} + e^- \rightarrow Cu$$ (1 mark)
b In the reaction, iron atoms are changed into iron(II) ions.
Write down a balanced ionic equation for this reaction. (2 marks)

5 Aluminium powder burns in air to produce aluminium oxide:
$$4Al + 3O_2 \rightarrow 2Al_2O_3$$

a Calculate the mass of one mole of oxygen, O_2.
(The molar mass of oxygen, O_2, is 32.0 g/mol) (1 mark)
b Use your answer to a to calculate the mass of one molecule of oxygen.
Give your answer to 3 significant figures.
(The Avogadro constant is 6.02×10^{23}/mol) (2 marks)
c Calculate the mass of aluminium needed to make 2.4 g of aluminium oxide.
(The relative atomic mass of Al is 27.0 and of O is 16.0) (3 marks)

6* A student carries out an investigation to see what happens to solid elements when they react with oxygen gas in the air. This is the method he uses:
i Put a piece of magnesium in a crucible.
ii Record the mass of the crucible and its contents.
iii Heat the magnesium in the open crucible to produce magnesium oxide powder.
iv Repeat step ii.
The student then uses the same method to make carbon dioxide gas from carbon. The table shows his results.

Element used	Mass of crucible and contents at start (g)	Mass of crucible and contents at end (g)
magnesium	20.24	20.40
carbon	20.24	20.10

Explain the observed changes in mass using the particle model. In your answer, include balanced symbol equations with state symbols. (6 marks)

C3.2 Energetics

C3.2.1 Exothermic and endothermic reactions

Learning outcomes

After studying this lesson you should be able to:

- identify exothermic and endothermic reactions
- compare exothermic and endothermic reactions.

Specification reference: C3.2a

Figure 1 *Explosions are exothermic reactions.*

Figure 2 *This first aider is treating a sports injury. The pack gets cold when urea crystals inside it mix with water, causing an endothermic change.*

'Exo' and 'endo'

Many scientific words begin with 'exo' or 'endo'. These come from the Greek language: *exo* means 'outside' and *endo* means 'inside'.

Explosives are useful if you want to demolish an old building. When the explosives detonate, chemical reactions produce rapidly expanding hot gases that push the building apart (Figure 1). Explosions are exothermic reactions in which huge amounts of energy are transferred to the surroundings.

What are exothermic and endothermic reactions?

If the temperature of the reaction mixture increases during a chemical reaction, the reaction is an **exothermic** one. Examples of exothermic reactions include:

- **combustion**, for example a fuel burning to cook food
- **neutralisation**, for example sodium hydroxide solution reacting with dilute hydrochloric acid.

If the temperature of the reaction mixture decreases during a chemical reaction, the reaction is an **endothermic** one (Figure 2). Examples of endothermic reactions include:

- the reaction between citric acid and sodium hydrogencarbonate in sherbet sweets, which is why they feel cold on your tongue
- photosynthesis in plants, which involves carbon dioxide and water reacting to form sugars by using energy in light.

> A Describe one way to identify exothermic reactions.
>
> B Write down an example of an exothermic reaction and of an endothermic reaction.

Study tip

Exothermic and endothermic

If you confuse exothermic with endothermic reactions, think of a fire exit sign. You go out of an *exit* and fires are *hot* – think of "*ex*othermic reactions get *hot*".

Investigating temperature changes

To find out if a reaction involving a solution is exothermic or endothermic:

1   Measure the start temperature of a solution in an insulated container (Figure 3).

2   Add the other **reactant** (reacting substance) and stir.

3   Measure the end temperature of the mixture.

4   Calculate the difference between the start temperature and the end temperature.

If both reactants are solutions, you may need to check that their temperatures are the same before mixing them.

You should wear eye protection when carrying out this practical.

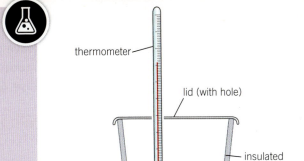

Figure 3 *Investigating temperature changes when magnesium reacts with acid.*

thermometer

lid (with hole)

insulated container

magnesium powder

dilute hydrochloric acid

C Explain why you could use a polystyrene cup, as in Figure 3, in the sort of experiment described in *Investigating temperature changes*.

---

1   Explain, in terms of energy transfers, why a lid is often used with the insulated container when investigating temperature changes (Figure 3).   *(2 marks)*

2   Some endothermic reactions only happen as long as they are heated. Suggest why it is difficult to determine if they are endothermic or exothermic by measuring temperature changes.   *(2 marks)*

3   Table 1 shows the results of three experiments involving solutions mixed in a polystyrene cup.

Table 1 *Experimental results.*

Experiment	Start temperature (°C)	End temperature (°C)
1	20.8	16.3
2	20.8	34.5
3	20.8	20.8

a Identify the exothermic and endothermic reactions, and explain your answer.   *(2 marks)*

b Suggest a reason for the result in Experiment 3.   *(1 mark)*

c Suggest why the temperature eventually returns to 20.8 °C in Experiments 1 and 2.   *(4 marks)*

Go further

Some endothermic reactions do not involve a decrease in temperature, but energy must be transferred continuously to the reaction mixture to keep them going. Decomposition reactions, in which one substance breaks down to form two or more products, are like these:

- Electrolysis is a type of decomposition reaction that only happens while an electric current is passed through the reaction mixture.

- Thermal decomposition is a type of decomposition reaction that only happens while the reaction mixture is being heated.

Find out how aluminium is produced from aluminium oxide, and why this is an expensive process.

C3.2.2 Reaction profiles

Learning outcomes

After studying this lesson you should be able to:

- draw and label reaction profiles for exothermic and endothermic reactions
- explain the meaning of 'activation energy'.

Specification reference: C3.2b, C3.2c

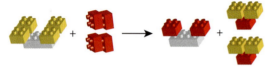

Figure 1 *A Lego® model of methane and oxygen reacting to produce carbon dioxide and water.*

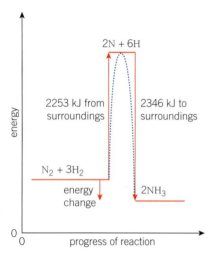

Figure 2 *A reaction profile for the reaction between nitrogen and hydrogen to form ammonia. The amount of energy stored in the reactants and products is represented as horizontal lines.*

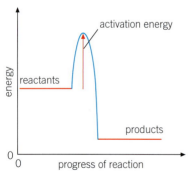

Figure 3 *The activation energy is shown as the difference between the energy of the reactants and the top of the peak.*

Have you ever played with Lego® bricks (Figure 1)? You can make a model, break it up, and then use the same bricks to make something else. Chemical reactions work in a similar way.

What are reaction profiles?

In a chemical reaction, bonds break in the reactants to produce separate atoms, and then new bonds form between the atoms to make the **products**. The **surroundings** include everything except the reacting particles themselves. During a chemical reaction, energy (usually measured in kilojoules, kJ) is transferred:

- from the surroundings to break bonds in the reactants to form separate atoms
- to the surroundings from the reacting particles when bonds form between atoms.

A **reaction profile** is a chart that shows the energy involved. For example, in the reaction between nitrogen and hydrogen to make ammonia (Figure 2):

- 2253 kJ is transferred to break bonds in the reactants
- 2346 kJ is transferred when new bonds form to make products.

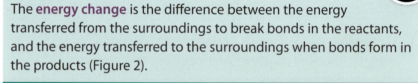

Energy change

The **energy change** is the difference between the energy transferred from the surroundings to break bonds in the reactants, and the energy transferred to the surroundings when bonds form in the products (Figure 2).

A Use data from Figure 2 to show that the energy change is 93 kJ when nitrogen reacts with hydrogen.

B Explain, in terms of energy transfers, what happens when bonds are made or broken.

What is activation energy?

Energy must be transferred to the reactant particles to start most chemical reactions. **Activation energy** is the minimum energy needed for a reaction to start (Figure 3). It is often provided by a flame or by heating, and it breaks bonds in the reactants.

All reactions have an activation energy. However, it can be so low for some reactions that the reactants already have enough energy to react and just need to be mixed together. Neutralisation reactions are like this. The acid and alkali react as soon as you mix them, without heating.

C Explain why a mixture of nitrogen and hydrogen needs heating for it to react.

How do you draw reaction profiles?

In an endothermic reaction, the amount of energy transferred to break bonds is more than the energy transferred when new bonds form. This means that the energy change is positive. Overall, energy is transferred from the surroundings, which cool down. This is shown by an upward arrow on the energy profile (Figure 4). It is the opposite situation for an exothermic reaction (Figure 5).

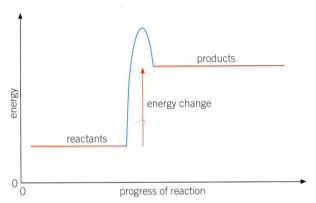

Figure 4 *In endothermic reactions, more chemical energy is stored in the products than in the reactants.*

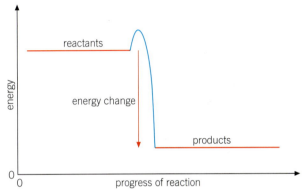

Figure 5 *In an exothermic reaction, less energy is transferred to break bonds than is transferred when new bonds form, so the energy change is negative and energy is transferred to the surroundings.*

Drawing a reaction profile

When you draw a reaction profile:

1 Draw both axes.

2 Draw two horizontal lines, representing the chemical energy stores in the reactants and products. Remember to draw the reactant line lower than the product line for an exothermic reaction, and higher than the product line for an endothermic reaction.

3 Draw a curve to represent the activation energy, and add labels to your diagram.

---

1 Explain why the energy change is negative in exothermic reactions, and overall energy is transferred to the surroundings. (3 marks)

2 Write down the meaning of 'activation energy'. (2 marks)

3 Sketch reaction profiles for these reactions. Include the formulae of the reactants and products. Label arrows for the activation energy and energy change.

   a The combustion of methane, an exothermic reaction:

   $$CH_4 + 2O_2 \rightarrow CO_2 + 2H_2O$$ (4 marks)

   b The thermal decomposition of calcium carbonate, an endothermic reaction:

   $$CaCO_3 \rightarrow CaO + CO_2$$ (4 marks)

Learning outcomes

After studying this lesson you should be able to:

- define 'bond energy'
- calculate energy changes in chemical reactions using bond energy values.

Specification reference: C3.2d

Figure 1 *Cutting steel pipe with the 3500 °C flame from an oxyacetylene torch.*

Study tip

You do not need to learn the values for bond energies.

Oxygen and ethyne, C_2H_2, react together in oxyacetylene torches to produce flames hot enough to cut steel (Figure 1). Using bond energies you can calculate the energy changes in reactions.

What are bond energies?

Energy must be transferred to molecules to break their covalent bonds. The energy needed to break 1 mol of a particular bond is the **bond energy**, and it is measured in kilojoules per mole (kJ/mol). Different bonds have different bond energies. For example, 347 kJ of energy must be transferred to break 1 mol of C–C bonds. This is also the amount of energy transferred to the surroundings when 1 mol of these bonds form.

Table 1 *Bond energies are measured in kilojoules per mole (kJ/mol).*

Bond	Bond energy (kJ / mol)
C–C	347
N–H	391
C–H	413
H–H	436
O–H	464
O=O	498
C=C	612
C=O	805
N≡N	945

A Compare the bond energies of double and triple bonds with those of single bonds. What do you notice?

B Calculate the energy transferred when 2 mol of C–C bonds form.

How do you calculate energy changes in reactions?

You can use bond energies to calculate energy changes in reactions. You need to consider:

- the energy transferred to break the bonds in the reactants, and
- the energy transferred when the bonds form in the products.

Mean bond energies

Bond energies are usually mean values for the same bonds in different substances. This means that the calculated energy change for a reaction may differ from the value found by experiment.

Combustion of ethene

Calculate the energy change when ethene burns completely in oxygen:

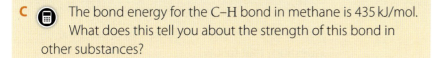

Step 1: Write down the number of bonds of each type in the reactants:

$(1 \times [C=C]) + (4 \times [C—H]) + (3 \times [O=O])$

Step 2: Calculate the total energy transferred to break all these bonds:

energy transferred $= (1 \times 612\,kJ/mol) + (4 \times 413\,kJ/mol) + (3 \times 498\,kJ/mol)$

$= 612 + 1652 + 1494\,kJ/mol$

$= 3758\,kJ/mol$

Step 3: Write down the number of bonds of each type in the products. Make sure you count all the bonds. For example, two O=C=O molecules contain four C=O bonds:

$(4 \times [C=O]) + (4 \times [O—H])$

Step 4: Calculate the total energy transferred when all these bonds are made:

energy transferred $= (4 \times 805\,kJ/mol) + (4 \times 464\,kJ/mol)$

$= 3220 + 1856\,kJ/mol$

$= 5076\,kJ/mol$

Step 5: Calculate the energy change:

energy change = (energy transferred to break bonds) − (energy transferred when bonds made)

$= 3758\,kJ/mol − 5076\,kJ/mol$

$= −1318\,kJ/mol$

The negative sign in this answer shows that the reaction is exothermic.

C The bond energy for the C–H bond in methane is 435 kJ/mol. What does this tell you about the strength of this bond in other substances?

1 Explain what a positive sign for an energy change tells you. *(1 mark)*

2 Water decomposes when electricity is passed through it:

$$2[H—O—H] \rightarrow 2[H—H] + O=O$$

   a Calculate the energy transferred:

     i to break all the bonds in the water molecules   *(1 mark)*

     ii when all the bonds form in the products.   *(1 mark)*

   b Use your answers to **a** to calculate the energy change for the reaction.   *(1 mark)*

   c State whether the reaction is exothermic or endothermic, and explain your answer.   *(2 marks)*

3 Nitrogen reacts with hydrogen to produce ammonia, NH_3:

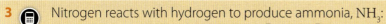

Calculate the energy change for the reaction. Explain what your answer shows.   *(5 marks)*

Summary questions

1 Write down simple definitions of the following terms.
   a exothermic
   b endothermic
   c activation energy.

2 During chemical reactions, chemical bonds are broken and formed. Copy each of these sentences, including only the correct word in each pair of words shown in bold.
   a In an **exothermic | endothermic** reaction, the temperature of the surroundings goes up.
   b Energy must be transferred **to | from** the surroundings **from | to** the reactant particles to break bonds.
   c In an **exothermic | endothermic** reaction, more energy is transferred from the surroundings than is transferred to the surroundings.

3 Two powders are soluble in water.
   a Describe a simple experiment you could carry out to see whether an exothermic change or an endothermic change happens when each powder dissolves in water. Include any essential apparatus, methods, and observations you would need.
   b One of these powders produces an alkaline solution when it dissolves. Explain two precautions you could take to reduce the risk of harm from this hazard.

4 Figure 1 is a reaction profile for a reaction.

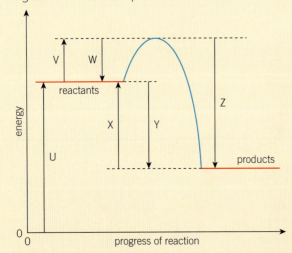

Figure 1 *Reaction profile.*

Write down which of the letters U–Z represent:
   a the energy transferred to break the bonds in the reactants

   b the energy transferred when bonds form in the products
   c the activation energy for the reaction
   d the overall energy change in the reaction.

5 Draw a reaction profile to model the changes in energy stored in the particles during an endothermic reaction.

6 Ethanol, C_2H_5OH, is used as a fuel. It burns in a plentiful supply of oxygen to produce carbon dioxide and water.
   a Write down a balanced equation for the reaction.
   b The reaction is exothermic. Draw a reaction profile to model this reaction.

7 Explain the term *bond energy*. **H**

8 Methane burns in a plentiful supply of oxygen to produce carbon dioxide and water vapour. The reaction can be modelled using displayed formulae as shown below.

$$H\!-\!\overset{\displaystyle H}{\underset{\displaystyle H}{C}}\!-\!H + 2[O\!=\!O] \longrightarrow O\!=\!C\!=\!O + 2[H\!-\!O\!-\!H]$$

Use the bond energies in Table 1 to answer the questions that follow.

Table 1 *Bond energies.*

Bond	Bond energy (kJ/mol)
C–H	413
O–H	464
O=O	498
C=O	805

   a Calculate the total energy transferred to break the bonds in the reactants.
   b Calculate the total energy transferred when the bonds form in the products.
   c Use your answers to calculate the overall energy transferred in the reaction.
   d Use your answer to c to explain whether the reaction is exothermic or endothermic.

Revision questions

1 The diagram shows the reaction profile for a chemical reaction.

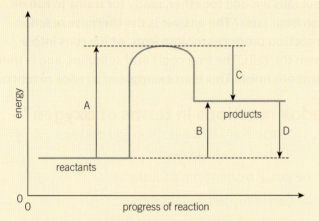

a Which arrow (**A**, **B**, **C**, or **D**) shows the energy change in the reaction? (*1 mark*)

b Which arrow (**A**, **B**, **C**, or **D**) shows the activation energy for the reaction? (*1 mark*)

2 Describe what is meant by the term *activation energy*. (*1 mark*)

3 A student adds three different powdered solids to beakers containing dilute hydrochloric acid. She measures the temperature of the acid before and after adding the powders.
The following table shows her results.

Powder	Temperature at start (°C)	Temperature at end (°C)	Temperature change (°C)
A	18.1	24.2	
B	17.8	14.2	
C	18.2	18.2	

a Calculate the temperature change for each reaction. (*3 marks*)

b Suggest a reason for the results obtained for substance **C**. (*1 mark*)

c Explain which substance gave an endothermic reaction with hydrochloric acid. (*2 marks*)

4 Hydrogen reacts with chlorine to produce hydrogen chloride:

$$H–H + Cl–Cl \rightarrow 2H–Cl$$

The following table shows the bond energies of the bonds involved.

Bond	Bond energy (kJ/mol)
H–H	436
Cl–Cl	243
H–Cl	432

a Calculate the energy transferred to break the bonds in the reactants. (*1 mark*)

b Calculate the energy transferred when bonds are made in the products. (*1 mark*)

c Use your answers to **a** and **b** to calculate the energy change in the reaction. (*1 mark*)

d Explain whether this reaction is exothermic or endothermic. (*2 marks*)

5* Zinc powder reacts with copper(II) sulfate solution to produce zinc sulfate solution and copper. Iron powder reacts with copper(II) sulfate solution to produce iron(II) sulfate solution and copper. Both reactions are exothermic.
Describe you how could carry out an investigation to determine which reaction produces the greatest temperature change. In your answer, include essential apparatus and precautions necessary for valid results. (*6 marks*)

C.3.3.1 Redox reactions

Learning outcomes

After studying this lesson you should be able to:

- explain redox reactions in terms of transfer of oxygen
- identify oxidising agents and reducing agents
- explain redox reactions in terms of transfer of electrons. **H**

Specification reference: C3.3a, C3.3b

Figure 1 *Molten iron from the thermite reaction joining two rail tracks.*

How are two steel rails welded together, ready for trains to run on them less than an hour later? The answer is the thermite reaction (Figure 1). This reaction produces molten iron, which runs into a small gap between the rails. The iron cools and solidifies, and is then machined to a smooth finish. This is an example of a redox reaction.

What are redox reactions in terms of oxygen?

A **redox** reaction is a reaction in which reduction and oxidation happen at the same time. In terms of transfer of oxygen:

- **reduction** is the loss of oxygen from a substance
- **oxidation** is the gain of oxygen by a substance.

The thermite reaction is the reaction between aluminium and iron(III) oxide:

$$\text{aluminium} + \text{iron(III) oxide} \rightarrow \text{aluminium oxide} + \text{iron}$$

$$2Al(s) + Fe_2O_3(s) \rightarrow Al_2O_3(s) + 2Fe(l)$$

Aluminium gains oxygen and is **oxidised** to aluminium oxide. At the same time, iron oxide loses oxygen and is **reduced** to iron. Overall, oxygen is transferred from iron oxide to aluminium.

Aluminium is acting as a **reducing agent** because it reduces iron oxide to iron. Iron oxide is acting as an **oxidising agent** because it oxidises aluminium to aluminium oxide.

> **A**   Magnesium and copper oxide react together when heated (Figure 2):
>
> $$Mg(s) + CuO(s) \rightarrow MgO(s) + Cu(l)$$
>
> Explain why this is a *redox* reaction.

Figure 2 *The reaction between magnesium and copper oxide is violent and exothermic.*

What are redox reactions in terms of electrons? H

A **half equation** shows the change that happens to one reactant in a reaction. These are the two half equations for the thermite reaction:

$$Al \rightarrow Al^{3+} + 3e^-$$

$$Fe^{3+} + 3e^- \rightarrow Fe$$

The oxide ions, O^{2-}, are unchanged in the reaction. However, aluminium atoms lose electrons to become aluminium ions, and iron(III) ions gain electrons to form iron atoms. In terms of electrons:

- reduction is the gain of electrons
- oxidation is the loss of electrons.

So aluminium is oxidised and iron is reduced. Overall, electrons are transferred from aluminium atoms to iron(III) ions.

B Explain, in terms of electrons and half equations, why the reaction between magnesium and copper(II) oxide is a redox reaction.

Redox reactions do not have to involve oxygen at all. For example, zinc reacts with copper(II) chloride solution to form zinc chloride solution and copper:

$$Zn(s) + CuCl_2(aq) \rightarrow ZnCl_2(aq) + Cu(s)$$

The chloride ions, Cl^-, are unchanged in the reaction. However, zinc atoms are oxidised to zinc ions, and copper(II) ions are reduced to copper atoms.

Copper(II) ions are acting as an oxidising agent because they accept electrons from zinc atoms, which oxidises the zinc atoms. Zinc atoms are acting as a reducing agent because they donate electrons to copper(II) ions, which reduces the copper(II) ions.

> **Study tip**
>
> 'OIL RIG' is a useful memory aid: **o**xidation **i**s **l**oss of electrons, **r**eduction **i**s **g**ain of electrons.

1. Zinc reacts with copper oxide to form zinc oxide and copper.
   a Write a balanced equation for the reaction. (*1 mark*)
   b Explain, in terms of oxygen, why this is a redox reaction. (*4 marks*)
   c Identify the reducing agent. Explain your answer. (*2 marks*)

2. Write half equations to model the reaction between zinc and copper(II) chloride solution. H (*2 marks*)

3. Iron reacts with chlorine: H
   $$2Fe(s) + 3Cl_2(g) \rightarrow 2FeCl_3(s)$$
   a Write half equations to model the reaction. (*4 marks*)
   b Explain, in terms of electrons, why this is a redox reaction. (*6 marks*)
   c Identify the oxidising agent and reducing agent. Explain your answer. (*3 marks*)

Learning outcomes

After studying this lesson you should be able to:

- recall the ions formed by acids and by alkalis in solution
- recall how acidity and alkalinity are measured using the pH scale
- describe how to measure pH.

Specification reference: C3.3c, C3.3h, C3.3k

Figure 1 *Using a pH meter to measure the pH of soil.*

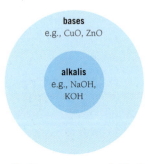

Figure 2 *Alkalis are water-soluble bases.*

Do you like crisps and sweets? Potatoes grow best in weakly acidic soil, while sugar beet grows best in weakly alkaline soil. Farmers check the pH of their soil to make sure it is suitable for their crops (Figure 1).

What are acids and alkalis?

An **acid** is a substance that releases hydrogen ions, $H^+(aq)$, when it dissolves in water to make an **aqueous solution**. For example, hydrogen chloride gas, $HCl(g)$, dissolves in water to form hydrochloric acid, $HCl(aq)$. This solution contains hydrogen ions and chloride ions:

$$HCl(aq) \rightarrow H^+(aq) + Cl^-(aq)$$

The hydrogen ions make the solution acidic.

A **base** is a substance, usually a metal oxide or metal hydroxide, that can neutralise acids. If a base can dissolve in water, it is also an **alkali** (Figure 2). An alkali releases hydroxide ions, $OH^-(aq)$, when it dissolves in water. For example, when sodium hydroxide forms an aqueous solution:

$$NaOH(aq) \rightarrow Na^+(aq) + OH^-(aq)$$

The hydroxide ions make the solution alkaline.

> **A** Write down the names and formulae of the ions that make solutions acidic or alkaline.

What is the pH scale?

The **pH** of a solution describes its relative acidity or alkalinity. On the **pH scale**:

- pH < 7 means acidic
- pH = 7 means **neutral**
- pH > 7 means alkaline.

The scale normally goes from pH 0 (most strongly acidic) to pH 14 (most strongly alkaline). You can use a universal indicator (Figure 3) or a pH meter to show the pH of a solution.

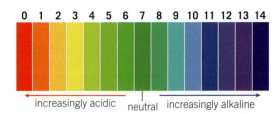

Figure 3 *The pH scale with approximate colours of universal indicator.*

> ## The symbols '<' and '>'
> Remember that '<' means 'less than' and '>' means 'greater than'.

B A weakly acidic solution has a pH of 5. Explain what a pH of 9 tells you.

C Universal indicator solution is green before use. Explain what this tells you.

Measuring pH with a pH meter

You may need to **calibrate** the pH meter (Figure 4) to get accurate measurements:

1 Wash the pH probe with water, then put it into a calibration **buffer**.

2 Adjust the reading to match the pH of the buffer solution.

Some pH meters need to be calibrated against two or three different buffers.

To measure the pH of a solution, wash the probe with water and then put it into the solution. Record the reading on the meter, which is usually precise to two decimal places.

You should wash the probe with water between each measurement to make sure that the probe is not contaminated with the previous sample, as this would affect the reading.

Synoptic link

You will have used a universal indicator paper and solution at Key Stage 3 to estimate pH values. Remember to use a colour chart to match the colour to the pH.

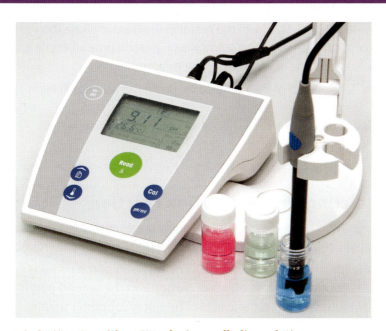

Figure 4 *A pH meter with a pH probe in an alkaline solution.*

Go further

Universal indicator is a mixed indicator. It contains several indicators that change colour over different pH ranges. Phenolphthalein is a single indicator. It is colourless below pH 8.2 and pink above this pH value.

Methyl orange is another single indicator. Find out what colours it has at different pH values.

1 Explain why:
   a nitric acid, $HNO_3(aq)$, is acidic (2 marks)
   b limewater, $Ca(OH)_2(aq)$, is alkaline. (2 marks)

2 Explain why all alkalis are bases, but not all bases are alkalis. (2 marks)

3 Compare the advantages and disadvantages of determining the pH of solution using an indicator or a pH meter. (4 marks)

Learning outcomes

After studying this lesson you should be able to:

- describe neutralisation in terms of reactants and products
- describe neutralisation in terms of ions
- write equations predicting products from given reactants.

Specification reference: C3.3d, C3.3e

Figure 1 *A farmer liming a field with powdered calcium oxide.*

Table 1 *You can predict the second part of the name of a salt formed in a neutralisation reaction if you know which acid has been used.*

Acid used	Type of salt made
hydrochloric acid, $HCl(aq)$	chloride
nitric acid, $HNO_3(aq)$	nitrate
sulfuric acid, $H_2SO_4(aq)$	sulfate
phosphoric acid, $H_3PO_4(aq)$	phosphate

Crops generally grow well at pH 6.8 to 7.0, but many soils are naturally more acidic than this. Acid rain caused by air pollution makes soil even more acidic. The solution is to 'lime' the fields, spreading powdered calcium oxide to neutralise excess acidity (Figure 1). How does this work?

What is neutralisation?

Neutralisation is the reaction between an acid and a base, or an alkali, to form a **salt** and water only. The salt made depends upon the acid and base used. In general:

$$acid + base \rightarrow salt + water$$

The pH changes when neutralisation happens. It increases if a base or an alkali is added to an acid. If enough base or alkali is added to the acid, the pH may increase from less than 7 to more than 7. The opposite happens if an acid is added to a base or an alkali. These changes can be sudden or gradual, depending on the reactants used.

How do you predict the salt made?

You can predict the salt produced in a neutralisation reaction if you know the base or alkali used, and the acid used. The names of salts have two parts:

1 The first part comes from the metal in the base or alkali. The exception to this is when ammonia or ammonium carbonate is used, when the name of the salt starts with 'ammonium'.

2 The second part of the name comes from the acid used (Table 1).

A Name the salt formed when ammonia solution neutralises hydrochloric acid.

Predicting a neutralisation reaction

Copper(II) oxide powder neutralises sulfuric acid. Name the soluble salt produced and write a balanced equation. Include state symbols.

Step 1: Identify the salt produced. The first part of the name is copper(II) (from copper(II) oxide), and the second part is sulfate (from sulfuric acid). The salt is copper(II) sulfate.

Step 2: Write the word equation:

copper(II) oxide + sulfuric acid → copper(II) sulfate + water

Step 3: Write the formulae underneath, balancing if necessary, and adding state symbols when asked:

$$CuO(s) + H_2SO_4(aq) \rightarrow CuSO_4(aq) + H_2O(l)$$

B Sodium hydroxide solution neutralises nitric acid. Name the salt produced, and write a balanced equation for the reaction.

What happens during neutralisation in solution?

Acidic solutions contain hydrogen ions, H^+(aq), and alkaline solutions contain hydroxide ions, OH^-(aq). These react together when neutralisation happens, producing water. This ionic equation models the reaction:

$$H^+(aq) + OH^-(aq) \rightarrow H_2O(l)$$

The salt produced depends upon the other ions present. For example, sodium hydroxide solution contains Na^+(aq) ions, and hydrochloric acid contains Cl^-(aq) ions. They form a solution of sodium chloride, NaCl(aq) (Figure 2).

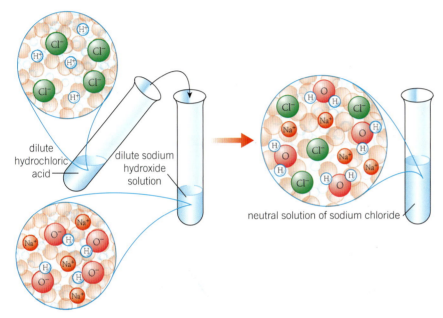

dilute hydrochloric acid

dilute sodium hydroxide solution

neutral solution of sodium chloride

Figure 2 *H^+ and OH^- ions combine to make water when sodium hydroxide reacts with hydrochloric acid. This leaves Na^+ ions from the alkali and Cl^- ions from the acid.*

C Phosphoric acid contains phosphate ions, PO_4^{3-}(aq). Explain why sodium phosphate solution, Na_3PO_4(aq), forms when sodium hydroxide solution neutralises phosphoric acid.

1 Name the salt formed when potassium hydroxide solution reacts with dilute sulfuric acid. *(1 mark)*

2 Describe neutralisation in solutions in terms of reactants and products, and in terms of reacting ions. *(3 marks)*

3 Nitric acid is a substance present in acid rain. It is neutralised by calcium oxide spread on fields.
   a Name the soluble salt formed in the reaction. *(1 mark)*
   b Write a balanced equation for the reaction, including state symbols. *(3 marks)*

Learning outcomes

After studying this lesson you should be able to:

- recall that acids react with carbonates and with some metals
- write equations predicting products from given reactants.

Specification reference: C3.3f

Figure 1 *A huge limestone cavern created by reactions with acids in running water.*

Figure 2 *Limestone, marble, and chalk are mostly calcium carbonate. Geologists test rocks with dilute hydrochloric acid to see if they contain carbonates.*

It can take thousands or millions of years for a limestone cave to form (Figure 1). Rainwater trickles through the ground into the rocks below. Carbonic acid produced naturally in the water reacts with the calcium carbonate in the limestone, dissolving the rock and forming spaces that grow larger over time.

What happens when acids react with carbonates?

Carbonates are ionic compounds that contain the carbonate ion, CO_3^{2-}. Carbonates react with acids to form a salt, plus water and carbon dioxide. In general:

$$acid + carbonate \rightarrow salt + water + carbon\ dioxide$$

The salt made depends upon the acid and carbonate used. For example, this equation models the reaction between calcium carbonate and dilute hydrochloric acid:

$$CaCO_3(s) + 2HCl(aq) \rightarrow CaCl_2(aq) + H_2O(l) + CO_2(g)$$

The carbon dioxide is released as bubbles in the acid during the reaction. Calcium carbonate is insoluble in water but calcium chloride, the salt formed, is soluble in water. This is why a lump of calcium carbonate appears to dissolve if you add acid to it (Figure 2).

> **Study tip**
>
> Most carbonates are insoluble in water. Sodium carbonate, the carbonates of other Group 1 metals, and ammonium carbonate are soluble in water.

A Sodium carbonate solution reacts with dilute hydrochloric acid. Name the salt formed, and write a balanced equation for the reaction, including state symbols.

Effervescence

Chemists often describe bubbling or fizzing as **effervescence**.

What happens when acids react with metals?

If a metal reacts with a dilute acid, the reaction produces a salt and hydrogen. In general:

$$\text{acid} + \text{metal} \rightarrow \text{salt} + \text{hydrogen}$$

The salt made depends upon the acid and metal used. For example, this equation models the reaction between magnesium and dilute hydrochloric acid:

$$Mg(s) + 2HCl(aq) \rightarrow MgCl_2(aq) + H_2(g)$$

The hydrogen is released as bubbles in the acid (Figure 3). As the reaction goes on, soluble magnesium chloride forms. This is why a piece of magnesium ribbon appears to dissolve if you put it into acid, even though magnesium itself is insoluble.

Figure 3 *Hydrogen bubbles leave the surface of magnesium as it reacts with dilute hydrochloric acid.*

B Name the salt formed when calcium reacts with hydrochloric acid.

C Sodium reacts violently with dilute sulfuric acid. Name the salt formed, and write a balanced equation for the reaction, including state symbols.

1 Name the salt formed when ammonium carbonate reacts with dilute hydrochloric acid. *(1 mark)*

2 Magnesium carbonate reacts with dilute hydrochloric acid.
  a Name the soluble salt formed in the reaction. *(1 mark)*
  b Write a balanced equation for the reaction, including state symbols. *(3 marks)*

3 Aluminium reacts with dilute sulfuric acid, forming a soluble salt. Describe what you expect to observe in the reaction, and explain your answer with the help of a balanced equation. *(6 marks)*

Synoptic link

Whether or not a metal will react with an acid is linked to its position in the reactivity series. You can learn more about this in C4.1.6 *Reactivity of elements*.

Study tip

Carbon dioxide is given off from **carbon**ates reacting with acid. Hydrogen is produced by *any* metal *if* it reacts with water or with an acid.

Go further

Gold does not react with dilute acids, but it is oxidised by a mixture of concentrated strong acids called 'aqua regia'. Find out which acids aqua regia contains, and what its name means.

Learning outcomes

After studying this lesson you should be able to:

- explain the difference between dilute and concentrated acids
- explain the difference between weak and strong acids
- describe the link between pH and hydrogen ion concentration.

Specification reference: C3.3g, C3.3i, C3.3j

Figure 1 *Vinegar and ketchup both contain ethanoic acid, a weak acid.*

Synoptic link

You can learn more about how to calculate the concentration of solutions in moles per cubic decimetre (mol/dm^3) in C5.1.4 *Concentration of solution.*

Synoptic link

You can learn more about the $\rightleftharpoons$ symbol and what it means in C5.3.1 *Reversible reactions.*

Do you like vinegar or ketchup on your chips (Figure 1)? If you do, you are putting a dilute solution of a weak acid onto your food!

What is the difference between a dilute and a concentrated acid?

A solution consists of a solute dissolved in a solvent. For a given volume of solution, the greater the amount of solute it contains, the greater its **concentration**. For acids:

- a **dilute** acid contains a low ratio of acid to volume of solution
- a **concentrated** acid contains a high ratio of acid to volume of solution.

Schools usually buy concentrated acid, then add it to water to make it dilute acid. This reduces the concentration, making it safer for you to use.

What are weak acids and strong acids?

Acids release hydrogen ions, H$^+$(aq), in aqueous solution. A covalent bond in the acid molecule breaks, producing a hydrogen ion and a negatively charged ion:

- **weak acids** are partially ionised (only a small fraction of their molecules release H$^+$ ions)
- **strong acids** are fully ionised (all of their molecules release H$^+$ ions).

Ethanoic acid is a weak acid. This equation models how it reacts in aqueous solution:

$$CH_3COOH(aq) \rightleftharpoons CH_3COO^-(aq) + H^+(aq)$$

The $\rightleftharpoons$ symbol here shows that the reaction does not go to completion.

Nitric acid is a strong acid:

$$HNO_3(aq) \rightarrow H^+(aq) + NO_3^-(aq)$$

The arrow shows that the reaction goes to completion.

> **A** Propanoic acid, CH$_3$CH$_2$COOH(aq), is a weak acid. Write an equation to show how it reacts in aqueous solution.
>
> **B** Hydrochloric acid, HCl(aq), is a strong acid. Write an equation to show how it reacts in aqueous solution.

How is pH linked to hydrogen ion concentration?

In aqueous solution, as the concentration of $H^+(aq)$ ions increases by a factor of 10, the pH of a solution decreases by 1.

This means:

- An acid has a lower pH (its solution is more acidic) when it is concentrated than when it is dilute.
- A strong acid has a lower pH than a weak acid at the same concentration.

C Explain why a dilute strong acid may have the same pH as a concentrated weak acid.

A **pH titration curve** (Figure 2) shows the effect on pH of changing the hydrogen ion concentration during a neutralisation reaction.

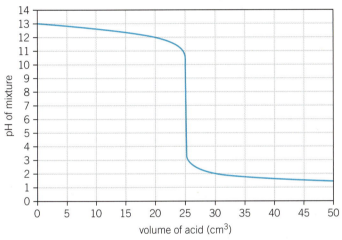

Figure 2 *A pH titration curve for a strong acid added to a strong alkali. The pH decreases rapidly just before and after the acid is completely neutralised.*

Synoptic link

You can learn more about titrations in C5.1.5 *Titrations*.

pH titration curves

You can investigate neutralisation reactions for different combinations of acid and alkali.

1 Wearing eye protection, transfer 25 cm³ of dilute alkali to a beaker.

2 Estimate its pH using universal indicator solution and a colour chart, or measure it with a pH meter.

3 Add 1 cm³ of dilute acid, stir, and record the pH.

4 Continue until you have added an excess of acid (more than enough to neutralise the alkali).

5 Plot a graph of pH against volume of acid added.

You need to stir the mixture in the beaker before each measurement of the pH, to make sure the acid and alkali are completely mixed.

1 Explain the difference in pH between dilute ethanoic acid and concentrated ethanoic acid. (*3 marks*)

2 Explain why hydrochloric acid is a strong acid but propanoic acid is a weak acid. (*2 marks*)

3 A student adds water to some acid, reducing its concentration by a factor of 100. Calculate the change in its pH. (*3 marks*)

C3.3 Types of chemical reaction

Summary questions

1 Substances dissolve in water to form acidic, neutral, or alkaline solutions. Copy and complete each of these sentences, including only the correct word in each pair of words shown in bold.

   a In an **acidic** | **alkaline** solution, the pH is less than 7.

   b If the pH is 7, the solution is **neutral** | **alkaline**.

   c When they dissolve in water, acids form **H⁺(aq)** | **OH⁻(aq)** ions, and bases form **H⁺(aq)** | **OH⁻(aq)** ions.

   d Alkalis are **soluble** | **insoluble** bases.

2 Aluminium powder reacts with chromium(III) oxide powder to produce aluminium oxide and chromium:
$$2Al + Cr_2O_3 \rightarrow Al_2O_3 + 2Cr$$

   a Explain, in terms of oxygen, which substance is oxidised, and which substance is reduced.

   b Explain why the reaction is an example of a redox reaction.

   c Write down the name of the reducing agent in this reaction, and explain your answer.

   d Describe, in terms of oxygen, what an oxidising agent does.

3 Magnesium reacts with chlorine to form magnesium chloride: **H**
$$Mg(s) + Cl_2(g) \rightarrow MgCl_2(s)$$

   a Explain, in terms of electrons, which substance is:

     i reduced in this reaction

     ii oxidised in this reaction

     iii acting as a reducing agent

     iv acting as an oxidising agent.

   b Write half equations to model the changes that happen to:

     i magnesium

     ii chlorine.

4 a Write down word equations to model the reaction between the following pairs of reactants:

     i potassium hydroxide solution and hydrochloric acid

     ii copper(II) oxide powder and sulfuric acid

     iii calcium lumps and hydrochloric acid

     iv sodium carbonate powder and nitric acid.

   b Write down balanced word equations for the reactions in **a**. State symbols are not required.

5 Explain what the following terms mean: **H**

   a dilute acid and concentrated acid

   b weak acid and strong acid.

6 Hydrochloric acid is a strong acid. Its pH is 0.52 at a concentration of $0.30 \, mol/dm^3$.

Write down the pH of hydrochloric acid at $0.030 \, mol/dm^3$, and explain your answer.

7 Figure 1 is a pH curve. It was obtained by adding $50 \, cm^3$ of $0.10 \, mol/dm^3$ sodium hydroxide solution to $25 \, cm^3$ of hydrochloric acid.

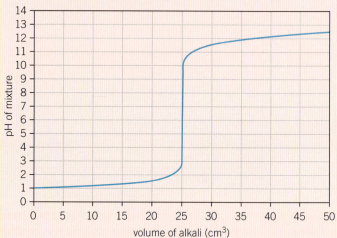

Figure 1 *A pH curve.*

   a Write down the pH when:

     i $20 \, cm^3$ of alkali has been added

     ii $30 \, cm^3$ of alkali has been added.

   b i Write down the volume of alkali needed to neutralise the acid.

     ii Use your answer to **i** to suggest the concentration of the hydrochloric acid at the start.

   c Explain, in terms of hydrogen ion concentration, why the pH increases from 1 to 3.

   d Explain, in terms of ions, why the pH is 12.5 at the end of the experiment.

Revision questions

1 Zinc powder reacts with copper(II) oxide. Zinc oxide and copper are produced.
   Which of the following describes what happens in this reaction?
   A Zinc is reduced.
   B Zinc oxide is reduced.
   C Copper oxide is reduced.
   D Copper is reduced. *(1 mark)*

2 Which ions are released into solution when acids and alkalis dissolve in water?

	Solutions of acids	Solutions of alkalis
A	hydrogen ions, H^+	hydroxide ions, OH^-
B	hydrogen ions, H^-	hydroxide ions, OH^+
C	hydroxide ions, OH^-	hydrogen ions, H^+
D	hydroxide ions, OH^+	hydrogen ions, H^-

(1 mark)

3 Dilute sulfuric acid, H_2SO_4, reacts with sodium hydroxide solution, NaOH. Water and sodium sulfate solution, Na_2SO_4, are produced.
   a Write down a balanced symbol equation for the reaction, including state symbols. *(2 marks)*
   b The reaction is an example of neutralisation. Describe what is meant by *neutralisation*. *(2 marks)*
   c Write down a general equation for neutralisation, including state symbols. *(2 marks)*

4 Copper(II) carbonate powder, $CuCO_3$, reacts with dilute hydrochloric acid, HCl. Copper(II) chloride, $CuCl_2$, water, and carbon dioxide are produced.
   a Write down a balanced symbol equation for the reaction. *(2 marks)*
   b i Name the salt produced if dilute nitric acid is used instead. *(1 mark)*
   ii Write down the formula for the salt named in **i**. *(1 mark)*

5 Hydrochloric acid is a strong acid but ethanoic acid, CH_3COOH, is a weak acid. **H**
   a Explain, in terms of ionisation, the difference between a strong acid and a weak acid. *(2 marks)*
   b 1 mol/dm^3 hydrochloric acid has a pH of 0. Write down the pH of 0.01 mol/dm^3 hydrochloric acid, and explain your answer. *(3 marks)*

6 Calcium reacts with dilute hydrochloric acid to produce calcium chloride solution and hydrogen. The ionic equation for this reaction is:
$$Ca + 2H^+ \rightarrow Ca^{2+} + H_2$$
   a Balance this half equation: $H^+ + e^- \rightarrow H_2$ *(1 mark)*
   b Write down a balanced half equation for the change that happens to calcium. *(1 mark)*
   c Identify which species is oxidised and which species is reduced in the reaction. Explain your answer in terms of the gain or loss of electrons. *(2 marks)*

7* A student investigates the acidity and alkalinity of some household substances. She uses an indicator and a pH probe to measure the pH value of each substance.
   Describe the two methods the student should use in her investigation. Include in your answer essential apparatus and substances, and suitable precautions to reduce the risk of harm. *(6 marks)*

C3.4.1 Electrolysis of molten salts

Learning outcomes

After studying this lesson you should be able to:

- describe electrolysis in terms of the ions present and reactions at the electrodes
- predict the products of electrolysis of molten ionic compounds.

Specification reference: C3.4b, C3.4d

Figure 1 *A nineteenth century engraving of Sir Humphry Davy (right) using electrolysis to isolate reactive metals.*

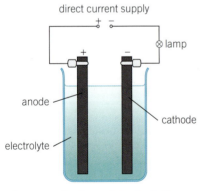

direct current supply

lamp

anode

cathode

electrolyte

Figure 2 *The lamp is included so you can see if a current is flowing.*

Cations

The word cation is pronounced 'cat ion' not 'cayshun'.

Imagine it is 1807. You are a respected chemist who has just become the first person to see potassium (Figure 1). What would do you do next? Sir Humphry Davy decided to dance around his laboratory when this happened to him. Health and Safety, and electrolysis, have come on a long way since then!

What is electrolysis?

Electrolysis is a process in which an electric current is passed through a compound, causing a chemical change. You need three components for electrolysis to work, arranged as shown in Figure 2:

1. An **electrolyte**, a compound in its liquid state or in solution, which contains mobile ions and conducts electricity.

2. Two **electrodes**, made from a metal or graphite, which conduct electricity to the electrolyte.

3. An electrical supply such as a power pack or battery.

There are two electrodes:

- the negative electrode, the **cathode**
- the positive electrode, the **anode**.

During electrolysis:

- positive ions gain electrons at the cathode and become atoms
- negative ions lose electrons at the anode, and become atoms.

If the atoms formed are non-metal atoms, covalent bonds may form between them, making molecules. In electrolysis, positive ions are called **cations** and negative ions are called **anions**.

> **A** Suggest where the terms 'cation' and 'anion' come from.

How do you predict the products of electrolysis?

A **binary ionic compound** contains just two elements. **Molten** lead bromide, $PbBr_2(l)$, consists of lead ions and bromide ions. You can model what happens at the electrodes using half equations:

- lead is produced at the cathode: $Pb^{2+} + 2e^- \rightarrow Pb$
- bromine is produced at the anode: $2Br^- \rightarrow Br_2 + 2e^-$

When ions become atoms or molecules at an electrode, you say that the ions have been **discharged**.

Overall, ions move to an oppositely charged electrode during electrolysis (Figure 3). The concentration of ions close to each electrode goes down as ions gain or lose electrons to become atoms. Other ions in the electrolyte can move to replace them by **diffusion** and **convection**. This could not happen in ionic compounds in the solid state, which is why ionic compounds in the solid state cannot conduct electricity.

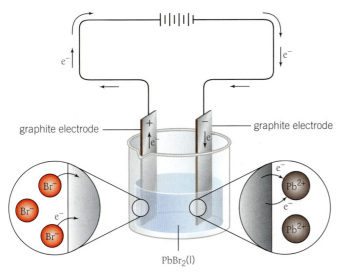

Figure 3 *Lead bromide must be heated so that it melts to form an electrolyte.*

B Suggest why the electrolysis of molten lead bromide is usually carried out in a fume cupboard.

C Predict the products formed during the electrolysis of molten sodium chloride.

The importance of new technology

Alessandro Volta invented the battery in 1800. This allowed chemists to explore new ways to carry out chemical reactions. Just seven years later, Humphry Davy used an electric current to isolate potassium. Molten potassium hydroxide was decomposed to molten potassium at the cathode, and oxygen gas at the anode. Using electrolysis, Davy also became the first person to isolate sodium, magnesium, boron, and barium.

1 Explain why molten potassium hydroxide is an electrolyte but potassium hydroxide in the solid state is not. *(3 marks)*

2 Name the products formed at the anode and cathode during the electrolysis of the following substances, justifying your answers:
   a molten potassium iodide *(3 marks)*
   b molten calcium chloride. *(3 marks)*

3 Explain, in terms of electron transfer, why the electrolysis of molten aluminium oxide is a redox reaction. In your answer, identify the products formed and include half equations to model the reactions at the electrodes. *(6 marks)*

Learning outcomes

After studying this lesson you should be able to:

- describe the technique of electrolysis using inert electrodes
- predict and explain the products of electrolysis of ionic compounds in solution.

Specification reference: C3.4a, C3.4c, C3.4d, C3.4e

Figure 1 *Solar panels supply electricity to the International Space Station, some of which is used for the electrolysis of water.*

How do you get oxygen in space? Electrolysis comes to the rescue in the International Space Station (Figure 1), producing hydrogen and oxygen from water.

What are inert electrodes?

Inert electrodes are not changed during electrolysis. They are usually made from unreactive metals such as copper or platinum, or from graphite. Inert electrodes can be used for the electrolysis of many ionic compounds in their liquid state or in aqueous solution (dissolved in water).

> **A** Suggest why schools usually use copper or graphite electrodes, rather than platinum electrodes.

What happens during the electrolysis of water?

Water is naturally partially ionised. It contains small concentrations of hydrogen ions and hydroxide ions:

$$H_2O(l) \rightleftharpoons H^+(aq) + OH^-(aq)$$

During electrolysis, hydrogen ions are discharged at the cathode as hydrogen in its gas state, and hydroxide ions are discharged at the anode, forming water and oxygen in its gas state (Figure 2):

$$4H^+(aq) + 4e^- \rightarrow 2H_2(g) \quad \text{(reduction)}$$

$$4OH^-(aq) \rightarrow 2H_2O(l) + O_2(g) + 4e^- \quad \text{(oxidation)}$$

Figure 2 *This apparatus is used for the electrolysis of water, and ionic compounds, in aqueous solution.*

> **B** Suggest why the volume of hydrogen produced during the electrolysis of water is double the volume of oxygen produced.

Go further

You can add together two half equations to make a balanced equation, provided there are equal numbers of electrons in both half equations.

Combine the two half equations shown above Figure 2 to produce the overall equation for the electrolysis of water. Remember that $H^+(aq) + OH^-(aq) \rightarrow H_2O(l)$, and you can cancel out substances that appear on both sides.

What happens during the electrolysis of solutions?

An aqueous solution contains ions from a dissolved ionic compound. Unlike a molten electrolyte, it also contains hydrogen ions and hydroxide ions from water. The reactions of each ion compete at the electrodes and only one ion is discharged at each electrode:

- Hydrogen is produced at the cathode, but if ions from a less reactive metal than hydrogen are present, that metal is produced instead.

- Oxygen is produced at the anode, but if ions from an element in Group 7 (IUPAC Group 17) element are present at a high enough concentration then the Group 7 element is produced instead.

You can use a **reactivity series** to decide if a metal is less reactive than hydrogen, see Figure 3.

Table 1 shows some examples of products formed by electrolysis.

Table 1 *Some products formed by electrolysis*

Electrolyte	Cations present	Anions present	Cathode product	Anode product
$CuCl_2(aq)$	Cu^{2+}, H^+	Cl^-, OH^-	copper	chlorine
$CuSO_4(aq)$	Cu^{2+}, H^+	SO_4^{2-}, OH^-	copper	oxygen
$KBr(aq)$	K^+, H^+	Br^-, OH^-	hydrogen	bromine
$NaOH(aq)$	Na^+, H^+	OH^-	hydrogen	oxygen
$H_2SO_4(aq)$	H^+	SO_4^{2-}, OH^-	hydrogen	oxygen

Electrolysis of solutions

You can easily test your predictions for the electrolysis of solutions.

1 Wearing eye protection, add some test solution to a beaker and dip two graphite electrodes into it.

2 Connect the electrodes to the terminals on a battery, and observe what happens.

C Explain why bubbles should be produced at both electrodes during the electrolysis of concentrated sodium chloride solution. Predict the products formed at each electrode.

1 Predict what you would observe at each electrode during the electrolysis of copper sulfate solution. *(2 marks)*

2 Potassium iodide dissolves in water to form potassium iodide solution.
   a Identify the ions present in this solution. *(2 marks)*
   b Predict the products formed at each electrode. *(2 marks)*

3 Explain why electrolysis provides evidence for the existence of ions, and their movement when ionic compounds are in the solid state and liquid state, and when dissolved. *(6 marks)*

Synoptic link

You can learn more about the reactivity series of metals in C4.1.6 *Reactivity of elements*.

potassium	most reactive
sodium	
calcium	
magnesium	
aluminium	
zinc	
iron	
tin	
lead	
(hydrogen)	
copper	
silver	
gold	
platinum	least reactive

Figure 3 *A reactivity series of metals, with hydrogen shown for comparison.*

Go further

Multiple factors can affect which anion discharges at the anode. For example, if the concentration of hydroxide and halide ions is similar, hydroxide ions preferentially discharge. If halide ion concentration is much higher, the halide ions will preferentially discharge. These effects are explained in terms of a property called electrode potential. Find out what this property is.

Study tip

Nitrates do not produce nitrogen at the anode during electrolysis of solutions, and sulfates do not produce sulfur. They produce oxygen there instead.

C3.4.3 Electroplating

Learning outcomes

After studying this lesson you should be able to:

- describe how electrolysis is used to electroplate metals
- describe how electrolysis is used to purify copper.

Specification reference: C3.4e

Ordinary steel objects rust in air and water. A rusted steel bathroom tap would not turn smoothly and would not look very appealing. One solution is to use electrolysis to coat the tap with a layer of chromium (Figure 1). This keeps air and water from the steel, so the tap stays shiny.

What do you need for electroplating?

Electroplating is a type of electrolysis that uses **non-inert electrodes**; these electrodes change during electrolysis. In electroplating:

- The cathode is the object you want to coat.
- The anode is a piece of the metal you want to coat the object with.
- The electrolyte is a solution containing ions of the coating metal.

A solid silver piece of jewellery would be expensive, so electroplating is often used to coat copper or nickel jewellery with silver (Figure 2):

- The cathode is the piece of jewellery.
- The anode is a piece of silver.
- The electrolyte is silver nitrate solution.

Silver nitrate solution is used because it contains silver ions and all nitrate salts are soluble in water, so you know that silver nitrate will dissolve.

Figure 1 *Shiny chromium-plated taps are common in bathrooms and kitchens.*

> **A** Describe suitable substances for the cathode, anode, and electrolyte when chromium-plating a tap.

How does electroplating work?

During electroplating, metal ions from the electrolyte are discharged on the surface of the object (the cathode). These ions are replaced by metal ions leaving the surface of the anode. Overall, metal leaves the anode and is deposited on the object. The process continues until the anode is used up.

During electroplating with silver:

- silver atoms lose electrons at the anode: $Ag(s) \rightarrow Ag^+(aq) + e^-$
- silver ions gain electrons at the cathode: $Ag^+(aq) + e^- \rightarrow Ag(s)$

Silver ions move through the electrolyte and electrons move through the wires between the two electrodes.

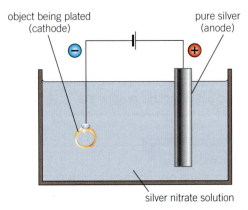

object being plated (cathode)   pure silver (anode)

silver nitrate solution

Figure 2 *How to electroplate jewellery with silver.*

> **B** One way of plating a tap with chromium involves chromium(III) ions, Cr^{3+}. Write half equations to show what happens at the anode and cathode.   **H**

How is copper purified?

The conducting part of electrical cables is made from copper. However, copper obtained straight from copper ores must be purified so it can conduct electricity well enough to be useful. Copper is purified in a similar way to electroplating, except that both electrodes are made from copper. The anode is impure copper and the cathode is very pure copper (Figure 3 and Figure 4). Copper(II) sulfate solution is commonly used as the electrolyte. During electrolysis, the cathode gains copper atoms and increases in mass, while the anode loses copper atoms and decreases in mass. Impurities from the anode fall off and collect underneath it.

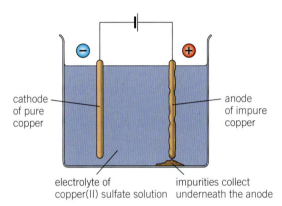

cathode of pure copper

anode of impure copper

electrolyte of copper(II) sulfate solution

impurities collect underneath the anode

Figure 3 *How to purify copper using electrolysis.*

Figure 4 *A cathode coated in copper being lifted from an electrolysis cell in a copper refinery.*

C Write half equations to show what happens at the anode and cathode during the purification of copper using electrolysis. **H**

Go further

Acidified copper(II) sulfate solution is used as the electrolyte for the industrial purification of copper. Find out which acid is used, and why.

1 Explain why, in electroplating, the object to be coated must be the cathode. *(2 marks)*

2 Food cans are steel-coated with a thin layer of tin. Describe how you could electroplate a piece of steel with tin. *(4 marks)*

3 Copper ores often contain small amounts of other metals such as silver and gold. These collect with the other impurities underneath the anode during the purification of copper.
   a Suggest why these other metals collect underneath the anode, rather than being deposited on the cathode. *(2 marks)*
   b Suggest how the overall cost of producing pure copper could be reduced. *(2 marks)*

C3 Chemical reactions

C3.4 Electrolysis

Summary questions

1 Write down simple definitions for the following terms.
   a electrolysis
   b electrolyte
   c electrode
   d cathode
   e anode

2 Figure 1 shows the apparatus used in the electrolysis of zinc chloride, $ZnCl_2$.

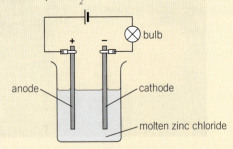

Figure 1 *Electrolysis of zinc chloride.*

   a Explain why the lamp lights when zinc chloride is molten (in the liquid state), but not when it is in the solid state.
   b Write down the names of the ions discharged at each electrode.
   c Write down the names of the elements released at each electrode.
   d Write half equations to model the reactions at: **H**
   i the anode
   ii the cathode.
   e Use your answers to **d** to explain which substance has been oxidised, and which has been reduced.

3 Copy and complete Table 1 for the electrolysis of molten salts by adding the correct formulae. Five have been completed already.

Table 1 *Electrolysis of molten salts.*

Electrolyte	Positive ion present	Negative ion present	Product at cathode	Product at anode
lead bromide, $PbBr_2(l)$	Pb^{2+}			Br_2
sodium chloride, $NaCl(l)$		Cl^-	Na	
calcium oxide, $CaO(l)$				
potassium iodide, $KI(l)$			K	

4 Figure 2 shows the apparatus used in the electrolysis of dilute sulfuric acid, $H_2SO_4(aq)$.

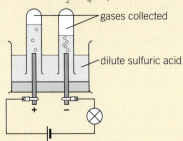

Figure 2 *Electrolysis of dilute sulfuric acid.*

   a Write down the formulae for the two ions produced by the sulfuric acid.
   b Explain why the dilute sulfuric acid also contains hydroxide ions, OH^-.
   c Name the gas collected in the test tube:
   i above the positive electrode
   ii above the negative electrode.
   d Explain why sulfur is not one of the products formed.
   e Write half equations to model the reactions at each electrode. **H**

5 Figure 3 shows apparatus used to electroplate a nickel spoon with silver.

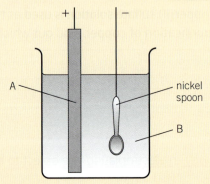

Figure 3 *Electroplating a nickel spoon with silver.*

   a Write down:
   i The name of the metal that should be used for the electrode labelled A.
   ii The formula of the ion that must be present in the electrolyte solution labelled B.
   b Suggest a suitable compound to dissolve in water to make the electrolyte solution labelled B.
   c Describe how the nickel spoon becomes electroplated with silver.

Revision questions

1 Which row about cations is correct?

	Charge on ion	Electrode to which ion is attracted
A	positive	positive
B	positive	negative
C	negative	positive
D	negative	negative

(1 mark)

2 During the electrolysis of molten lead bromide, what is made at the anode?
   A lead
   B bromide
   C hydrogen
   D bromine   (1 mark)

3 a Describe what is meant by *electrolysis*.   (1 mark)
   b Sodium chloride conducts electricity when it is molten or dissolved in water, but not when it is solid. Explain these observations.   (2 marks)

4 A student investigates the electrolysis of dilute sulfuric acid, H_2SO_4. She uses graphite electrodes. The diagram shows the apparatus she uses.

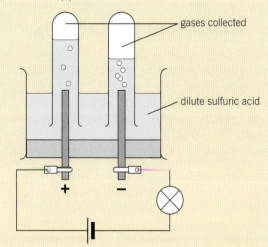

gases collected

dilute sulfuric acid

a Write down the formulae of the ions present in the acid due to:
   i the acid itself   (1 mark)
   ii the water the acid is dissolved in.   (1 mark)
b Name the two gases produced in the experiment.
        (2 marks)

5 Copper and oxygen are produced during the electrolysis of copper(II) sulfate solution using platinum electrodes.
   a Describe what you would see at each electrode.
        (3 marks)
   b Explain the differences, if any, you would see if carbon electrodes are used instead.   (2 marks)

6 Sodium chloride solution contains these ions: H^+, Na^+, OH^-, Cl^-
Hydrogen and chlorine are produced during the electrolysis of sodium chloride solution using platinum electrodes.
   a Explain why hydrogen is produced instead of sodium.   (1 mark)
   b During electrolysis, the solution gradually becomes alkaline.
      i Describe one way to show that the solution becomes alkaline.   (2 marks)
      ii Name the ion responsible for an alkaline solution.   (1 mark)

7* Copper is purified industrially using electrolysis. Copper(II) sulfate solution and two copper electrodes are used. During electrolysis, copper leaves the impure copper anode and is deposited as pure copper on the cathode.
A student investigates the changes in mass at each electrode. The diagram shows the apparatus he uses.

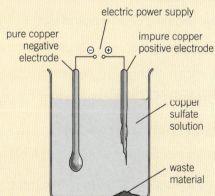

electric power supply

pure copper negative electrode

impure copper positive electrode

copper sulfate solution

waste material

The student expected the loss in mass at the anode to equal the increase in mass at the cathode. Describe how he could investigate this prediction. Include in your answer essential additional apparatus, and precautions to obtain valid results.   (6 marks)

C3.1 Introducing chemical reactions

- Write down the formulae of elements and simple covalent compounds.
- Use the formulae of common ions to work out the formulae of ionic compounds.
- State and use the law of conservation of mass.
- Explain mass changes in reactions using the particle model.
- Write, and balance, chemical equations including state symbols.
- Write half equations and balanced ionic equations. **H**
- Explain the meanings of the terms *mole* and *Avogadro constant*.
- Calculate the mass of a mole of a substance.
- Use a balanced equation to calculate masses of reactants or products.
- Explain the effect of a limiting amount of a reactant.
- Calculate the stoichiometry of an equation using given data.

C3.2 Energetics

- Identify and compare exothermic and endothermic reactions.
- Draw and label reaction profiles for exothermic and endothermic reactions.
- Explain the meaning of *activation energy*.
- Define *bond energy*. **H**
- Calculate energy changes in chemical reactions using bond energy values.

C3.3 Types of chemical reaction

- Explain redox reactions in terms of transfer of oxygen.
- Explain redox reactions in terms of transfer of electrons. **H**
- Identify oxidising agents and reducing agents.
- Recall how acidity and alkalinity are measured using the pH scale.
- Describe how to measure pH.

- Describe neutralisation in terms of reactants and products, and in terms of the ions produced by acids and alkalis.
- Describe the reactions of acids with carbonates and with some metals.
- Explain the difference between dilute and concentrated acids. **H**
- Explain the difference between weak and strong acids.
- Describe the link between pH and hydrogen ion concentration.

C3.4 Electrolysis

- Describe how electrolysis is carried out.
- Explain electrolysis in terms of the ions present and reactions at the electrodes.
- Predict the products of electrolysis of molten ionic compounds.
- Predict and explain the products of electrolysis of ionic compounds in solution.
- Describe how electrolysis is used to electroplate metals and to purify metals.

Chemical equations

- model chemical reactions
- show rearrangement of atoms in reaction and relative amounts of substances involved
- balanced equations show equal number of atoms of each element on either side of reaction arrow (law of conservation of mass)
- state symbols show physical state of substances in reaction

- half equations model change to one reactant in a reaction **H**
- complete ionic equation shows ions in a reaction mixture
- net ionic equations omit spectator ions and are used to model precipitation

The mole **H**

- mole (mol) = unit for amount of substance
- Avogadro constant = number of entities in 1 mol
- molar mass (g/mol) = mass of 1 mol of a substance

  mass (g) = molar mass (g/mol) × amount (mol)

- one reactant usually in excess (some left over), the other limiting (determines amount of product formed)

Chemical formulae

- Periodic Table shows element names and chemical symbols
- Group 0 (IUPAC Group 18) formulae = chemical symbol
- Group 7 (IUPAC Group 17) formulae contain subscript 2 (diatomic molecules)
- simple covalent compound formulae show element symbols and number of atoms of each element in one molecule
- ionic compound molecular formulae must have equal positive and negative charges

Redox reactions

reduction	oxidation
gain of oxygen	loss of oxygen

redox = reduction and oxidation happening together

magnesium oxidised to magnesium oxide	copper oxide reduced to copper

$$Mg + CuO \rightarrow MgO + Cu$$

oxidising agent ← → reducing agent

OIL RIG **H**

- oxidation is loss of electrons
- reduction is gain of electrons

Electrolysis

Molten ionic compounds

Cathode (−)	Anode (+)
metal or hydrogen formed (e.g., $Pb^{2+} + 2e^- \rightarrow Pb$)	non-metal formed (e.g., $2Br^- \rightarrow Br_2 + 2e^-$)

Ionic compounds in aqueous solution

Cathode (−)	Anode (+)
hydrogen formed (unless metal is more reactive than hydrogen)	oxygen formed (unless halide ions Cl^-, Br^-, or I^- present)

Electroplating	Purifying copper
object being plated (cathode) ⊖ pure silver (anode) ⊕ / silver nitrate solution	⊖ ⊕ / cathode of pure copper / anode of impure copper / electrolyte of copper(II) sulfate solution / impurities collect underneath the anode

C3 Chemical reactions

Energetics

	Exothermic reactions	Endothermic reactions
Temperature change of surroundings	↑	↓
Examples	• combustion • neutralisation	• thermal decomposition • electrolysis • photosynthesis
Energy of products	less than energy of reactants	more than energy of reactants
Energy change	− (energy transferred to surroundings)	+ (energy transferred from surroundings)

Activation energy

- minimum energy needed to start reaction
- can be supplied by sparks, flames, heating, electrical energy

CALCULATED USING **H**

Bond energies
- bond breaking = endothermic (energy in)
- bond making = exothermic (energy out)

energy change = energy in − energy out

The pH scale

- pH = relative acidity/alkalinity
- measured using indicators or pH meters

Acids	Alkalis
form H^+ ions when dissolved in water	form OH^- ions when dissolved in water
$HCl(aq) \rightarrow$ $H^+(aq) + Cl^-(aq)$	$NaOH(aq) \rightarrow$ $Na^+(aq) + OH^-(aq)$
pH < 7	pH > 7

SOLUBLE BASES = ALKALIS

acid + base → salt + water
acid + alkali → salt + water
acid + carbonate → salt + water + carbon dioxide
acid + metal → salt + hydrogen

- neutralisation: $H^+(aq) + OH^-(aq) \rightarrow H_2O(l)$
- pH decreases by 1 if $H^+(aq)$ concentration increases 10 times

Learning outcomes

After studying this lesson you should be able to:

- recall some physical and chemical properties of Group 1 elements
- explain the reactions of Group 1 elements
- predict properties from given trends.

Specification reference: C4.1a, C4.1b

Figure 1 *Gloves are worn when handling sodium to stop it reacting with the skin.*

Sodium floats on water, so it might seem like a good material for making a boat. Unfortunately, sodium is soft enough to cut with a knife (Figure 1) and it can burst into flames on water. Sodium, and the other elements in Group 1, have some fascinating properties.

What are the Group 1 elements like?

Group 1 elements are placed in the vertical column on the far left of the Periodic Table (Figure 2). They are called the **alkali metals** because they react with water to form alkaline solutions. Group 1 elements have some typical properties of metals. They are shiny when freshly cut and are good conductors of electricity, but they also have some unusual properties.

The alkali metals are in the solid state at room temperature, but they are all soft enough to cut with a knife. Lithium is the hardest and each alkali metal is softer as you go down the group. A general pattern or direction of change is a **trend**. The alkali metals show other trends in their physical properties. Going down the group:

- their **density** increases, although sodium is denser than expected (Figure 3)
- their melting point decreases (Figure 4).

(1)

Figure 2 *Francium, at the bottom of Group 1, is radioactive and extremely rare.*

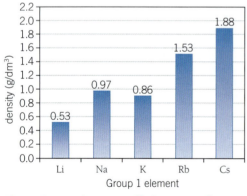

Figure 3 *Density generally increases down Group 1.*

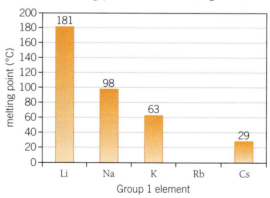

Figure 4 *Melting point decreases down Group 1.*

Density

A substance will float on water if its density is less than 1.00 g/cm³, the density of water.

A From Figure 3 predict and explain which of the elements in Group 1 will float on water, and which will sink.

B Use Figure 4 to predict the melting point of rubidium, Rb.

What are the reactions of Group 1 elements?

The only Group 1 metals you are likely to see in school are lithium, sodium, and potassium. They react rapidly with oxygen in the air and with water, so they are stored in oil (Figure 5).

The alkali metals react with water to produce the metal hydroxide and hydrogen. For example, sodium reacts with water to produce sodium hydroxide and hydrogen:

$$2Na(s) + 2H_2O(l) \rightarrow 2NaOH(aq) + H_2(g)$$

The **reactivity** of alkali metals with water increases down the group:

- lithium fizzes steadily and slowly disappears
- sodium melts to form a silvery ball, fizzes vigorously, and quickly disappears
- potassium immediately ignites, burns with a lilac flame, and very quickly disappears.

Figure 5 *The Group 1 elements are stored in oil to prevent them from reacting with oxygen in the air and with water.*

How do you explain the trend in reactivity?

The Group 1 elements have similar chemical properties because all their atoms have one electron in their outer shell. They lose these outer electrons in reactions, leaving a complete outer shell and forming ions with a single positive charge. This general ionic equation models the change (M stands for the symbol of the metal):

$$M \rightarrow M^+ + e^-$$

The easier it is for an atom of a Group 1 element to lose its outer electron, the more reactive the elements is. For example, potassium is more reactive than lithium because its atoms lose their outer electron more easily than lithium atoms do.

> **C** Explain how the general equation above shows why Group 1 metals are oxidised in their reactions with water and non-metals such as oxygen. **H**

1 Explain the colour you expect universal indicator to be in potassium hydroxide solution. (*2 marks*)

2 Caesium reacts explosively on contact with water.
   a Write a balanced equation to model the reaction, including state symbols. (*3 marks*)
   b Explain why rubidium is less reactive than caesium, but more reactive than potassium. (*3 marks*)

3 Predict the physical and chemical properties of francium, and explain your answers. (*6 marks*)

Figure 1 *Chlorine compounds are added to swimming pool water to kill microorganisms.*

Table 1 *Densities of Group 7 (IUPAC Group 17) elements.*

Element	Density (g/cm³) at room temperature and pressure
fluorine, F_2	0.001 55
chlorine, Cl_2	0.002 90
bromine, Br_2	3.10
iodine, I_2	4.93

Have you ever had stinging eyes after swimming (Figure 1)? You might think it is caused by chlorine, but in fact it is caused by chloroamines. These are compounds formed when chlorine in the water reacts with sweat and other body substances. Chlorine is a reactive non-metal, like the other elements in Group 7 (IUPAC Group 17).

What are the Group 7 (IUPAC Group 17) elements like?

The elements in **Group 7** (IUPAC Group 17) are placed in a vertical column on the right of the Periodic Table (Figure 2). They have typical properties of non-metals, including being brittle in the solid state and poor conductors of electricity.

Group 7 (IUPAC Group 17) elements exist as diatomic molecules with weak intermolecular forces. They are coloured or form coloured vapours (see Figure 5), and they occur in different states at room temperature:

- fluorine, F_2, is a pale yellow gas
- chlorine, Cl_2, is a green gas
- bromine, Br_2, is an orange-brown liquid that vaporises easily
- iodine, I_2, is a shiny grey-black **crystalline** solid that sublimes to form a purple vapour.

Going down Group 7 (IUPAC Group 17):

- density increases (Table 1)
- melting points and boiling points increase (Figure 3).

Figure 2 *Astatine at the bottom of Group 7 (IUPAC Group 17) is an extremely rare, radioactive element.*

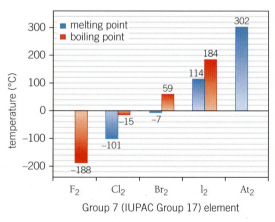

Figure 3 *The melting points and boiling points of the halogens increase going down the group.*

> **A** Use the graph in Figure 3 to predict the melting point of fluorine.

What are the reactions of Group 7 (IUPAC Group 17) elements?

Group 7 (IUPAC Group 17) elements are also called the **halogens** because they react with metals to produce salts. They react vigorously with Group 1 metals, particularly if the metal is heated first. For example, sodium reacts with chlorine to produce sodium chloride:

$$2Na(l) + Cl_2(g) \rightarrow 2NaCl(s)$$

The reactivity of the halogens decreases down the group. This is the opposite trend in reactivity to the alkali metals.

> **B** Iron reacts with chlorine to produce orange-brown iron(III) chloride (Figure 4). Write a balanced equation for the reaction between iron and chlorine.

How do you explain the trend in reactivity?

The halogens have similar chemical properties because all their atoms have seven electrons in their outer shell. They gain an electron in reactions, completing their outer shell and forming ions with a single negative charge. This general ionic equation models the change (X stands for the symbol of the halogen):

$$X_2 + 2e^- \rightarrow 2X^-$$

The easier it is for a halogen atom to gain an outer electron, the more reactive the element is. For example, chlorine is more reactive than iodine because its atoms gain an outer electron more easily than iodine atoms do.

> **C** Explain how the general equation above shows that Group 7 (IUPAC Group 17) elements are reduced in their reactions with metals. **H**

> **1** Use the graph of melting and boiling points in Figure 3 to predict the boiling point of astatine. Explain your answer. (*2 marks*)
>
> **2** Two scientists reacted a small piece of caesium with fluorine.
> **a** Explain what you would expect to see in the reaction. (*3 marks*)
> **b** Write a balanced equation to model the reaction, including state symbols. (*3 marks*)
> **c** Explain why chlorine is less reactive than fluorine, but more reactive than bromine. (*3 marks*)
>
> **3** Predict the physical and chemical properties of astatine, and explain your answers. (*4 marks*)

Halogens

The word 'halogen' comes from Greek word *halos* meaning 'salt'. The 'gen' part means 'maker'.

Figure 4 *Hot iron reacts vigorously with chlorine to make Fe(III) chloride.*

Figure 5 *Flasks containing chlorine, bromine, and iodine.*

Learning outcome

After studying this lesson you should be able to:

- describe and explain displacement reactions involving halogens and halides.

Specification reference: C4.1a

Figure 1 *These scientists are testing a new flame retardant.*

Figure 2 *Chlorine gas displacing bromine from potassium bromide.*

Some bromine compounds are useful as flame retardants. These substances are used to make buildings and materials more difficult to ignite (Figure 1). Without them, fires would spread more quickly. Bromine is obtained using a halogen displacement reaction.

What is a halogen displacement reaction?

Halogens can react with **halides** in solution. A halide is a compound containing a Group 7 (IUPAC Group 17) element and one other element, usually hydrogen or a metal.

For example, chlorine reacts with sodium bromide solution to form sodium chloride and bromine (Figure 2):

$$Cl_2(g) + 2NaBr(aq) \rightarrow 2NaCl(aq) + Br_2(aq)$$

The reaction mixture turns orange-brown as bromine is produced. Chlorine can **displace** or 'push out' bromine from sodium bromide in a **displacement** reaction.

A halogen will displace a less reactive halogen from its **halide ions** in solution. This means that:

- Chlorine displaces bromine from bromides, and iodine from iodides.
- Bromine displaces iodine from iodides, but it cannot displace chlorine from chlorides.
- Iodine cannot displace chlorine from chlorides, or bromine from bromides.

It is safer in school to use aqueous solutions of chlorine, bromine, and iodine, rather than the pure elements.

A Explain why bromine can displace iodine from sodium iodide solution.

B Explain why potassium iodide solution turns darker brown when chlorine solution is added to it.

Halides and halide ions

A 'halide' is a compound of a Group 7 (IUPAC Group 17) element and one other element. Hydrogen chloride and sodium chloride are halides. A 'halide ion' is a negative ion formed by a Group 7 (IUPAC Group 17) element, for example chloride ions from chlorine.

Halogen displacement reactions

You can use halogen displacement reactions to confirm the order of reactivity for chlorine, bromine, and iodine.

1 Wearing eye protection, place a small volume of potassium chloride solution in a spotting tile well (Figure 3).

2 Add a few drops of bromine water.

3 Note your observations in a table.

4 You will then need to repeat for other pairs of solutions.

Asthmatics should take particular care not to breathe in any chemical fumes.

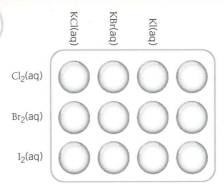

Figure 3 *A spotting tile for the halogen displacement reactions.*

C In a practical like that described in *Halogen displacement reactions*, which other pairs of solutions would you need to test to confirm the halogen order of reactivity? Use the spotting tile diagram in Figure 3 to help you draw a suitable table for your results.

D In a practical like that described in *Halogen displacement reactions*, which halogen should take part in two reactions, which halogen should take part in one reaction, and which halogen should not take part in any reactions at all?

Go further

Find out how bromine is extracted industrially using chlorine and seawater.

How can ionic equations model displacement reactions? **H**

You can write a half equation to model what happens to each reactant. For example, chlorine displaces bromine from bromide ions:

$$Cl_2 + 2e^- \rightarrow 2Cl^- \qquad \text{reduction}$$
$$2Br^- \rightarrow Br_2 + 2e^- \qquad \text{oxidation}$$

You can combine these half equations to make the ionic equation:

$$Cl_2 + 2Br^- \rightarrow 2Cl^- + Br_2$$

Notice that halogen displacement reactions are also redox reactions.

E Write the ionic equation for chlorine displacing iodine from iodide ions. **H**

1 Write down the word equation for the displacement reaction between chlorine and lithium iodide. (*1 mark*)

2 Predict whether or not iodine will displace astatine from sodium astatide, and explain your prediction. (*2 marks*)

3 Fluorine reacts with water, so displacement reactions involving fluorine in the gas state are carried out on filter paper dampened with a halide solution. **H**

   a Explain why fluorine can displace chlorine from sodium chloride solution. (*1 mark*)

   b Write a balanced equation to show the reaction, including state symbols. (*3 marks*)

   c Write an ionic equation for the reaction, and explain why it represents a redox reaction. (*4 marks*)

C4.1.4 Group 0 – the noble gases

Learning outcomes

After studying this lesson you should be able to:

- recall and explain the properties of Group 0 (IUPAC Group 18) elements
- explain the lack of reactivity of Group 0 (IUPAC Group 18) elements
- predict properties from given trends.

Specification reference: C4.1a, C4.1b

How do you measure atmospheric conditions 1 km above the ground? Meteorologists, who are scientists that study the weather and climate, use weather balloons filled with helium (Figure 1). The balloons carry scientific instruments, and a radio transmitter sends data back down to receivers on the ground.

What are the Group 0 (IUPAC Group 18) elements like?

The elements in **Group 0** (IUPAC Group 18) are placed in a vertical column on the far right of the Periodic Table (Figure 2). They are called the **noble gases** because they are so unreactive. Rather like the 'noble' men and women of the past who did not take part in ordinary everyday activities, the noble gases take part in very few chemical reactions. The Group 0 (IUPAC Group 18) elements are non-metals, and all are in the gas state at room temperature.

Why are the noble gases so unreactive?

The atoms of Group 0 (IUPAC Group 18) elements have complete outer electron shells (Figure 3). This means that they have no tendency to lose or gain electrons to form ions in reactions, or to share electrons to form molecules in reactions. As a result, the noble gases are very unreactive.

Figure 2 *Radon at the bottom of Group 0 (IUPAC Group 18) is a dense, radioactive non-metal.*

Figure 1 *These meteorologists are preparing a helium-filled weather balloon for its ascent.*

> **Study tip**
>
> All the noble gases have eight electrons in their outer shell except for helium, which completes its outer shell with two electrons.

He   Ne   Ar

helium   neon   argon

Figure 3 *Electron diagrams for the first three noble gases.*

A   Write down the electronic structures for helium, neon, and argon.

What trends in properties do the noble gases show?

The noble gases are **monatomic**. They exist as single atoms with very weak forces of attraction between them. These forces are easily overcome by heating. This gives the noble gases very low boiling points. Going down Group 0 (IUPAC Group 18):

- the attractive forces between atoms get stronger
- the boiling point increases.

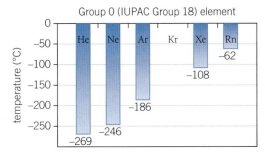

Figure 4 *Boiling point increases down Group 0 (IUPAC Group 18).*

B Use the bar chart in Figure 4 to predict the boiling point of krypton.

The noble gases have very low densities. This is because their atoms are far apart in the gas state, so there is very little mass in a given volume. As you go down Group 0 (IUPAC Group 18), the density increases.

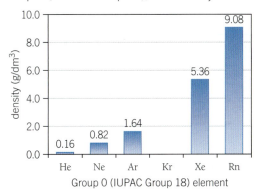

Figure 5 *Density increases down Group 0 (IUPAC Group 18).*

C Use the bar chart in Figure 5 to predict the density of krypton.

1 Explain why the Group 0 (IUPAC Group 18) elements are unreactive.
(2 marks)

2 The density of air at room temperature and pressure is $1.20\,g/dm^3$. Explain, in terms of density, why helium balloons and neon balloons can rise in the air, but balloons filled with other noble gases sink. *(2 marks)*

3 Three atoms of ununoctium, Uuo, were made early this century. Uuo is placed below radon in the Periodic Table. Scientists believe that it may be a metal, rather than a non-metal. Predict its chemical and physical properties, and justify your answer. *(6 marks)*

Noble not inert

The Group 0 (IUPAC Group 18) elements used to be called the 'inert gases' because chemists thought that they could not react at all. Then in 1962 Neil Bartlett made the first noble gas compound, $XePtF_6$. Several noble gas compounds have been made since then, but none involving neon or helium.

Learning outcome

After studying this lesson you should be able to:

● recall the general properties of transition metals.

Specification reference: C4.1c

Figure 1 *A 10 km diameter asteroid contains enough iridium to account for the worldwide iridium layer.*

What caused the dinosaurs to become extinct? One theory is that a massive asteroid may have hit the Earth 65 million years ago, drastically changing the climate (Figure 1). An iridium-rich layer in the Earth's crust provides evidence of this impact, as asteroids have an unusually high iridium–iron ratio. Iridium and iron are just two examples of transition metals.

What are the transition metals like?

The **transition metals** are placed between Groups 2 and 3 (IUPAC Groups 2 and 13) in the Periodic Table, occupying IUPAC Groups 3 to 12 (Figure 2). They are all metals and their properties are typical of metals. In general, they are:

● shiny when freshly cut

● good conductors of electricity

● strong

● malleable (they can be bent or hammered into shape).

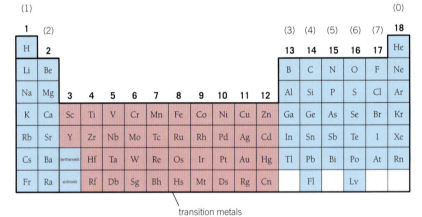

Figure 2 *The transition metals in the Periodic Table.*

Compared to the alkali metals of Group 1, the transition metals:

● are stronger and harder

● have higher densities

● have higher melting points (except for mercury, which is in the liquid state at room temperature).

These properties mean the transition metals are useful for making everyday objects (Figure 3). For example, copper is a good conductor of electricity and can be made into strong, flexible electrical wires.

Figure 3 *Steel nails and screws, chromium-plated stopper, copper pipes, iron magnet, copper–nickel alloy coins, and copper-plated steel coins.*

> **A** Suggest why iron in the form of steel is suitable for making ships, but sodium from Group 1 is not.

What are the chemical properties of transition metals?

The transition metals are less reactive than the alkali metals. They react slowly, if at all:

- iron reacts slowly with water and oxygen to produce rust, hydrated iron(III) oxide
- gold, platinum, and iridium do not react with water or oxygen at all.

The alkali metals produce white or colourless ionic compounds, but the transition metals produce coloured ionic compounds (Figure 4).

B Explain why sodium sulfate solution is colourless, but the other solutions in Figure 4 are coloured.

Transition metal names

Many of transition metal names reflect the colours of their ions. Iridium is named from 'iris', the Greek word for 'rainbow', and rhodium is named from 'rhodon', the Greek word for 'rose'.

Figure 4 *Aqueous solutions of some different metal compounds.*

The alkali metals can only form ions with a single positive charge, but many transition metals can form more than one type of ion. For example, iron(II) oxide, FeO, contains Fe^{2+} ions, and iron(III) oxide, Fe_2O_3, contains Fe^{3+} ions. The Roman numbers in brackets tell you the charge on the metal ions.

Transition metals are often good **catalysts**. Catalysts are substances that increase the rates of chemical reactions without being used up. Platinum, rhodium, and palladium are used in devices called catalytic converters. These convert harmful gases in vehicle exhaust fumes into less harmful ones.

C Use Figure 4 to suggest how to tell iron(II) sulfate from iron(III) chloride.

Synoptic link

You can learn more about catalysts in C5.2.5 *Catalysts and reaction rate.*

Go further

Vanadium forms several different ions in solution. Find the origin of the metal's name, and the formulae and colours of some of its ions.

1 Describe two differences between Group 1 metal ions and transition metal ions. *(2 marks)*

2 A vanadium compound, V_2O_5, is used as a catalyst for one stage in the manufacture of sulfuric acid.
  a Explain what a catalyst is. *(2 marks)*
  b Write down the symbol for the vanadium ion in this compound. *(1 mark)*
  c Suggest why this compound is in the solid state at room temperature, and yellow-brown in colour. *(3 marks)*

3 Explain, in terms of their properties, why platinum and iridium are suitable for making electrodes for electrolysis. *(4 marks)*

C4.1.6 Reactivity of elements

Learning outcomes

After studying this lesson you should be able to:

- explain the reactivity of different metals
- use experimental data to put metals into order of reactivity
- predict the reactions and reactivity of elements from their position in the Periodic Table.

Specification reference: C4.1c, C4.1d, C4.1e, C4.1f

Figure 1 *Before and after photographs showing copper displacing silver from silver nitrate solution.*

Figure 2 *A reactivity series of metals, with hydrogen shown for comparison.*

A teacher put a spiral of copper wire into silver nitrate solution. The solution gradually turned pale blue and silver crystals formed on the copper. These observations are the result of a metal displacement reaction (Figure 1).

How do metals react with water and dilute acids?

Metals form positive ions in reactions. The more easily this happens, the more reactive the metal. A metal can react with water or dilute acids if it is more reactive than hydrogen (Figure 2). For example, calcium reacts with water and dilute acids, but copper does not. Metals react with:

- water to produce a metal hydroxide and hydrogen, for example
$$Ca + 2H_2O \rightarrow Ca(OH)_2 + H_2$$
- acids to produce a salt and hydrogen, for example
$$Ca + 2HCl \rightarrow CaCl_2 + H_2$$

What does the rate of reaction tell you?

The reactions of metals with water, or with dilute hydrochloric acid, can be used to put metals in order of their reactivity. In general, the more reactive the metal, the greater the rate of hydrogen production. This means that there is more vigorous bubbling. Eye protection should be worn when you do this.

You could place about $2 \, cm^3$ of water in a test tube, add a small piece of metal, and then note the rate of bubbling. If there was no bubbling, or very little, you could then gently warm the water to see if bubbling starts or increases in rate. You would repeat this process with other metals, and with dilute hydrochloric acid instead of water.

If you were carrying out this test, you should should not boil the water or dilute hydrochloric acid because it would bubble when it boiled, so you would not be able to tell if the bubbles were due to a reaction. It would also be unsafe to boil the dilute hydrochloric acid.

What are metal displacement reactions?

A more reactive metal can displace a less reactive metal from solutions of its compounds. Copper is more reactive than silver. It can displace silver from silver nitrate solution:

$$Cu(s) + 2AgNO_3(aq) \rightarrow Cu(NO_3)_2(aq) + 2Ag(s)$$

Copper nitrate solution is blue, which explains the colour change in Figure 1. Displacement reactions are examples of redox reactions. These half equations model the changes:

$$Cu(s) \rightarrow Cu^{2+}(aq) + 2e^- \quad \text{oxidation (loss of electrons)}$$
$$Ag^+(aq) + e^- \rightarrow Ag(s) \quad \text{reduction (gain of electrons)}$$

A Explain why iron can displace copper from copper(II) nitrate solution, but copper cannot displace iron from iron(III) nitrate solution.

How can you predict reactions from the Periodic Table?

There are some useful patterns of reactivity in the Periodic Table (Figure 3):

- Elements in Group 0 (IUPAC Group 18) do not react.
- Reactive non-metals may form covalent compounds with each other.
- Metals may form ionic compounds with reactive non-metals.
- Metals in Groups 1 and 2 are more reactive than transition metals and other metals.
- Metals in Groups 1 and 2 become more reactive down the group.
- Non-metals in Group 7 (IUPAC Group 17) become less reactive down the group.

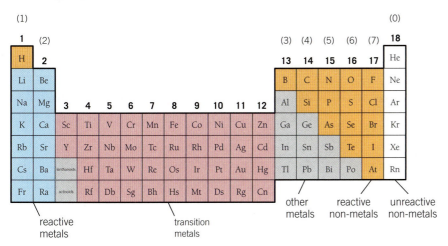

Figure 3 *Patterns in the Periodic Table.*

Displacement reactions

You can use displacement reactions to confirm the order of reactivity for metals such as magnesium, zinc, and copper.

1. Place a small volume of copper(II) sulfate solution in a spotting tile well.
2. Add a piece of magnesium.
3. Note your observations in a table.
4. You will then need to repeat for other pairs of solutions and metals.

Eye protection should be worn throughout this practical.

B In a practical like that described in *Displacement reactions*, which other pairs of solutions would you need to test? Use Figure 4 to help you draw a suitable table for your results.

C In a practical like that described in *Displacement reactions*, which metal should cause two reactions, which should cause one reaction, and which should not cause any reaction at all?

D Name the most reactive metal and the most reactive non-metal.

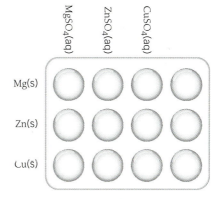

Figure 4 *A spotting tile for metal displacement reactions.*

1 In the reactivity series, identify metals from Group 1, Group 2, Group 3 (IUPAC Group 13), and the transition metals. Describe the trend you see. *(5 marks)*

2 Suggest why you cannot compare the reactivity of sodium and potassium using displacement reactions in solution. *(2 marks)*

3 Outline experiments that could be used to determine the reactivity of an unknown metal. Describe what you might observe in each experiment. *(6 marks)*

C4.1 Predicting chemical reactions

Summary questions

1 Copy and complete each of these sentences about Group 1 elements, including only the correct word in each pair of words shown in **bold**.

 a The Group 1 elements are all **metals | non-metals**.

 b They react with water to form **hydrogen | oxygen** and **acidic | alkaline** solutions.

 c The Group 1 elements are **poor | good** conductors of electricity.

 d The atoms of Group 1 elements **lose | gain** electrons when they react with non-metal elements.

2 Table 1 shows some information about Group 1 elements.

Table 1 *Group 1 elements.*

Element	Atomic number	Melting point (°C)	Boiling point (°C)	Density (g/cm³)	Relative hardness
lithium	3	181	1330	0.53	0.6
sodium	11	98	883	0.97	0.5
potassium	19	63	759	0.86	0.4
rubidium	37	39			
caesium	55	29	671	1.88	0.2

 a Use your knowledge of trends in physical properties of Group 1 elements, and the data in Table 1, to predict the missing values for rubidium.

 b Write down the names of the Group 1 elements that will be in the solid state at:

  i 30 °C

  ii 100 °C.

 c Write down the electronic structures of lithium, sodium, and potassium.

 d Explain, in terms of electrons, why Group 1 elements have similar chemical properties.

3 Copy and complete each of these sentences about Group 7 (IUPAC Group 17) elements, including only the correct word in each pair of words shown in **bold**.

 a Apart from astatine, the Group 7 (IUPAC Group 17) elements are all **metals | non-metals**.

 b The reactivity of Group 7 (IUPAC Group 17) elements **increases | decreases** going down the group.

 c The melting points of Group 7 (IUPAC Group 17) elements **increase | decrease** going down the group.

 d The atoms of Group 7 (IUPAC Group 17) elements **lose | gain** electrons when they react with Group 1 elements.

4 Group 7 (IUPAC Group 17) elements react with their ions in solution. A student carried out an experiment in which they added solutions of chlorine, bromine, or iodine to solutions of sodium chloride, sodium bromide, or sodium iodide.

 a Copy and complete Table 2 by adding a tick (✓) in each box where you would expect to see a reaction.

Table 2 *Group 7 (IUPAC Group 17) elements and ions.*

	Chlorine solution	Bromine solution	Iodine solution
sodium chloride			
sodium bromide			
sodium iodide			

 b Write down a balanced equation to model the reaction between fluorine and sodium chloride. State symbols are not needed.

5 Explain, in terms of electronic structures, why Group 0 (IUPAC Group 18) elements are unreactive.

6 Copy and complete each of these sentences about transition metals, including only the correct word in each pair of words shown in **bold**.

 a Compared to Group 1 elements, most transition metals are **weaker | stronger**, and **more | less** reactive.

 b Transition metal compounds are usually **coloured | colourless**.

7 A more reactive metal will displace a less reactive metal from its oxide.

Table 3 shows the results of similar experiments with four metals (A, B, C, and D). A tick (✓) shows where a reaction happened.

Table 3 *Metals and their oxides.*

	A oxide	B oxide	C oxide	D oxide
metal A		✓		
metal B				
metal C	✓	✓		✓
metal D	✓	✓		

 a Use the results to place the metals in order of decreasing reactivity.

 b Explain which metal (A, B, C, or D) forms positive ions most easily.

Revision questions

1 Elements may exist in different physical states. Which row about the elements in Groups 1, 7 and 0 at room temperature is correct?

	Group 1	Group 7	Group 0
A	all in solid state	all in solid state	all in gas state
B	some in liquid state	some in solid state	some in gas state
C	some in liquid state	some in gas state	all in gas state
D	all in solid state	some in gas state	all in gas state

(1 mark)

2 What happens when Group 1 elements are added to water?

A They all float, and hydrogen gas and an alkaline solution are produced.

B Some float, and carbon dioxide gas and an alkaline solution are produced.

C Some float, and hydrogen gas and an alkaline solution are produced.

D Some float, and hydrogen gas and an acidic solution are produced. *(1 mark)*

3 Iron is an element placed in the block of metals in the middle of the Periodic Table.

a Write down the name of this block of elements. *(1 mark)*

b Iron has a higher density than the metals in Group 1 of the Periodic Table.
Describe three other properties of iron that are different from those of elements in Group 1. *(3 marks)*

4 The table shows the atomic radius for some elements in Group 7.

Element	Atomic number	Atomic radius (pm)
lithium	3	145
sodium	11	180
potassium	19	to determine
rubidium	37	235
caesium	55	260
francium	87	not known

a Using graph paper, plot a line graph of these data. Show atomic number on the horizontal axis and atomic radius on the vertical axis.
Draw the line of best fit. *(4 marks)*

b Use your graph to predict the atomic radius for:
i potassium *(1 mark)*
ii francium. *(1 mark)*

5* The elements in Group 1 become more reactive down the group, but the elements in Group 7 become less reactive down the group. The elements in Group 0 are unreactive.
Explain these observations in terms of atomic structure and the transfer of electrons. *(6 marks)*

6 A student investigates the reactions of some metals with dilute hydrochloric acid. **H**

a Predict and explain what the student will observe when he adds zinc to dilute hydrochloric acid.
Include a balanced symbol equation in your answer. *(3 marks)*

b i Write a half equation to model what happens when the student adds magnesium to dilute hydrochloric acid. *(1 mark)*

ii Use your half equation in **i** to help you to explain whether magnesium is oxidised or reduced when it reacts with dilute hydrochloric acid. *(1 mark)*

c Magnesium reacts more vigorously than zinc with dilute hydrochloric acid. Copper does not react with dilute hydrochloric acid. Explain these observations. *(2 marks)*

Learning outcome

After studying this lesson you should be able to:

● describe laboratory tests to detect oxygen, hydrogen, carbon dioxide, and chlorine.

Specification reference: C4.2a

Figure 1 *Lake Nyos is an African lake formed in the crater of an extinct volcano.*

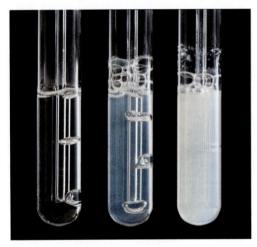

Figure 2 *A white precipitate forms in limewater when carbon dioxide is bubbled through it.*

In 1986, more than 1700 people and their animals died around Lake Nyos in Cameroon (Figure 1). Over 80 million cubic metres of an invisible, odourless gas suddenly escaped from the lake. Such a huge volume in a very short time suffocated them. How could you identify the gas?

How do you detect carbon dioxide?

You test for carbon dioxide using **limewater**. Limewater is calcium hydroxide solution. Limewater turns cloudy white when carbon dioxide is bubbled through it (Figure 2). The change is caused by the calcium hydroxide reacting with carbon dioxide to form water and a white precipitate of calcium carbonate.

> **A** Write down the balanced equation for the reaction between calcium hydroxide and carbon dioxide, including state symbols.

How do you detect chlorine?

Chlorine dissolves in water to form an acidic solution. It also bleaches dyes – changes them from coloured to colourless. These properties are the basis of a simple laboratory test for chlorine:

1 Use a drop of tap water to dampen a piece of blue **litmus paper**.

2 Hold the paper near to a container that holds the substance.

3 If chlorine is present, the paper turns red then white (Figure 3).

> **B** Explain why the litmus paper must be damp for the chlorine test to work.

Figure 3 *Chlorine water releases chlorine in the gas state, which bleaches damp litmus paper.*

How do you smell substances in the laboratory?

You can often use your sense of smell to detect substances. For example, chlorine has a distinctive 'swimming pool' smell. However, you must not just breathe in a lungful of what might be a harmful or toxic substance! To safely smell substances in the laboratory you should always follow these two steps:

1   With the container well away from your nose, breathe in enough air to almost fill your lungs.

2   Hold the container a few centimetres away from your nose, and waft any smell towards you. Take a cautious sniff.

Holding the container well away from your nose makes sure you do not breathe in a lot of the substance. This also ensures that, should any of the substance escape, it will not go up your nose.

How do you detect hydrogen and oxygen?

You test for hydrogen by placing a lighted splint near the mouth of the container of gas:

● If hydrogen is present, it should ignite with a squeaky pop.

You test for oxygen by placing a glowing splint near the mouth of the container of gas:

● If oxygen is present, the splint should relight (Figure 4).

> **C**   In the test for hydrogen, oxygen in the air reacts with hydrogen to produce water. Write down the balanced equation for this reaction.

Go further

There is another test for chlorine. Damp **starch–iodide paper** turns blue-black in the presence of chlorine. Chlorine displaces iodine from iodide ions in the damp paper. The starch in the paper then turns blue-black.

Write an equation for the reaction between chlorine and potassium iodide.

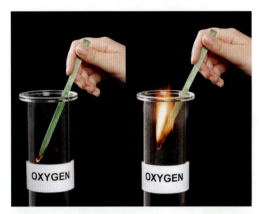

Figure 4   *A glowing wooden splint relights in oxygen.*

1   Describe how to detect chlorine and oxygen. State what you would do, and what you would observe.   *(4 marks)*

2   Hydrogen and carbon dioxide are both invisible and odourless (do not smell). Describe how you could confirm that the gas which escaped from Lake Nyos (Figure 1) was carbon dioxide, and not hydrogen.   *(4 marks)*

3   If excess carbon dioxide is bubbled through limewater, the calcium carbonate precipitate eventually re-dissolves.

    a   Write down the balanced equation, including state symbols, for the reaction between water and carbon dioxide to form carbonic acid, H_2CO_3.   *(2 marks)*

    b   Use your answer to **a** to suggest why the precipitate re-dissolves.   *(2 marks)*

Learning outcomes

After studying this lesson you should be able to:

- describe how to carry out and interpret flame tests
- describe hydroxide precipitate tests to identify aqueous metal ions.

Specification reference: C4.2b

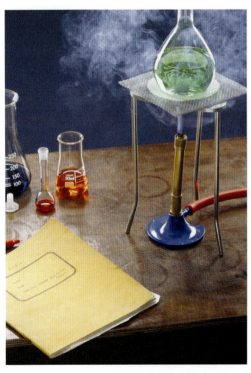

Figure 1 *Chemists have used the Bunsen burner since 1855.*

Figure 2 *A flame test with a copper compound.*

Who invented the Bunsen burner (Figure 1)? Surprisingly, it was not the German chemist Robert Bunsen, although he was involved in its development. Bunsen needed a clean, hot flame for his experiments with metal compounds. Peter Desaga, his technician, developed an earlier design by Humphry Davy's assistant, Michael Faraday.

What is a flame test?

When metal ions are heated, energy is transferred to their electrons. This makes their electrons move into higher electron shells. When they move back to their normal electron shells, energy is transferred to the surroundings as radiation, which you see as light. Different metal ions produce different colours of light. This is the basis of **flame tests** (Figure 2), and it is also what makes fireworks so exciting to watch. Table 1 shows the flame colours for common metal cations.

Table 1 *Flame test colours.*

Metal	Ion	Flame test colour
lithium	Li^+	red
sodium	Na^+	yellow
potassium	K^+	lilac
calcium	Ca^{2+}	orange-red
copper	Cu^{2+}	green-blue

Doing a flame test

You can carry out flame tests using a nichrome wire loop attached to a handle. You may need to clean the loop several times by dipping it in hydrochloric acid, then rinsing it with distilled water. Once it is clean, there will be no change in the colour of the Bunsen burner flame when you hold the loop in it.

To test a substance:

1   Dip the clean loop into the test powder or solution.

2   Use the handle to hold the loop in the edge of a roaring blue flame.

3   Record the flame colour.

Cleaning the wire loop between each test ensures that it is not contaminated with the previous test substance so you will not confuse flame colours or mix them.

A   Flame tests can be carried out using disposable damp wooden splints. Suggest an advantage of using these instead of wire loops.

What are hydroxide precipitate tests?

Group 1 metal hydroxides are soluble in water but most other metal hydroxides are insoluble. This is why you often use sodium hydroxide solution (and not other metal hydroxides) in experiments. Copper(II) hydroxide is an insoluble metal hydroxide formed when copper(II) sulfate solution reacts with sodium hydroxide solution:

$$CuSO_4(aq) + 2NaOH(aq) \rightarrow Cu(OH)_2(s) + Na_2SO_4(aq)$$

Different metals produce different coloured precipitates (Figure 3). This is the basis of hydroxide precipitate tests. You add a few drops of sodium hydroxide solution to a solution containing metal ions, and note the colour of any precipitate formed (Table 2).

Table 2 *Colours of different hydroxide precipitates.*

Name of ion	Formula of ion	Colour of hydroxide precipitate
iron(II)	Fe^{2+}	green
iron(III)	Fe^{3+}	orange-brown
copper(II)	Cu^{2+}	blue
calcium	Ca^{2+}	white
zinc	Zn^{2+}	white

Calcium hydroxide and zinc hydroxide are both white. However, if you add an excess of sodium hydroxide solution, zinc hydroxide dissolves to form a colourless solution but calcium hydroxide does not dissolve.

B Describe how you could distinguish between iron(II) chloride solution and iron(III) chloride solution.

Go further

Robert Bunsen and Gustav Kirchhoff discovered two elements in 1860 using a device they had invented to analyse colours from flame tests. Find out what these elements were and what their device was.

Figure 3 *Iron(II) hydroxide, iron(III) hydroxide, and copper(II) hydroxide.*

1 Outline two ways to detect copper(II) compounds. (*4 marks*)

2 Explain why you must use flame tests to detect lithium, sodium, and potassium ions, rather than hydroxide precipitate tests. (*3 marks*)

3 Calcium chloride, $CaCl_2$, and zinc chloride, $ZnCl_2$, are both soluble white solids at room temperature. Describe the tests you would use to distinguish between them. Include what you would do, what you would see, and equations for the formation of any precipitates. (*6 marks*)

C4.2.3 Detecting anions

Learning outcomes

After studying this lesson you should be able to:

- describe tests to detect sulfates, carbonates, and halides
- identify compounds from test results.

Specification reference: C4.2b, C4.2d

Figure 1 *This patient is having a 'barium meal', a sort of milkshake containing barium sulfate.*

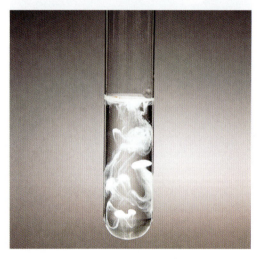

Figure 2 *A white precipitate of barium sulfate.*

X-rays are often used to make images of broken bones, but they can also be used to make images of the digestive system. The patient swallows some barium sulfate given to them by the radiographer. Barium ions absorb X-rays. This makes the digestive system appear white on X-ray photographs, ready for doctors to study (Figure 1). Barium ions can also be used in the laboratory for simple chemical tests.

How do you detect sulfate ions?

Barium ions react with sulfate ions to produce white, insoluble barium sulfate:

$$Ba^{2+}(aq) + SO_4^{2-}(aq) \rightarrow BaSO_4(s)$$

This forms the basis of a simple laboratory test to detect sulfate ions in a solution:

- add a few drops of dilute hydrochloric acid
- then a few drops of barium chloride solution, $BaCl_2(aq)$.

If sulfate ions are present, you get a white precipitate (Figure 2).

A Write down the balanced equation, including state symbols, for the reaction between barium chloride solution and sodium sulfate solution, $Na_2SO_4(aq)$.

How do you detect carbonate ions?

Hydrogen ions from dilute acids react with carbonate ions to produce carbon dioxide and water:

$$2H^+(aq) + CO_3^{2-}(aq) \rightarrow CO_2(g) + H_2O(l)$$

This forms the basis of a simple laboratory test to detect carbonate ions. It works whether the carbonate compound is in the solid state or in a solution. This is useful because most carbonates, such as magnesium carbonate, $MgCO_3$, are insoluble. To detect carbonate ions:

- add a few drops of dilute hydrochloric acid.

If carbonate ions are present, bubbles of gas will be produced (Figure 3). You can confirm that the gas is carbon dioxide by bubbling it through limewater. You could use dilute nitric acid, but dilute hydrochloric acid is used more often.

B Write down the balanced equation, including state symbols, for the reaction between magnesium carbonate powder and dilute hydrochloric acid.

How do you detect halide ions?

Silver fluoride is soluble in water, but the other silver halides are insoluble. This forms the basis of a simple laboratory test to detect chloride, bromide, and iodide ions in a solution:

- add a few drops of dilute nitric acid
- then a few drops of silver nitrate solution, $AgNO_3(aq)$.

If one of these ions is present, you get a precipitate. Table 1 and Figure 4 show their colours.

Table 1 *Colours of silver halide precipitates.*

Name of halide ion	Formula of ion	Colour of silver halide precipitate
chloride	Cl^-	white
bromide	Br^-	cream
iodide	I^-	yellow

Figure 4 *$AgCl(s)$ is white, $AgBr(s)$ is cream coloured, and $AgI(s)$ is yellow.*

Figure 3 *Bubbles of gas from the reaction between magnesium carbonate and hydrochloric acid.*

C Write a balanced equation (include state symbols) for the reaction between silver nitrate solution and sodium chloride solution.

1 Soluble barium salts are toxic but insoluble ones are not.
   a Suggest a suitable precaution to take when carrying out tests for sulfate ions. *(2 marks)*
   b Explain why barium sulfate is used for medical imaging but barium chloride is not. *(2 marks)*

2 Silver carbonate is insoluble in water.
   a Describe how you would carry out a test to detect the carbonate ions in silver carbonate. *(2 marks)*
   b Suggest why dilute nitric acid is added first when testing for the presence of halide ions using silver nitrate solution. *(3 marks)*

3 Describe the chemical tests you would use to distinguish between sodium iodide, potassium sulfate, and potassium carbonate. In your answer, outline what you would do and the results you expect. *(6 marks)*

Learning outcomes

After studying this lesson you should be able to:

- describe the advantages of using scientific instruments to analyse substances
- interpret results from an instrumental analysis.

Specification reference: C4.2f, C4.2g

How do you analyse the soil and atmosphere on another planet? It would be exciting to go there yourself, but that is too difficult and expensive at the moment. Instead, space scientists use automated probes and rovers to do the job for them (Figure 1). These carry scientific instruments, similar to ones used on Earth, to analyse samples.

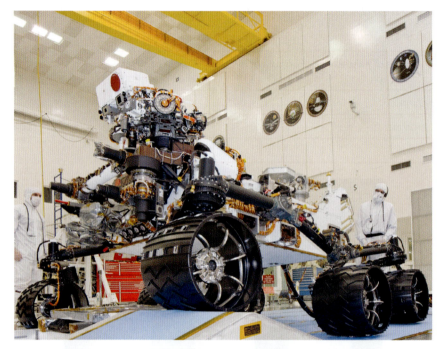

Figure 1 *The Mars Curiosity rover, launched in 2011, carries scientific instruments such as a gas chromatograph and mass spectrometer.*

What are instrumental methods of analysis?

An **instrumental method of analysis** relies on a machine to carry out an analysis of a substance. Figure 2 shows an example of an instrumental method of analysis. This is different to simple chemical methods such as hydroxide precipitate tests. Instrumental methods have several advantages. These include:

- Sensitivity – instruments can analyse very small amounts of substances. This is useful if the substance is expensive or difficult to obtain.
- Accuracy – instruments are very accurate. They can be calibrated using internationally accepted standards.
- Speed – instruments can carry out analyses quickly and they can run all the time.

Figure 2 *A 'carousel' of samples in a laboratory, with a robot arm to carry them to the instrument.*

A Suggest why instrumental methods of analysis are useful for analysing samples from a crime scene.

How do you interpret gas chromatograms?

A gas chromatogram is a chart that represents the different substances in a mixture:

- each peak represents a substance present in the mixture
- the areas under each peak show the relative amount of each substance in the mixture
- the **retention time**, the time taken for a substance to travel through the chromatography column, is different for different substances.

> **B** The police suspect that a car driver is over the legal limit for alcohol. The gas chromatogram in Figure 3 is from a sample of the driver's urine. Explain how the chromatogram could be used to determine whether the driver was over the legal limit.

How do you interpret mass spectra?

A **mass spectrometer** can measure the masses of atoms and molecules. It is used to analyse the relative amounts of different isotopes of an element, and the structure of molecules. Mass spectrometry is used in many fields of science, such as in environmental science, for which scientists use it to detect toxic substances.

The mass spectrometer is particularly useful for analysing molecules. The sample molecules are ionised by the machine to form **molecular ions**. These may break up to form fragments, which the machine can separate and detect.

In a mass spectrum (Figure 4):

- each peak represents a fragment of the molecule
- the peak on the far right represents the molecular ion.

The mass to charge ratio of the **molecular ion** peak is equal to the relative formula mass, M_r, of the molecule.

> **C** Explain why the mass spectrum in Figure 4 could be for butane, C_4H_{10}.

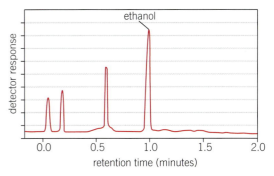

Figure 3 *A gas chromatogram of a sample of urine.*

> **Synoptic link**
>
> You can learn more about how gas chromatography works C2.1.6 *Chromatography*.

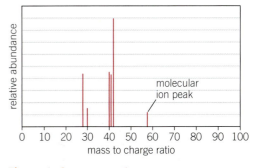

Figure 4 *A mass spectrum.*

Table 1 *Wavenumbers for different bonds in an infrared spectrum.*

Bond	Wavenumber
C–H	2850–3300
C–C	750–1100
C=C	1620–1680
C=O	1680–1750
C–O	1000–1300
N–H	3360–3500
O–H	3230–3550

> **1** Describe two advantages of instrumental methods of analysis. *(2 marks)*
>
> **2** A gas chromatogram can be coupled to a mass spectrometer. As each substance leaves the chromatography column, it passes to the mass spectrometer. Suggest a reason for combining these two instruments. *(2 marks)*
>
> **3** An instrument called an infrared spectrometer can detect covalent bonds in a molecule. Table 1 shows the 'wavenumber' for different bonds in an infrared spectrum. Ethanol produces a spectrum with peaks at 3358, 2974, 1274, and 880. Use Table 1 to identify the bonds present. *(4 marks)*

Predicting and identifying reactions and products

C4.2 Identifying the products of chemical reactions

Summary questions

1 Hydrogen and oxygen can be detected using a wooden splint. Copy and complete each of these sentences, including only the correct word in each pair of words shown in **bold**.
   a A **lighted | glowing** splint **pops | ignites** in hydrogen.
   b A **lighted | glowing** splint **pops | ignites** in oxygen.

2 Describe simple laboratory tests for the following substances in the gas state. In your answers, write down any essential materials and what you expect to observe.
   a carbon dioxide
   b chlorine.

3 Some metal ions can be detected using flame tests.
   a Complete Table 1 (some boxes are already completed for you).

   Table 1 *Flame tests.*

Metal	Formula of ion	Flame test colour
		orange–red
	K^+	
sodium		
	Li^+	

   b Outline how to carry out a flame test.
   c Suggest why magnesium ions cannot be detected using a flame test.

4 Group 1 metal hydroxides are soluble in water. Their hydroxide ions react with ions of other metals to form insoluble hydroxide precipitates. The colour of each precipitate can be used to identify the metal ion it contains.
   a Write down the colour of the hydroxide precipitate formed by:
      i copper(II) ions, Cu^{2+}
      ii iron(II) ions, Fe^{2+}
      iii iron(III) ions, Fe^{3+}.
   b Calcium hydroxide, $Ca(OH)_2(s)$, and zinc hydroxide, $Zn(OH)_2(s)$, are both white. Explain how you would use sodium hydroxide solution to distinguish between them.

5 Halide ions (Cl^-, Br^-, or I^-) can be detected in a simple laboratory test in which precipitates are formed.
   a Copy and complete each of these sentences, including only the correct word in each pair of words shown in **bold**.
   A few drops of dilute **nitric | hydrochloric** acid are added to the test solution.
   A few drops of **barium | silver** nitrate solution are then added, and the colour of the precipitate observed.
   b Complete Table 2 with the correct precipitate colours.

   Table 2 *Halide precipitates.*

Halide ion	Colour of precipitate formed
bromide	
chloride	
iodide	

6 Sulfate ions, SO_4^{2-}, can be detected in a simple laboratory test in which a precipitate is formed.
   a Copy and complete each of these sentences, including only the correct word in each pair of words shown in bold.
   A few drops of dilute **sulfuric | hydrochloric** acid are added to the test solution.
   A few drops of **silver | barium** chloride solution are then added.
   A **white | yellow** precipitate forms if sulfate ions are present.
   b Write down the name or formula of the precipitate.

7 Carbonate ions, CO_3^{2-}, can be detected using acids. Write down a balanced equation, including state symbols, for the reaction between calcium carbonate powder and hydrochloric acid.

8 Mass spectrometry, gas chromatography, and infrared spectroscopy are examples of instrumental methods of analysis. Describe three advantages of instrumental methods over simple methods such as flame tests.

154

Revision questions

1 Which of the following describes the correct results of a laboratory test for chlorine?
   A Damp blue litmus paper turns red, then white.
   B Damp blue litmus paper turns white, then red.
   C Damp red litmus paper turns blue, then white.
   D Damp red litmus paper turns white, then blue.
   (1 mark)

2 A student carries out a flame test on a sample of sodium chloride.
   What colour does the flame become during the test?
   A yellow
   B red
   C green–blue
   D lilac *(1 mark)*

3 A student adds sodium hydroxide solution to a small sample of iron(III) sulfate solution.
   A precipitate is made.
   What is the colour of the precipitate?
   A orange–brown
   B green
   C blue
   D white *(1 mark)*

4 Some gases can be identified using different laboratory tests.
   For each of the following, describe how you carry out the test and what you observe.
   a hydrogen *(2 marks)*
   b oxygen *(2 marks)*
   c carbon dioxide *(2 marks)*

S 5 A student wants to identify the ions contained in a solid, **X**.
   She dissolves it in a little water, then carries out some tests on the solution.
   The following table shows her results.

Test	Method used	Observations
Test 1	Carry out a flame test	Orange–red flame seen
Test 2	Add dilute sodium hydroxide solution	White precipitate forms, which redissolves in excess sodium hydroxide solution
Test 3	Add dilute nitric acid, then silver nitrate solution	Cream coloured precipitate seen
Test 4	Add dilute hydrochloric acid, then barium chloride solution	No change – mixture stays clear and colourless

   a Use the table to identify the ion that produces:
      i the flame colour seen in Test 1 *(1 mark)*
      ii the white hydroxide precipitate in Test 2 *(1 mark)*
      iii the yellow precipitate in Test 3. *(1 mark)*
   b Identify the ion **not** present in compound **X**, as shown in Test 4. *(1 mark)*
   c The student could have carried out a flame test on **X** when it was solid.
      Describe how she should do this. *(3 marks)*

S **H** 6* a Describe in detail how you would carry out simple laboratory tests to distinguish between the following four compounds:
   sodium iodide
   sodium carbonate
   potassium sulfate
   iron(III) chloride
   Include the observations you would expect to make. *(6 marks)*
   b Write balanced symbol equations, with state symbols, for four of the reactions you described in **a**. *(4 marks)*

C4.1 Predicting chemical reactions

- Describe some physical and chemical properties of Group 1 elements.

- Explain the reactions of Group 1 elements in terms of electronic structure.

- Explain the trend in reactivity of Group 1 elements in terms of how easily their atoms lose electrons in reactions.

- Describe physical and chemical properties of Group 7 (IUPAC Group 17) elements.

- Explain the reactions of Group 7 (IUPAC Group 17) elements in terms of electronic structure.

- Explain the trend in reactivity of Group 7 (IUPAC Group 17) elements in terms of how easily their atoms gain electrons in reactions.

- Describe and explain displacement reactions between Group 7 (IUPAC Group 17) elements and their ions.

- Describe and explain the properties of Group 0 (IUPAC Group 18) elements.

- Explain the lack of reactivity of Group 0 (IUPAC Group 18) elements in terms of the electronic structure of their atoms.

- Predict properties of elements in Groups 1, 7, and 0 (IUPAC Groups 1, 17, and 18) from given trends.

- Describe the general properties of transition metals. **S**

- Explain the reactivity of different metals in terms of how easily their atoms lose electrons in reactions.

- Use experimental data to put metals into their order of reactivity.

- Predict the reactions and reactivity of elements from their position in the Periodic Table.

C4.2 Identifying the products of chemical reactions

- Describe laboratory tests to detect oxygen, hydrogen, carbon dioxide, and chlorine.

- Describe how to carry out and interpret flame tests. **S**

- Describe hydroxide precipitate tests to identify aqueous metal ions.

- Describe tests to detect sulfates, carbonates, and halides.

- Identify compounds from test results.

- Describe the advantages of using scientific instruments to analyse substances.

- Interpret results from an instrumental analysis.

Reactivity series

potassium — most reactive
sodium
calcium
magnesium
aluminium
zinc
iron
tin
lead
(hydrogen)
copper
silver
gold
platinum — least reactive

Gas tests

Hydrogen
- lighted splint ignites gas with a pop

Oxygen
- glowing splint relights

Chlorine
- damp blue litmus paper turns red then white

Carbon dioxide
- turns limewater cloudy

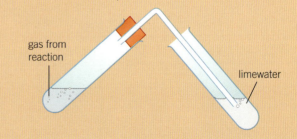

gas from reaction

limewater

Detecting cations

Flame tests

- carried out using nichrome wire loop and Bunsen burner
- different metal cations produce different flame colours

Flame colour	red	yellow	lilac	orange-red	green-blue
Metal ion	Li^+ lithium	Na^+ sodium	K^+ potassium	Ca^{2+} calcium	Cu^{2+} copper

Hydroxide precipitate tests

TELL APART AS Ca^{2+} PRECIPITATE WILL NOT DISSOLVE IN EXCESS SODIUM HYDROXIDE

- carried out using sodium hydroxide solution
- different metals produce different coloured precipitates

Precipitate colour	green	orange-brown	blue	white	white
Metal ion	Fe^{2+} iron(II)	Fe^{3+} iron(III)	Cu^{2+} copper	Ca^{2+} calcium	Zn^{2+} zinc

C4 Predicting and identifying reactions and products

Detecting anions

Halide ion test

1 Add a few drops of dilute nitric acid
2 Add a few drops of silver nitrate solution
3 Observe colour of precipitate formed

Precipitate colour	white	cream	yellow
Halide ion	Cl^- chloride	Br^- bromide	I^- iodide

Sulfate ion test

1 Add a few drops of dilute hydrochloric acid
2 Add a few drops of barium chloride solution
3 Look for white precipitate of barium sulfate

Carbonate ion test

1 Add a few drops of dilute nitric acid
2 Look for bubbles
3 Confirm gas is carbon dioxide using limewater

Group 1 – Alkali metals

- solid at room temperature
- float on water
- reactive metals (lithium fizzes, sodium melts, and potassium ignites with water)

Going down the group

- melting point ↓
- reactivity ↑ (atoms lose electrons)
- hardness ↓
- density ↑

Group 7 – Halogens
IUPAC Group 17

- reactive non metals

Going down the group

- melting point ↑
- reactivity ↓ (atoms gain electrons)
- boiling point ↑
- density ↑

Halogen displacement

- a more reactive halogen displaces a less reactive halogen from its halide ions in solution:
- e.g. $Cl_2 + 2KI \rightarrow 2KCl + I_2$ (mixture darkens)

Group 0 – Noble gases
IUPAC Group 18

- gas at room temperature
- unreactive (inert) non-metals

Going down the group

- melting point ↑
- boiling point ↑
- density ↑

Transition metals

- form coloured ions with different charges
- used as catalysts

Compared to Group 1

- higher melting points
- less reactive

Instrumental methods of analysis

Advantages

- Sensitivity (instruments can analyse very small amounts of substances)
- Accuracy (instruments are very accurate)
- Speed (instruments can carry out analyses quickly and they can run all the time)

C5 Monitoring and controlling chemical reactions
C5.1 Monitoring chemical reactions
C5.1.1 Theoretical yield
GCSE CHEMISTRY ONLY

Learning outcome

After studying this lesson you should be able to:

- calculate the theoretical yield of a product from a given mass of reactant.

Specification reference: C5.1g

Figure 1 *Agronomists study ways to improve the quality and yield of crops.*

Synoptic link

You can learn more about the law of conservation of mass in C3.1.3 *Conservation of mass*.

Figure 2 *Ammonia is used in some household cleaning fluids.*

Agronomists are scientists who specialise in producing and improving food crops. They work with farmers to increase crop yields, the total mass of food produced in a field (Figure 1). Chemists are also interested in yields, but in chemistry the yield is the mass of chemical product made in a reaction.

How can you calculate the theoretical yield?

The **yield** of a product is the mass of that product made in a chemical process. The **theoretical yield** is the maximum mass it is possible to make from a given mass of reactants. You can calculate a theoretical yield using the law of conservation of mass if you know:

- the mass of the **limiting reactant**
- the relative formula masses of reactants and products
- the balanced equation for the reaction.

Theoretical yield

Nitrogen reacts with hydrogen to make ammonia (Figure 2):

$$N_2 + 3H_2 \rightarrow 2NH_3$$

Calculate the theoretical yield of ammonia when 12.0 g of hydrogen reacts with excess nitrogen.

(A_r of N = 14.0, A_r of H = 1.0)

Step 1: Calculate the relative formula masses of the limiting reactant and the product, if you are not given them.

Hydrogen is the limiting reactant here because nitrogen, the only other reactant, is in excess.

M_r of H_2 = (2 × 1.0) = 2.0   (limiting reactant)

M_r of NH_3 = 14.0 + (3 × 1.0) = 17.0   (product)

Step 2: Use the numbers from the balanced equation to calculate the sum of the M_r for each substance in Step 1.

sum of M_r for H_2 = (3 × 2.0) = 6.0

sum of M_r for NH_3 = (2 × 17.0) = 34.0

Step 3: Use the mass of limiting reactant and ratios from Step 2 to calculate the theoretical yield.

$$\text{theoretical yield} = \frac{\text{mass of limiting reactant}}{\text{sum of } M_r \text{ for limiting reactant}} \times \text{sum of } M_r \text{ for products}$$

$$\text{theoretical yield} = \frac{12.0\,g}{6.0} \times 34.0 = 68.0\,g$$

A Write down the meaning of the term 'theoretical yield'.

B Methane burns in oxygen to produce carbon dioxide and water:

$$CH_4 + 2O_2 \rightarrow CO_2 + 2H_2O$$

Calculate the theoretical yield of water when 4.80 g of methane burns completely.

(A_r of C = 12.0, A_r of H = 1.0, A_r of O = 16.0)

C Aluminium reacts with iodine to produce aluminium iodide (Figure 3):

$$2Al + 3I_2 \rightarrow 2AlI_3$$

Calculate the theoretical yield of aluminium iodide when 10.8 g of aluminium reacts with excess iodine.

(A_r of Al = 27.0, M_r of AlI_3 = 407.7)

Figure 3 *Aluminium reacts vigorously with iodine in the presence of water.*

Study tip

If the products in a reaction have complex formulae, it may be easier to calculate the sum of M_r values for the reactants, rather than for the products. Since mass is conserved in chemical reactions, you could use this value for the sum of M_r values for the products.

1 Carbon burns in oxygen to produce carbon dioxide:

$$C + O_2 \rightarrow CO_2$$

  a Calculate the M_r of carbon dioxide. *(1 mark)*
  b Calculate the theoretical yield of carbon dioxide when 6.0 g of carbon burns completely. *(2 marks)*

2 Copper(II) carbonate decomposes when heated to produce copper(II) oxide and carbon dioxide:

$$CuCO_3 \rightarrow CuO + CO_2$$

Calculate the theoretical yield of copper(II) oxide from 24.7 g of copper(II) carbonate (A_r of Cu = 63.5). *(3 marks)*

3 1.35 g of aluminium reacts with hydrochloric acid, HCl. The products are aluminium chloride, $AlCl_3$, and hydrogen, H_2.
  a Write the balanced equation for the reaction, including state symbols. *(2 marks)*

  b Calculate the amount, in moles, of aluminium. *(2 marks)* **H**
  c Calculate the amount of hydrogen, in moles, that could be made. *(1 mark)*
  d Use your answer to **c** to calculate the theoretical yield of hydrogen. *(2 marks)*

C5.1.2 Percentage yield and atom economy

Learning outcomes

After studying this lesson you should be able to:

- calculate the percentage yield of a product
- define the atom economy of a reaction
- calculate the atom economy of a reaction.

Specification reference: C5.1h, C5.1i, C5.1j

Figure 1 *It is difficult to dispose of toxic and hazardous waste.*

Figure 2 *Purifying blue copper(II) sulfate crystals using filtration.*

Sustainable development is the idea that you should meet your needs without damaging the ability of future generations to meet their needs. Chemists play a part in this by developing processes that have high percentage yields and atom economies, reducing the use of raw materials and production of waste (Figure 1).

How do you calculate percentage yield?

The **actual yield** is the mass of product you actually make in a chemical reaction. It is usually less than the theoretical yield. You calculate the **percentage yield** of a reaction product using this equation:

$$\text{percentage yield} = \frac{\text{actual yield}}{\text{theoretical yield}} \times 100$$

Percentage yield

The theoretical yield of a product was 8.00 g but the actual yield was only 7.10 g. Calculate the percentage yield, giving your answer to 3 significant figures.

$$\text{percentage yield} = \frac{\text{actual yield}}{\text{theoretical yield}} \times 100 = \frac{7.10\,\text{g}}{8.00\,\text{g}} \times 100$$

$$= 88.8\% \text{ to three significant figures}$$

A The theoretical yield of a product was 3.24 g and its actual yield was 2.97 g. Calculate the percentage yield, giving your answer to three significant figures.

What affects percentage yield?

Percentage yield varies from 0% to 100%, depending upon the mass of product you actually obtain. It may be less than 100% because:

- The reactants may react in a different way than expected. For example, if you burn lithium in air to make lithium oxide, you might also make lithium nitride.

- The reaction may not go to completion – some of the reactants present do not react. This often happens in **reversible reactions**, but it happens in other reactions too.

- You may lose some of the product when you separate it from the reaction mixture and purify it – for example, during filtration or when you transfer substances between containers.

B Explain how Figure 2 shows that some copper(II) sulfate was lost during filtration.

How do you calculate atom economy?

The **atom economy** of a reaction is a measure of how many atoms in the reactants form a desired product. You calculate it using this equation:

$$\text{atom economy} = \frac{\text{sum of } M_r \text{ of the desired product}}{\text{sum of } M_r \text{ of all products}} \times 100$$

Atom economy

Hydrogen can be produced by reacting methane with steam:

$$CH_4 + H_2O \rightarrow CO + 3H_2$$

Calculate the atom economy of the process, giving your answer to three significant figures.

Step 1: Calculate the M_r for each product.

M_r of CO = 12.0 + 16.0 = 28.0

M_r of H_2 = (2.0 × 1.0) = 2.0

Step 2: Calculate the atom economy using the values from Step 1 and numbers from the balanced equation.

$$\text{atom economy} = \frac{\text{sum of } M_r \text{ of the desired product}}{\text{sum of } M_r \text{ of all products}} \times 100$$

$$= \frac{(3 \times 2.0)}{28.0 + (3 \times 2.0)} \times 100$$

$$= \frac{6.0}{34.0} \times 100$$

= 17.6% to three significant figures

Figure 3 *Modern chemical processes have high percentage yields or atom economies.*

C Hydrogen can also be produced by reacting carbon with steam:

$$C + 2H_2O \rightarrow CO_2 + 2H_2$$

Calculate the atom economy of the process, giving your answer to three significant figures.

Study tip

Percentage yield and atom economy cannot be greater than 100%.

---

1 The theoretical yield of a product was 16.3 kg and its actual yield was 12.7 kg. Calculate the percentage yield, giving your answer to three significant figures. *(2 marks)*

2 Hydrogen can be produced by the electrolysis of water:

$$2H_2O \rightarrow 2H_2 + O_2$$

Calculate the atom economy of the process, giving your answer to three significant figures. *(2 marks)*

3 Evaluate the three processes to make hydrogen described on this page. Explain which one may be best for sustainable development. *(6 marks)*

Learning outcome

After studying this lesson you should be able to:

- explain why a particular reaction pathway is chosen to produce a specified product, using appropriate data.

Specification reference: C5.1k

Figure 1 *Ethanediol is an ingredient of car screenwash.*

Synoptic link

You can learn more about rate of reaction in C5.2.1 *Rate of reaction*.

Car screenwash contains ethanediol, an antifreeze made from epoxyethane, in order to stop it freezing in winter (Figure 1). Now, epoxyethane is made in a different way to how it used to be made. The original method for making epoxyethane was abandoned in the 1970s.

What factors determine how you make a substance?

Manufacturers in the chemical industry must be able to make a profit when producing substances, otherwise they will go out of business. Chemical engineers consider many factors when choosing a **reaction pathway**, that is the reaction or series of reactions for making a particular substance. These include the:

- yield of the product
- atom economy of the reaction
- usefulness or otherwise of by-products
- rate of the reaction
- equilibrium position, if it is a reversible reaction.

A Suggest why chemical engineers consider the cost of raw materials when choosing a reaction pathway.

What happened with epoxyethane?

The manufacture of epoxyethane, C_2H_4O, illustrates how manufacturers choose a reaction pathway. Epoxyethane (Figure 2) is made from ethene, a compound obtained from crude oil. The original pathway was complex and used two steps:

Step 1: ethene + chlorine + water → chloroethanol + hydrogen chloride

Step 2: chloroethanol + calcium hydroxide → calcium chloride + water + epoxyethane

The overall reaction was:

ethene + chlorine + calcium hydroxide → calcium chloride + water + epoxyethane

$$C_2H_4 + Cl_2 + Ca(OH)_2 \rightarrow CaCl_2 + H_2O + C_2H_4O$$

The yield of this process was about 80%, but its atom economy was just 25.4%.

B Use your knowledge and understanding of Group 7 (IUPAC Group 17) elements, and their compounds, to suggest why the first reaction (Step 1) may have been hazardous.

The modern process has just one step, using silver as a catalyst:

$$\text{ethene} + \text{oxygen} \rightarrow \text{epoxyethane}$$
$$2C_2H_4 + O_2 \rightarrow 2C_2H_4O$$

The yield of this process is also about 80%, but its atom economy is 100%. Although both processes waste some reactants, the higher atom economy of the modern process means that it is more efficient overall.

What can you do with by-products?

A **by-product** is a substance formed in a reaction in addition to the desired product. Sometimes a by-product is useful and can be sold. This improves the atom economy of a process because the by-product becomes a desirable product. Calcium chloride is one of the by-products of the original epoxyethane process. It has many uses, including as an additive in concrete and polymers and to trap moisture in de-humidifiers (Figure 3). On the other hand, by-products may be toxic or be of little use, making a reaction pathway involving them less desirable.

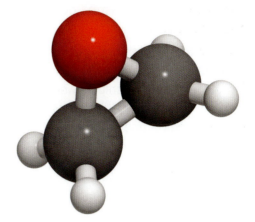

Figure 2 *A ball-and-stick model of an epoxyethane molecule.*

Figure 3 *Calcium chloride is used to trap moisture in de-humidifiers.*

1 The atom economy of the original epoxyethane process improved when calcium chloride could be sold. Show by calculation that it became 89.6%. *(3 marks)*

2 Write down three factors that affect the choice of reaction pathway for making a substance. *(3 marks)*

3 Ethanol can be manufactured by reacting ethene with steam, or by fermentation of plant sugars.
   a Explain, using data from Table 1, which process appears to be more desirable. *(3 marks)*
   b Carbon dioxide is the by-product of fermentation.
      i Explain what the term 'by-product' means. *(1 mark)*
      ii Carbon dioxide can be sold to fizzy drinks manufacturers. Explain how this may affect the choice of process for making ethanol. *(3 marks)*

Table 1 *Comparison of methods of manufacturing ethanol*

Process	Atom economy	Yield	Rate
hydration of ethene	100%	95%	fast
fermentation	51.1%	15%	slow

Learning outcomes

After studying this lesson you should be able to:

- calculate concentration of solution in mol/dm^3
- explain the relationship between concentration of solution, mass of solute, and volume of solution.

Specification reference: C5.1a, C5.1f

Figure 1 *A saline drip contains 9 g/dm^3 or 0.154 mol/dm^3 sodium chloride.*

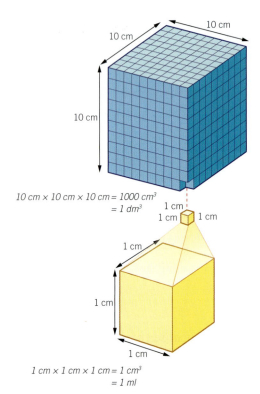

10 cm

10 cm

10 cm

10 cm

10 cm × 10 cm × 10 cm = 1000 cm^3
= 1 dm^3

1 cm

1 cm 1 cm

1 cm

1 cm

1 cm

1 cm × 1 cm × 1 cm = 1 cm^3
= 1 ml

Figure 2 *A cubic decimetre (blue) contains one thousand cubic centimetres (yellow).*

Hospital patients are often given an intravenous drip. This is a sterile solution delivered slowly into a vein (Figure 1). It is very important that the solutes in the drip are at the right concentrations. How do manufacturers work out these concentrations?

How do you calculate volumes?

When you buy a bottle of drink in a shop, the label usually gives the volume in millilitres, ml, or litres, l. However, chemists measure volumes in cubic centimetres, cm^3, and cubic decimetres, dm^3. One litre is one cubic decimetre, so one millilitre is one cubic centimetre. This useful to know if you see laboratory apparatus graduated (marked) in ml rather than in cm^3.

To carry out concentration calculations, you must be able to convert between cm^3 and dm^3:

- divide by 1000 to convert from cm^3 to dm^3
- multiply by 1000 to convert dm^3 to cm^3 (Figure 2).

A Explain why 25 cm^3 is equal to 0.025 dm^3.

How do you calculate concentrations?

There are two common units used to give concentrations: grams per cubic decimetre and moles per cubic decimetre. You use this expression to calculate concentration in g/dm^3:

$$\text{concentration in g/dm}^3 = \frac{\text{mass of solute in g}}{\text{volume of solution in dm}^3}$$

Concentration in g/dm^3

Calculate the concentration, in g/dm^3, of a solution containing 4.5 g of solute in 250 cm^3 of solution.

Step 1: Convert the volume to dm^3 if necessary.

$$\text{volume} = 250 \text{ cm}^3 \div 1000$$
$$= 0.25 \text{ dm}^3$$

Step 2: Calculate the concentration from the mass and volume.

$$\text{concentration in g/dm}^3 = \frac{4.5 \text{ g}}{0.25 \text{ dm}^3}$$
$$= 18 \text{ g/dm}^3$$

B Calculate the concentration, in g/dm^3, of a solution containing 0.75 g of solute in 150 cm^3 of solution.

You use this expression to calculate concentration in moles per cubic decimetre, mol/dm³:

$$\text{concentration in mol/dm}^3 = \frac{\text{amount of solute in mol}}{\text{volume of solution in dm}^3}$$

Concentration in mol/dm³

Calculate the concentration, in mol/dm³, of a solution containing 5.85 g of sodium chloride in 125 cm³ of solution.

Step 1: Convert the volume to dm³ if necessary.

$$\text{volume} = 125\,\text{cm}^3 \div 1000$$
$$= 0.125\,\text{dm}^3$$

Step 2: Calculate the amount of solute from its mass and molar mass.

$$\text{molar mass of NaCl} = (23.0 + 35.5)$$
$$= 58.5\,\text{g/mol}$$

$$\text{amount of NaCl in mol} = \frac{\text{mass in g}}{\text{molar mass in g/mol}}$$
$$= \frac{5.85\,\text{g}}{58.5\,\text{g/mol}}$$
$$= 0.100\,\text{mol}$$

Step 3: Calculate the concentration from the amount and volume.

$$\text{concentration in mol/dm}^3 = \frac{0.100\,\text{mol}}{0.125\,\text{dm}^3}$$
$$= 0.8\,\text{mol/dm}^3$$

C Calculate the concentration, in mol/dm³, of a solution containing 0.450 g of sodium chloride in 50.0 cm³ of solution. Give your answer to three significant figures.

Figure 3 *Like sheep in a flock, the more crowded a solution is with solute particles, the higher its concentration.*

Study tip

Converting between concentration units

To convert a concentration in mol/dm³ to g/dm³, multiply by the molar mass of the solute. To convert a concentration in g/dm³ to mol/dm³, divide by the molar mass.

Go further

Doctors and nurses may use milligrams per decilitre, mg/dl, for units of concentration. Work out how you convert mg/dl to g/dm³.

1 Calculate the concentration, in g/dm³, of a solution containing 36.5 g of solute in 500 cm³ of solution. *(2 marks)*

2 Rearrange the equation for calculating concentration in mol/dm³ to show how you calculate:
   a amount of solute, in mol, in a solution *(1 mark)*
   b volume of solution, in dm³. *(1 mark)*

3 **a** Calculate the amount of solute dissolved in 2 dm³ of a 0.5 mol/dm³ solution. *(2 marks)*
   b Calculate the volume of a 2 mol/dm³ solution that contains 0.5 mol of solute. *(2 marks)*
   c Calculate the mass of potassium iodide, KI, which must be dissolved to make 0.500 dm³ of a 0.125 mol/dm³ solution. Give your answer to three significant figures. *(3 marks)*

C5.1.5 Titrations

Learning outcome

After studying this lesson you should be able to:

- describe how to carry out an acid–alkali titration.

Specification reference: C5.1b

Figure 1 *These students are carrying out a titration using methyl orange indicator.*

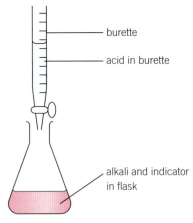

- burette
- acid in burette
- alkali and indicator in flask

Figure 2 *Phenolphthalein is pink in alkaline solutions, and colourless in acidic solutions.*

Go further

In your future chemistry studies after GCSE you might use titrations to study redox reactions. Find an example of a redox reaction that is suitable for titration.

Titration is a technique that uses a neutralisation reaction to find the concentration of an acid or an alkali (Figure 1). It involves several important laboratory skills.

What is a titration?

In a **titration** you usually add a known volume of alkali to a conical flask, and a few drops of a single indicator such as phenolphthalein (Figure 2). You then add acid to the flask from a **burette**, a long graduated glass tube with a tap at the bottom (Figure 3). At the **end-point**, when the alkali is neutralised and the indicator first changes colour, you stop adding acid.

> **A** Name the apparatus that is used in titrations and has a tap and graduated scale.

Standard solutions

A **standard solution** with a known, accurate concentration is often used in titrations. For example, an alkaline standard solution can be made by dissolving 1.00 g of $NaOH$ in about 150 cm^3 of water in a beaker. This solution is then added to a 250 cm^3 **volumetric flask**, and adding some more water to reach the 250 cm^3 mark on the flask.

> **B** Show that this standard solution has a concentration of 0.100 mol/dm^3.

How do you prepare to carry out a titration?

You need to take precautions to ensure you are ready to carry out an accurate titration.

- You could use a **measuring cylinder** to measure 25 cm^3 of alkali solution for the conical flask, but a **volumetric pipette** is more **accurate**. You need a **pipette filler** to use a volumetric pipette safely.
- The burette must be clamped vertically so you obtain accurate readings.
- When you have filled the burette with acid, and added a few drops of indicator to the alkali in the conical flask, you are ready.

Eye protection should be worn throughout this procedure.

> **C** Explain why a volumetric pipette is preferable to a measuring cylinder in titrations.

Doing a titration

When doing a titration you need to take an initial burette reading and a final burette reading at the end-point. The difference between these two readings is the **titre**, the volume of acid added to the alkali in the flask. Record your readings to two decimal places, ending in 0 if the bottom of the **meniscus** is on a burette line, or 5 if the meniscus is between two lines.

To obtain a **repeatable** titre you need to use the following techniques.

1 Swirl the flask during a titration to mix its contents.

2 Your first titration is usually a rough run, done quickly so that you get an idea of what the titre is. In later runs, you can quickly add the acid to within a few cm³ of the rough titre, then add the acid drop by drop.

Repeat the titration until you obtain at least two **concordant titres**, titres that are within 0.10 cm³ of each other. This allows you to calculate a mean titre that has high **precision**.

Eye protection should be worn throughout this practical.

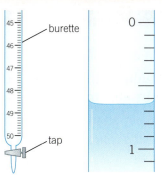

Figure 3 *The reading on this burette is 0.65 cm³, not 1.35 cm³.*

D When doing a titration, explain how you ensure a precise value for your mean titre.

E Explain why two burette readings are taken for each run in a titration.

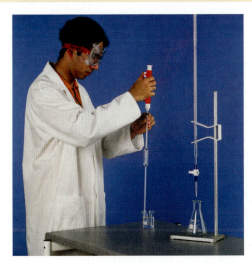

Figure 4 *This student is using a volumetric pipette with a pipette filler, and his burette is set up on the right.*

1 Describe one purpose of carrying out a titration. *(1 mark)*

2 Describe two precautions needed to obtain a repeatable titre. *(2 marks)*

3 Table 1 shows some results from repeating a titration.
   a Calculate the titre for Run 4. *(2 marks)*
   b Write down the concordant titres. *(1 mark)*
   c Use your answer to **b** to calculate the mean titre, giving your answer to two decimal places. *(2 marks)*

Table 1 *Results from a titration.*

	Run 1	Run 2	Run 3	Run 4
final burette reading (cm³)	25.90	49.30	24.35	48.90
initial burette reading (cm³)	Rough / .20	24.90	0.00	24.35
titre (cm³)	25.70	24.40	24.35	

Learning outcome

After studying this lesson you should be able to:

- carry out titration calculations involving concentrations and volumes.

Specification reference: C5.1c

Figure 1 *Samples of orange juice are tested during production.*

Study tip

This triangle may help you to rearrange the equation that links amount of substance, concentration, and volume.

n = amount in mol
c = concentration in mol/dm^3
V = volume in dm^3

Orange juice has a pH of around 3.5 because it contains citric acid and smaller amounts of other weak acids. Food scientists use titration to check the concentration of acid in the juice (Figure 1).

How can you use results from a titration?

When you have carried out a titration, you will know:

- the two reactants used (an acid and an alkali)
- the volume and concentration of one of the reactants
- the volume, but not the concentration, of the other reactant.

With this information, and the balanced equation, you can calculate the unknown concentration.

> **A** Write down an equation to show how to calculate the amount in moles from a concentration in mol/dm^3 and a volume in dm^3.

Titration calculation 1

$25.00\,cm^3$ of $0.100\,mol/dm^3$ sodium hydroxide solution was titrated with dilute hydrochloric acid. The mean titre of the acid was $20.00\,cm^3$. Calculate the concentration of the acid.

$$HCl(aq) + NaOH(aq) \rightarrow NaCl(aq) + H_2O(aq)$$

Step 1: Convert all volumes to dm^3.

volume of NaOH = $25.00\,cm^3 \div 1000 = 0.025\,dm^3$

volume of HCl = $20.00\,cm^3 \div 1000 = 0.020\,dm^3$

Step 2: You will know the concentration and volume of one of the reactants. Calculate its amount.

amount in mol = concentration in mol/dm^3 × volume in dm^3

amount of NaOH = $0.100\,mol/dm^3 \times 0.025\,dm^3$
$= 0.0025\,mol$

Step 3: Use the ratio in the balanced equation to calculate the amount of the other reactant.

1 mol of HCl reacts with 1 mol of NaOH

0.0025 mol of HCl reacts with 0.0025 mol of NaOH

Step 4: Use the amount of the reactant from Step 3, and its volume, to calculate its concentration.

$$\text{concentration in mol/}dm^3 = \frac{\text{amount in mol}}{\text{volume in }dm^3}$$

$$\text{concentration of HCl} = \frac{0.0025\,mol}{0.020\,dm^3}$$
$$= 0.125\,mol/dm^3$$

B 25.00 cm³ of 0.100 mol/dm³ sodium hydroxide solution was titrated with dilute hydrochloric acid. The mean titre of the acid was 29.40 cm³. Calculate the concentration of the acid, giving your answer to three significant figures.

Titration calculation 2

25.00 cm³ of 0.400 mol/dm³ sodium hydroxide solution was titrated with dilute sulfuric acid. The mean titre of the acid was 10.00 cm³. Calculate the concentration of the acid.

$$H_2SO_4(aq) + 2NaOH(aq) \rightarrow Na_2SO_4(aq) + 2H_2O(l)$$

Step 1: Convert all the volumes to dm³.

volume of $NaOH$ = 25.00 cm³ ÷ 1000 = 0.025 dm³

volume of H_2SO_4 = 10.00 cm³ ÷ 1000 = 0.010 dm³

Step 2: Calculate the amount of $NaOH$ (you know its concentration and volume).

amount of $NaOH$ = 0.400 mol/dm³ × 0.025 dm³
= 0.010 mol

Step 3: Calculate the amount of H_2SO_4 (you only know its volume).

1 mol of H_2SO_4 reacts with 2 mol of $NaOH$

amount of H_2SO_4 that reacts with 0.010 mol of $NaOH$ = (0.010 mol ÷ 2) = 0.0050 mol

Step 4: Use the amount of the reactant from Step 3, and its volume, to calculate its concentration.

$$\text{concentration of } H_2SO_4 = \frac{0.0050 \text{ mol}}{0.010 \text{ dm}^3}$$
$$= 0.50 \text{ mol/dm}^3$$

C 25.00 cm³ of 0.400 mol/dm³ sodium hydroxide solution was titrated with dilute sulfuric acid. The mean titre of acid was 20.00 cm³. Calculate the concentration of the acid.

1 25.00 cm³ of 0.200 mol/dm³ NaOH(aq) was titrated with 26.25 cm³ of HCl(aq). Calculate the concentration of the acid to three significant figures. *(5 marks)*

2 25.00 cm³ of 0.250 mol/dm³ NaOH(aq) was titrated with 16.40 cm³ H_2SO_4(aq). Calculate the concentration of the acid to three significant figures. *(5 marks)*

3 25.00 cm³ of orange juice was titrated with 27.75 cm³ of 0.0100 mol/dm³ NaOH(aq).
   a Calculate the amount, in mol, of sodium hydroxide. *(2 marks)*
   b 1 mol of acid in the orange juice reacts with 3 mol of NaOH. Calculate the amount of acid. *(2 marks)*
   c Calculate the concentration of acid in the orange juice. *(2 marks)*

C5.1.7 Gas calculations

Learning outcomes

After studying this lesson you should be able to:

● describe the relationship between molar amounts of gases and their volumes

● calculate the volumes of gases involved in reactions using the molar gas volume.

Specification reference: C5.1d, C5.1e

Figure 1 *Hydrogen gas cylinders on the roof of a hydrogen-powered bus.*

Figure 2 *A 36 cm diameter beach ball contains about 24 dm³ of air.*

Vehicles powered by hydrogen gas do not release carbon dioxide, a greenhouse gas, when in use (Figure 1). The hydrogen fuel is made in chemical reactions, and using mole calculations you can find the masses of reactants needed to make it.

What is the molar volume?

One mole of any substance in the gas state occupies the same volume at the same temperature and pressure. This volume is the **molar volume**, V_m. At room temperature and pressure (RTP):

$$V_m = 24\,dm^3/mol \ (24\,000\,cm^3/mol)$$

> **A** What volume does 1 mol of hydrogen occupy at RTP?

What is the relationship between amount and volume?

This equation links the amount and volume of a substance in the gas state at RTP:

$$\text{volume in } dm^3 = \text{amount in mol} \times 24\,dm^3/mol$$

You can use this equation to calculate the volume of a gas if you know its amount. For example, 2 mol of carbon dioxide occupies 48 dm³ at RTP (2 mol × 24 dm³/mol).

You can rearrange the equation so you can calculate the amount of gas if you know its volume:

$$\text{amount in mol} = \frac{\text{volume in } dm^3}{24\,dm^3/mol}$$

> **B** Calculate the amount, in mol, of 12 dm³ of oxygen at RTP.

How do you use the molar volume?

You can use the molar volume to calculate:

● the volume of a gaseous reactant or product, if you know its amount

● the amount of a gaseous reactant or product, if you know its volume.

Making hydrogen

1.31 g of zinc reacts with excess sulfuric acid:

$$Zn(s) + H_2SO_4(aq) \rightarrow ZnSO_4(aq) + H_2(g)$$

Calculate the volume of hydrogen produced at RTP.

Step 1: Calculate the amount of limiting product.

molar mass of zinc = 65.4 g/mol

$$\text{amount of zinc} = \frac{\text{mass in g}}{\text{molar mass in g/mol}}$$

$$= \frac{1.31\,g}{65.4\,g/mol}$$

$$= 0.0200\,mol$$

Step 2: Use the balanced equation to calculate the amount of hydrogen produced.

1 mol Zn produces 1 mol H_2

0.0200 mol of Zn produces 0.0200 mol of H_2

Step 3: Use your answer to Step 2, and V_m, to calculate the volume of hydrogen.

volume in dm^3 = amount in mol × 24 dm^3/mol

$$= 0.0200\,mol \times 24\,dm^3/mol$$

$$= 0.48\,dm^3$$

Go further

Molar volume increases if the pressure decreases or the temperature increases, so V_m is given for particular conditions. When temperature and pressure are not RTP, chemists use the ideal gas law equation, $pV = nRT$. Find out what the five terms in this equation are.

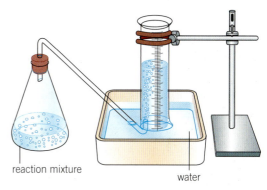

Figure 3 *Collecting gas using a measuring cylinder.*

reaction mixture

water

Measuring the volume of a gas

You can measure the volume of gas produced using a measuring cylinder.

1   Fill the measuring cylinder with water and, while keeping its mouth underwater, turn it upside down.

2   Clamp the measuring cylinder securely (Figure 3).

3   Mix the reactants in a conical flask and attach a delivery tube.

4   Measure the volume of gas produced in the reaction by recording the end reading on the measuring cylinder.

In an this investigation, you need to keep the mouth of the measuring cylinder underwater (Figure 3). This is in order to stop air getting in or water getting out. Either of these would affect the volume reading on the measuring cylinder.

Eye protection should be worn throughout this practical.

1   Calculate the volume of the following at RTP.
    a   0.25 mol of xenon   (1 mark)
    b   1.125 mol of nitrogen.   (1 mark)

2   A gas cylinder has a volume of 288 dm^3.
    a   Calculate the amount, in mol, of hydrogen that it could contain at RTP.   (1 mark)
    b   Use your answer to **a** to calculate the mass of hydrogen.   (2 marks)

3   The electrolysis of water produces hydrogen and oxygen:
    $$2H_2O(l) \rightarrow 2H_2(g) + O_2(g)$$
    a   1 cm^3 of water has a mass of 1 g. Calculate the amount, in mol, of water in 1 cm^3 of water.   (2 marks)
    b   Use your answer to **a** to calculate the volume of oxygen produced at RTP.   (2 marks)
    c   Calculate the total volume of products at RTP.   (2 marks)

Monitoring and controlling chemical reactions

C5.1 Monitoring chemical reactions — GCSE CHEMISTRY ONLY

Summary questions

1 A student added 3.60 g of magnesium, Mg, to dilute sulfuric acid, H_2SO_4(aq). No magnesium was left when the reaction was complete. The student transferred the solution to an evaporating basin, heated it using a Bunsen burner, and evaporated all the water. By calculation, the student should have made 18.0 g of magnesium sulfate, $MgSO_4$(s), but he only obtained 15.4 g.

a Explain how you can tell from the observations that sulfuric acid was in excess.

b Write a balanced equation to model this reaction, including state symbols.

c Use your equation to explain how the student could tell that the reaction was complete, other than by looking for magnesium in the reaction mixture.

d Calculate the percentage yield of magnesium sulfate, giving your answer to one decimal place.

e Write down one reason, other than a mistake, why the student may have obtained a percentage yield of less than 100%.

2 Iron is obtained from iron oxide, Fe_2O_3, by reduction with carbon monoxide:
$$Fe_2O_3 + 3CO \rightarrow 2Fe + 3CO_2$$

a Calculate the relative formula mass, M_r, of carbon dioxide (A_r of C = 12.0, A_r of O = 16.0, A_r of Fe = 55.8).

b Calculate the atom economy for the process, giving your answer to three significant figures.

3 a 1.20 g of copper(II) sulfate was dissolved in 500 cm³ of water. **[H]**

  i Calculate the volume of water in dm³.

  ii Use your answer to i to help you to calculate the concentration of the copper(II) sulfate solution in g/dm³.

b 5.05 g of potassium nitrate, KNO_3, was dissolved in 250 cm³ of water.

  i Calculate the molar mass of potassium nitrate (A_r of N = 14.0, A_r of O = 16.0, A_r of K = 39.1).

  ii Calculate the amount, in mol, of potassium nitrate that was used.

  iii Calculate the volume of water in dm³.

  iv Use your answers to ii and iii to calculate the concentration of the potassium nitrate solution in mol/dm³. Give your answer to two significant figures.

4 A school technician carried out a titration to check the concentration of the hydrochloric acid supplied to the students. He used 0.0800 mol/dm³ sodium hydroxide solution in the burette. The mean titre of sodium hydroxide solution needed to neutralise 25.00 cm³ of the hydrochloric acid was 27.40 cm³: **[H]**
$$NaOH(aq) + HCl(aq) \rightarrow NaCl(aq) + H_2O(l)$$
Give all your answers to three significant figures.

a Calculate the amount, in mol, of sodium hydroxide in the mean titre.

b Use the balanced equation and your answer to a to calculate the amount, in mol, of hydrochloric acid in the 25.00 cm³ sample.

c Use your answer to b to calculate the concentration of the hydrochloric acid.

d The concentration of sodium chloride at the end-point is the same as the concentration of hydrochloric acid. Calculate the concentration of the sodium chloride solution in g/dm³ (A_r of Na = 23.0, A_r of Cl = 35.5).

5 Baking powder contains sodium hydrogencarbonate, $NaHCO_3$. This decomposes when heated, forming carbon dioxide:
$$2NaHCO_3(s) \rightarrow Na_2CO_3(s) + H_2O(l) + CO_2(g)$$

a Calculate the amount of CO_2, in mol, that could be produced from 4.2 g of $NaHCO_3$.

b Use your answer to a to calculate the volume of CO_2 produced at RTP (molar volume at RTP = 24 dm³/mol).

Revision questions

1 Which row gives the correct names of the pieces of apparatus shown in the diagram? [S]

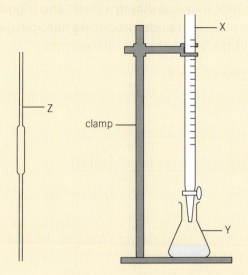

	Apparatus X	Apparatus Y	Apparatus Z
A	pipette	burette	beaker
B	conical flask	pipette	burette
C	burette	conical flask	pipette
D	beaker	pipette	burette

(1 mark)

2 Copper carbonate decomposes on heating to form copper oxide and carbon dioxide.
The following balanced equation models the reaction:
$$CuCO_3 \rightarrow CuO + CO_2$$
If 4 moles of copper carbonate are present at the start of the reaction, what is the theoretical amount of carbon dioxide that can be formed?
A 1 mole
B 2 moles
C 4 moles
D 8 moles *(1 mark)*

3 A student made sodium chloride from an acid and sodium hydroxide. [S]
The balanced equation below models the reaction:
$$HCl + NaOH \rightarrow NaCl + H_2O$$
a Write down the name of the acid the student used. *(1 mark)*
b The student used 0.1 mol of sodium hydroxide.
  i Calculate the theoretical yield of sodium chloride. Give your answer in grams. *(2 marks)*
  ii The actual yield of sodium chloride was 5.23 g. Calculate the percentage yield. *(2 marks)*

4 At high temperatures in car engines, nitrogen and oxygen from the air react together to make oxides of nitrogen, for example, nitrogen monoxide: [H]
$$N_2 + O_2 \rightarrow 2NO$$
1 mole of gas occupies 24 dm³ at room temperature and pressure.
A car produced 12 dm³ of nitrogen monoxide in a given time. The volume was measured at room temperature and pressure.
a Calculate the amount of nitrogen monoxide produced, in mol. *(1 mark)*
b Use your answer to **a** and data from the Periodic Table to calculate the mass of nitrogen gas used to make this volume of nitrogen dioxide. Give your answer in grams. *(2 marks)*

5 Hydrogen can be produced by reacting methane with steam:
$$CH_4 + H_2O \rightarrow CO + 3H_2$$
The atom economy for this process is 17.6%.
Hydrogen can also be produced by the decomposition of ammonia. The reaction requires a catalyst.
$$2NH_3 \rightarrow N_2 + 3H_2$$
a Calculate the atom economy for the production of hydrogen from ammonia. *(2 marks)*
b* Compare the atom economies for the two methods of producing hydrogen and suggest other factors that must be considered when deciding which reaction pathway to choose for the manufacture of hydrogen. *(6 marks)* [H]

C5.2 Controlling reactions
C5.2.1 Rate of reaction

Learning outcomes

After studying this lesson you should be able to:

- explain what is meant by rate of reaction
- suggest practical methods for determining rates
- interpret rate of reaction graphs.

Specification reference: C5.2a, C5.2b

Figure 1 *Testing car air bags with a crash test dummy.*

Synoptic link

You can learn more about measuring the volume of gas produced in a reaction using different apparatus in C5.1.7 *Gas calculations.*

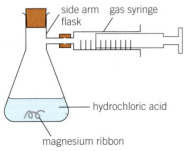

Figure 2 *This apparatus can be used to measure the rate of a reaction.*

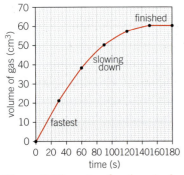

Figure 3 *The rate of a chemical reaction usually decreases as the reaction proceeds.*

Car air bags reduce the risk of injury in an accident (Figure 1). A sensor detects the rapid change in speed during a crash, and triggers the thermal decomposition of sodium azide, producing nitrogen gas. The nitrogen completely fills the bag within 30 milliseconds.

What is the rate of reaction?

The **rate of reaction** is a measure of how quickly reactants are used or products are formed:

$$\text{rate of reaction} = \frac{\text{amount of reactant used}}{\text{time taken}}$$

$$\text{rate of reaction} = \frac{\text{amount of product formed}}{\text{time taken}}$$

If a product is in the gas state, it is often easier to measure its volume rather than its mass.

> **A**   Explain whether a reaction with a high rate, or a low rate, will form a large amount of product in a short time.

How do you measure the volume of a gas?

You have already looked at measuring the volume of gas produced in a reaction using an upside-down measuring cylinder in water. Another method is to use a **gas syringe** (Figure 2). A gas syringe is made from glass, and has graduations marked in cm^3. As the syringe fills, the plunger moves outwards and you can record the volume of gas it contains. You are then able to calculate the rate of reaction.

Measuring the volume of gas

Magnesium reacts with dilute hydrochloric acid to form magnesium chloride solution and hydrogen:

$$Mg(s) + 2HCl(aq) \rightarrow MgCl_2(aq) + H_2(g)$$

You can measure how fast hydrogen is produced by carrying out the following steps.

1   Place dilute hydrochloric acid in a conical flask connected to a gas syringe.

2   Add a piece of magnesium ribbon into the acid, stopper the flask, and start a stop clock.

3   Draw a results table and record the time and volume of hydrogen at regular intervals.

4   Analyse your results by drawing a line graph as in Figure 3.

Before adding the stopper to the conical flask in this investigation, you should push the plunger all the way in, to make sure the reading starts at $0\,cm^3$.

You also need to make sure that you stopper the flask as soon as the reaction starts, in order to make sure that all the hydrogen is collected.

> **B**   Why is it easier to measure the volume of hydrogen produced rather than its mass?

Calculating mean rate of reaction

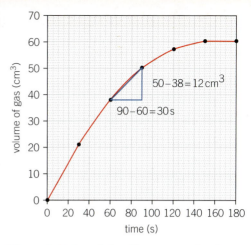

The graph in Figure 4 shows how the total volume of hydrogen changes with time during the reaction between magnesium and hydrochloric acid. Calculate the mean rate of reaction between 60 s and 90 s.

Step 1: Calculate the changes in volume and time.

change in volume = $(50\,cm^3 - 38\,cm^3) = 12\,cm^3$

change in time = $(90\,s - 60\,s) = 30\,s$

Step 2: Calculate the mean gradient (equal to the mean rate of reaction).

$$gradient = \frac{change\ in\ volume}{change\ in\ time} = \frac{12\,cm^3}{30\,s} = 0.40\,cm^3/s$$

Figure 4 *Determining the mean rate of reaction between two times.*

C Use Figure 4 to calculate the mean rate of reaction between 0 s and 30 s.

Instantaneous rate of reaction

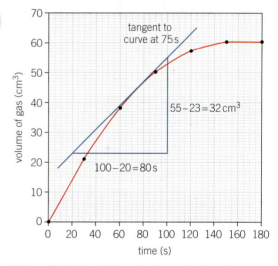

The graph in Figure 5 shows how you can calculate an instantaneous rate of reaction – the rate at a particular time. In this example, a tangent to the curve has been drawn at 75 s. The instantaneous rate of reaction at this time is equal to the gradient of the tangent.

D Use the information in Figure 5 to calculate the instantaneous rate of reaction at 75 s, including the units.

Figure 5 *Determining the instantaneous rate.*

1 Explain what is meant by the term 'rate of reaction' and how it is calculated. *(2 marks)*

2 Refer to the graph in Figure 3 for this question.
   a Calculate the mean rate for the whole reaction, using the results at 0 s and 150 s. *(3 marks)*
   b Explain why the result after 150 s was ignored in **a**. *(2 marks)*

3 Carbon dioxide is produced when calcium carbonate reacts with dilute hydrochloric acid. Outline how you could investigate the rate of this reaction, including how to analyse the results. *(6 marks)*

Air bags

The reaction in an air bag produces $60\,dm^3$ of nitrogen in 30 ms. Calculate the mean rate of reaction in m^3/s, expressing your answer in standard form.

C5.2.2 Temperature and reaction rate

Learning outcome

After studying this lesson you should be able to:

- describe, and explain, the effect of changes in temperature on the rate of reaction.

Specification reference: C5.2c, C5.2d

Figure 1 *Potatoes cook more slowly in boiling water at 100 °C than when fried in oil at 200 °C.*

Synoptic link

You can learn more about activation energy in C3.2.2 *Reaction profiles*.

Figure 2 *Before and after photos of the sodium thiosulfate 'disappearing cross' experiment.*

Cooking food by frying is quicker than by boiling (Figure 1). Chemical reactions, such as those that happen in cooking food, go faster at higher temperatures.

Why do reactions go faster at higher temperatures?

A reaction can only happen if:

- reactant particles **collide** with each other, and
- the colliding particles have enough energy to react.

A collision that leads to a reaction is a **successful collision**. A collision will not be successful if the particles have less energy than the activation energy.

As the temperature of a reaction mixture increases:

- the particles move more quickly, so they collide more often
- a greater proportion of the colliding particles have the activation energy or more.

The greater the rate of successful collisions, the greater the rate of reaction.

> **A** Explain, in terms of particles and energy, what is meant by a 'successful collision'.

Investigating the effect of temperature

The reaction between sodium thiosulfate solution and hydrochloric acid is often used to investigate the factors affecting rates of reaction:

$$Na_2S_2O_3(aq) + 2HCl(aq) \rightarrow 2NaCl(aq) + H_2O(l) + SO_2(g) + S(s)$$

This reaction produces lots of acidic fumes, especially at raised temperatures. In order to reduce the fumes, the beakers should be placed in an ice bath. Asthmatics should very careful not to breathe in chemical fumes, and the room should be well ventilated.

The equation above looks complicated, but you can easily measure how quickly a pale yellow precipitate of sulfur appears.

1 In a beaker, mix the reactants.

2 Look down through the mixture to a cross drawn on a piece of paper (Figure 2). The longer it takes for the cross to disappear, the lower the rate of reaction.

3 You can vary the temperature of the reaction mixture by warming up one of the solutions before mixing.

The reaction will start as soon as the sodium thiosulfate solution and dilute hydrochloric acid are mixed together. This means that you should start the stop clock as soon as you mix the solutions.

You should stop the clock as soon as the cross disappears, so that the timing will stop at the same point in all the reactions.

How can you use reaction times?

The rate of reaction is **inversely proportional** to the **reaction time**, the time taken for a reaction to happen. This means that 1/time is **directly proportional** to the rate:

$$\frac{1}{\text{reaction time}} \propto \text{rate of reaction}$$

You can plot a graph with rates of reaction instead of reaction times, if you calculate 1/time for each result. For example, a reaction time of 10 s gives a rate of:

$$\frac{1}{10\,\text{s}} = 0.10\,/\text{s}$$

Sometimes 1000/time is used instead of 1/time, as this can make the vertical axis scale easier to work out. The two graphs in Figures 3 and 4 represent the results of the same experiment carried out by a student.

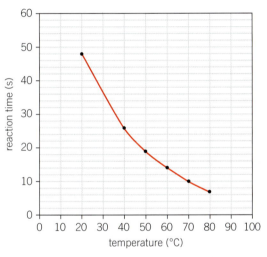

Figure 3 *Reaction time decreases as the temperature increases.*

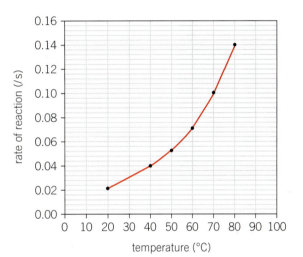

Figure 4 *The rate of reaction increases as the temperature increases.*

B Calculate 1/time for a reaction time of 26 s, giving your answer to two significant figures.

C Calculate 1000/time for a reaction time of 26.0 s, giving your answer to three significant figures.

1 Use your answers to questions **B** and **C** to explain whether the graph of rate against temperature in Figure 4 has been plotted using 1/time or 1000/time. *(2 marks)*

2 Explain what would happen to the rate of reaction if the temperature is decreased. *(3 marks)*

3 Refer to Figure 4 in this question.
　a Estimate the rate of reaction at 45 °C. *(1 mark)*
　b Explain, in terms of particles and energy, why the rate of reaction doubles between 60 °C and 80 °C. *(4 marks)*
　c Suggest why there are no results below 20 °C or above 80 °C. *(2 marks)*

Learning outcomes

After studying this lesson you should be able to:

● describe, and explain, the effect of changes in concentration of solutions on the rate of reaction

● describe, and explain, the effect of changes in pressure of reacting gases on the rate of reaction.

Specification reference: C5.2c, C5.2d

Figure 1 *Collisions are more likely in a crowded place.*

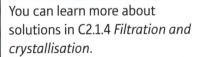
Synoptic link

You can learn more about solutions in C2.1.4 *Filtration and crystallisation*.

It can be difficult getting through a crowd without bumping into someone (Figure 1). Reactant particles also have a greater chance of colliding with one another if they are crowded close together.

Why do reactions go faster at higher concentrations?

The concentration of a solution is a measure of how much solute is dissolved in the solvent. The more concentrated a solution is, the more solute is dissolved in the solvent. If a reaction involves one or more reactants in solution, the rate of reaction increases as the concentration increases:

● the particles become more crowded, so they collide more often.

The energy stored in the particles does not change but, because the rate of collisions increases, the rate of successful collisions increases.

> **A** Magnesium reacts with hydrochloric acid. Explain why the reaction rate decreases when the acid is mixed with water.

How do you investigate the effect of concentration on reaction rate?

The reaction between magnesium ribbon and hydrochloric acid is often used to investigate rates of reaction:

$$Mg(s) + 2HCl(aq) \rightarrow MgCl_2(aq) + H_2(g)$$

You could use a gas syringe to measure the volume of hydrogen gas produced. However, it is easier to measure the time taken for a piece of magnesium ribbon to be used up in the reaction. You can add water to the hydrochloric acid to reduce its concentration. When you have recorded the reaction times at different concentrations of acid, you can calculate 1/time to obtain the reaction rates (Figures 2 and 3).

When carrying out this investigation, the length of the magnesium ribbon needs to be controlled. This is because the reaction time is affected by the mass (and therefore size) of the piece of magnesium, so this must be kept the same to make it a fair test.

> **B** Explain why the temperature should be kept the same when investigating the effect of concentration on reaction rate.

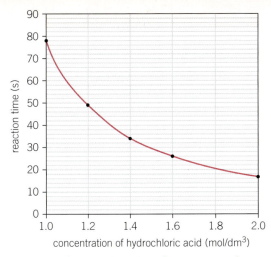

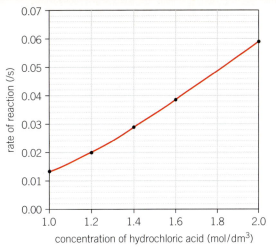

Figure 2 *The reaction time decreases as the concentration of hydrochloric acid increases.*

Figure 3 *The rate of reaction increases as the concentration of hydrochloric acid increases.*

Limiting reactants

In a reaction involving two reactants, one of the reactants will be limiting and the other in excess. The amount of product will be proportional to the amount of the limiting reactant.

C Compare the ratio of volume of gas produced to the ratio of magnesium used in Figure 4. Explain whether or not this shows that magnesium is the limiting reactant.

Why do reactions go faster at higher pressures?

If a reaction involves one or more reactants in the gas state, the rate of reaction increases as the **pressure** of the gas increases:

- the particles in the gas state become more crowded, so they collide more often (Figure 5).

The energy stored in the particles does not change but, because the rate of collisions increases, the rate of successful collisions increases. Industrial chemical processes often use high pressures to achieve high rates of reaction.

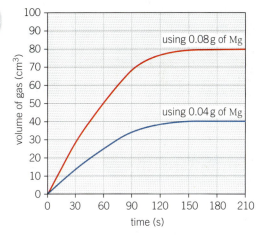

Figure 4 Two experiments involving the reaction between magnesium and hydrochloric acid.

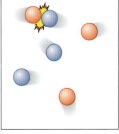
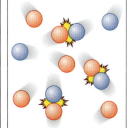

Figure 5 *Particles collide more frequently at high concentrations and/or pressures.*

1 Explain what happens to the rate of the reaction if the pressure is reduced for the following reaction:

$$N_2(g) + 3H_2(g) \rightarrow 2NH_3(g)$$ *(3 marks)*

2 Explain the change in the rate of reaction when the concentration of a reacting solution is increased. *(3 marks)*

3 Refer to the graph of rate against concentration in Figure 3.

a Estimate the rate of reaction at 1.8 mol/dm³. *(1 mark)*

b Compare the rate of reaction at 1.0 mol/dm³ and 2.0 mol/dm³. In this particular reaction, explain what happens when the concentration of hydrochloric acid is doubled. *(4 marks)*

Learning outcome

After studying this lesson you should be able to:

- describe, and explain, the effect on the rate of reaction of changes in the size of pieces of a reacting solid.

Specification reference: C5.2c, C5.2e

Figure 1 *Exploding powders can cause industrial accidents.*

Synoptic link

You can learn more about the surface area to volume ratio of nanoparticles in C2.3.4 *Nanoparticles*.

Powders such as flour and metal dust burn explosively in air. Factories that handle powders have strict rules to stop them escaping, and to avoid sparks or naked flames (Figure 1).

Why do reactions go faster with powders?

The particles in a substance in the solid state can only vibrate about fixed positions. They cannot move from place to place. This means that only the particles at the surface can take part in collisions. The rate of reaction increases as the surface area increases because:

- more reactant particles are available for collisions,
- collisions are more likely, so particles collide more often.

The energy stored in the particles does not change but, because the rate of collisions increases, the rate of successful collisions increases.

The total surface area available for collisions is larger when a big lump of a substance is broken down into smaller lumps (Figure 2). Powders have an even larger surface area, so their reactions are very fast. What matters is the surface area to volume ratio of the lumps:

- as the size of the lumps decreases, the surface area to volume ratio increases.

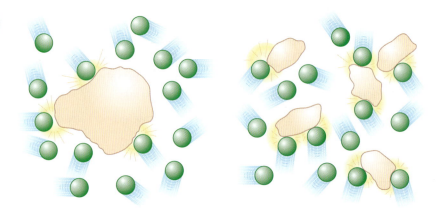

one big lump (slow reaction)   several small lumps (fast reaction)

Figure 2 *Small lumps have more frequent collisions than large lumps.*

A Suggest why an iron bar does not burn in a Bunsen burner flame, but iron filings do.

Investigating the effect of particle size

The reaction between calcium carbonate and dilute hydrochloric acid is often used to investigate rates of reaction:

$$CaCO_3(s) + 2HCl(aq) \rightarrow CaCl_2(aq) + H_2O(l) + CO_2(g)$$

You could use a gas syringe to measure the volume of carbon dioxide gas produced, but another method involves weighing the reaction mixture (Figure 3).

1 Add dilute hydrochloric acid to a conical flask and stopper it with cotton wool.

2 Place the flask on a balance next to a lump of calcium carbonate.

3 **Tare** the balance, remove the cotton wool, drop the calcium carbonate into the acid, start a stop clock, and replace the cotton wool (Figure 3).

The mass goes down as carbon dioxide escapes, so the balance reading must be recorded at regular intervals. You can repeat the experiment with different sized lumps of calcium carbonate or with calcium carbonate powder.

Figure 3 *Investigating the reaction between calcium carbonate and hydrochloric acid.*

B Explain why, if you carry out an investigation like *Investigating the effect of particle size*, you should start with the same mass of calcium carbonate each time.

C If you carry out an investigation like *Investigating the effect of particle size*, the cotton wool stops hydrochloric acid escaping as a fine spray during the reaction. Explain how the measured rate would be affected if the cotton wool were not used.

D Explain why you should keep temperature and concentration the same when investigating the effect of surface area on reaction rate.

Choosing appropriate techniques

The method you choose to measure the rate of a reaction depends on several factors. These include the properties of the reactants and products, and the ease of measuring their loss or production.

Refer to the graph in Figure 4 for questions **1** and **3**. Assume that 0.90 g of carbon dioxide was lost in each experiment.

1 **a** State the time taken for each reaction to finish. *(2 marks)*
   b Calculate the mean rates of reaction, giving your answers to two significant figures. *(4 marks)*
   c Explain the difference between your answers to **b**. *(4 marks)*

2 Explain why custard powder would be a hazard in a factory. *(3 marks)*

3 **a** Calculate the volume of carbon dioxide produced in the investigation, giving your answer to two significant figures ($V_m = 24\,000\ cm^3$/mol at RTP). *(3 marks)*
   b Suggest why a gas syringe was not used. *(1 mark)*

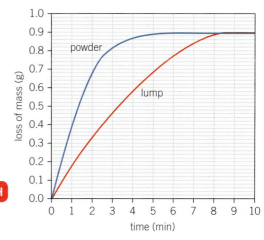

Figure 4 *The loss in mass when calcium carbonate and hydrochloric acid react together.*

C5.2.5 Catalysts and reaction rate

Learning outcomes

After studying this lesson you should be able to:

- describe the characteristics of catalysts and their effect on rates of reaction
- recall that enzymes act as catalysts in biological systems
- explain catalytic action in terms of activation energy.

Specification reference: C5.2f, C5.2g, C5.2h, C5.2i

Figure 1 *A catalytic converter cut open.*

Petrol and diesel vehicles produce harmful substances such as carbon monoxide and nitrogen oxides. To help solve this problem, their exhaust systems have catalytic converters (Figure 1). These speed up reactions that convert the harmful substances into less harmful ones.

What are catalysts like?

A catalyst is a substance that increases the rate of a reaction but is unchanged at the end of the reaction. If you add 1 g of catalyst to a reaction mixture, there will still be 1 g left when the reaction has finished. Catalysts are specific to particular reactions. A substance that acts as a catalyst for one reaction may not **catalyse** a different reaction.

A small amount of a catalyst will catalyse the reaction between large amounts of reactants. This is helpful because catalytic converters use platinum, rhodium, and palladium. These metals are very expensive, so they are coated onto an inert ceramic 'honeycomb'. This uses only a few grams of catalyst, provides a large surface area for the reactions, and allows exhaust gases through.

> **A** One stage in the manufacture of nitric acid involves reacting ammonia with oxygen in the gas state. Suggest why a fine mesh of platinum and rhodium is present in the reaction chamber.

Investigating a catalyst

Powdered manganese(IV) oxide, MnO_2, catalyses the decomposition of hydrogen peroxide:

$$2H_2O_2(aq) \rightarrow 2H_2O(l) + O_2(g)$$

1. Place hydrogen peroxide solution in a conical flask connected to a gas syringe.
2. Put a little manganese(IV) oxide into the flask, stopper it, and start a stop clock. Only small amounts should be added, as gas can be produced very quickly and violently.
3. Record the time and volume of oxygen at regular intervals.
4. Draw a graph to analyse your results.

You could vary the mass of catalyst (Figure 2), or try other catalysts. When investigating a catalyst, you should keep the volume, temperature, and concentration of hydrogen peroxide the same. This is because the volume of oxygen produced is affected by these variables, so they must be kept the same to make it fair test.

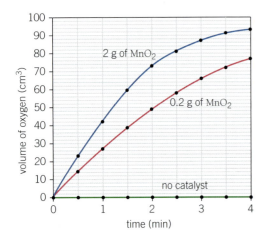

Figure 2 *Adding much more catalyst makes little difference to the reaction rate.*

What are biological catalysts?

Enzymes are proteins that act as catalysts in biological systems. Different enzymes catalyse different reactions. For example, liver cells contain an enzyme called catalase. This catalyses the decomposition of hydrogen peroxide (Figure 3).

B Describe what an enzyme is.

Catalytic activity

A 'catalyst' is a substance that 'catalyses' or speeds up a reaction. If a substance acts as a catalyst, it shows 'catalytic activity'.

How do catalysts work?

Catalysts work by providing an alternative reaction pathway with a lower activation energy (Figure 4). The amount of energy stored in the reactant particles does not change but, in the presence of a catalyst:

- a greater proportion of the colliding particles have the activation energy or more
- the rate of successful collisions increases compared to the rate in an uncatalysed reaction.

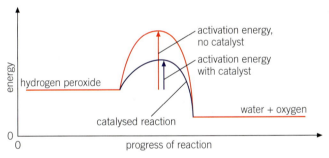

Figure 4 *Reaction profiles for a catalysed and an uncatalysed reaction.*

Figure 3 *Raw liver from an animal catalysing the decomposition of hydrogen peroxide.*

Go further

Find out what happens to enzymes, and the rate of a reaction catalysed by an enzyme, as temperature increases.

1 Explain what a catalyst is and how it works. (*4 marks*)

2 Refer to the graph in Figure 2 for this question.
  a Describe the effect of adding MnO_2 on the rate of reaction. (*2 marks*)
  b Calculate the catalysed rates of reaction over the first two minutes, expressing your answers to two significant figures. (*4 marks*)
  c Explain whether or not the reactions were complete after four minutes. (*2 marks*)

3 Zinc lumps react with dilute sulfuric acid to produce zinc sulfate solution and hydrogen:
$$Zn(s) + H_2SO_4(aq) \rightarrow ZnSO_4(aq) + H_2(g)$$
Outline how you could investigate the use of copper powder as a catalyst for this reaction. (*6 marks*)

C5.2 Controlling reactions

Summary questions

1 Write down simple definitions of the following terms:
  a rate of reaction
  b catalyst
  c enzyme.

2 Sodium thiosulfate solution reacts with dilute sulfuric acid to produce a yellow precipitate of sulfur:

$$Na_2S_2O_3(aq) + 2HCl(aq) \rightarrow 2NaCl(aq) + SO_2(g) + H_2O(l) + S(s)$$

A student investigated the effect of changing the concentration of sodium thiosulfate on the rate of reaction. She mixed $50\,cm^3$ of sodium thiosulfate solution with $10\,cm^3$ of dilute hydrochloric acid in a conical flask on top of a piece of paper marked with a cross. She timed how long it took for the mixture to become too cloudy to see the cross (Figure 1). She then repeated the experiment several times, but diluted the sodium thiosulfate solution each time.

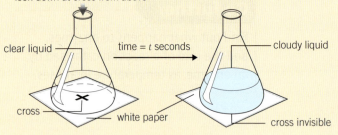

look down at cross from above

clear liquid — time = t seconds — cloudy liquid

cross — white paper — cross invisible

Figure 1 *Reacting sodium thiosulfate with hydrochloric acid.*

Table 1 shows the student's results.

Table 1 *Results from reacting sodium thiosulfate and hydrochloric acid.*

Volume of Na$_2$S$_2$O$_3$(aq) (cm³)	Volume of distilled water (cm³)	Volume of HCl(aq) (cm³)	Time for cross to disappear, t (s)	$1000 \div t$ (s)
50	0	10	104	9.6
40	10	10	55	
30	20	10	41	
20	30	10	29	
10	40	10	23	

  a Explain why the student added water to some of the mixtures.

  b Calculate $1000 \div t$ for each result, giving your answer to one decimal place. The first one has been done already.
  c i Plot a graph of $1000 \div t$ (vertical axis) against volume of Na$_2$S$_2$O$_3$(aq) (horizontal axis).
    ii Draw a straight line of best fit, ignoring any outliers.
  d i Write down the result that is an outlier.
    ii Use your graph to estimate the **time** this result should have been.
    iii Suggest a reason for the difference between your answer to **d ii** and the student's recorded time.
  e Describe, and explain, in terms of particles, the trend seen in the results.

3 Nitrogen and hydrogen react together to form ammonia:

$$N_2(g) + 3H_2(g) \rightarrow 2NH_3(g)$$

Describe, and explain, in terms of particles, the effect of changing the pressure on the rate of reaction.

4 Hydrogen peroxide decomposes slowly to form water and oxygen:

$$2H_2O_2(aq) \rightarrow 2H_2O(l) + O_2(g)$$

  a Manganese(IV) oxide powder acts as a catalyst for this reaction.
    i Describe the effect of manganese(IV) oxide powder on the rate of this reaction.
    ii Explain how manganese(IV) oxide has this effect.
  b Describe and explain the effect of increasing the temperature on the rate of this reaction.

Revision questions

1 Which statement about catalysts is true?

 A A catalyst slows down a reaction.

 B A catalyst increases the activation energy of a reaction.

 C A catalyst decreases the activation energy of a reaction.

 D A catalyst increases the time for a reaction to go to completion. (*1 mark*)

2 A student reacts sodium thiosulfate solution with hydrochloric acid.
 She places the reaction mixture over a cross.
 She measures the time for the cross to disappear.
 She does the experiment four more times. Each time, she changes one of the reaction conditions.
 Look at her table of results.

Experiment	Time for cross to disappear (s)
First experiment	31
A	9
B	24
C	31
D	62

 Give the letter of the experiment in which the student might have decreased the temperature compared to the first experiment.

3 Calcium carbonate reacts with sulfuric acid to make carbon dioxide.
 $CaCO_3(s) + H_2SO_4(aq) \rightarrow CaSO_4(aq) + H_2O(l) + CO_2(g)$
 A student sets up the apparatus in the following diagram.
 He measures the decrease in the mass of the reaction mixture every minute.

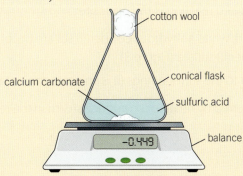

cotton wool

calcium carbonate

conical flask

sulfuric acid

−0.44g

balance

The student does the experiment twice with different sized pieces of calcium carbonate. He uses the same mass of calcium carbonate in both experiments, and acid of the same concentration.
Look at his table of results.

Decrease in mass since the start of the experiment (g)		
Time (min)	Experiment A	Experiment B
0	0.00	0.00
1	0.74	0.31
2	0.88	0.42

 a Explain why the mass of the flask and its contents decreases. (*1 mark*)

 b Write down the letter of the experiment in which the student used smaller pieces of calcium carbonate, and explain your answer. (*2 marks*)

 c Suggest how the student could change his experiment to investigate the effect of temperature on the rate of the reaction. In your answer, list two variables that the student must control. (*3 marks*)

4 Magnesium and hydrochloric acid react together to make hydrogen and magnesium sulfate. **H**
 $$Mg(s) + 2HCl(aq) \rightarrow MgCl_2(aq) + H_2(g)$$
 A student measures the rate of this reaction by measuring the time for a piece of magnesium ribbon to finish reacting.

 a Write down how she will know when the magnesium ribbon has finished reacting. (*1 mark*)

 b The student investigates how the concentration of acid affects the reaction rate.

 Look at her table of results.

Acid concentration (mol/dm³)	Time for magnesium to finish reacting (s)
0.50	240
0.75	79
1.00	40
1.50	21
2.00	12

 i Write down two variables the student should keep the same in her investigation. (*2 marks*)

 ii Plot the data for the acid concentrations of 1.50 and 2.00 mol/dm³ and draw a line of best fit. (*4 marks*)

C5.3 Equilibria

C5.3.1 Reversible reactions

Learning outcomes

After studying this lesson you should be able to:

- describe how some reactions are reversible
- describe some examples of reversible reactions
- describe the conditions under which dynamic equilibrium occurs.

Specification reference: C5.3a, C5.3b

Figure 1 *A see-saw can move in two directions.*

Figure 2 *When heated, blue hydrated copper(II) sulfate forms white anhydrous copper(II) sulfate and water.*

See-saws can be a lot of fun (Figure 1). Push off with your legs and you can make it move up or down. Some chemical reactions can also move in two directions. They can go forward or backward.

What are reversible reactions?

In many reactions, the products cannot easily be changed back into the reactants. However, in a reversible reaction, the products *can* react together to form the original reactants.

The reaction between copper(II) sulfate and water is a reversible reaction. Blue crystals of copper(II) sulfate are **hydrated** copper(II) sulfate, $CuSO_4.5H_2O$. Each Cu^{2+} ion is surrounded by five water molecules. If you remove these water molecules by heating, you get white **anhydrous** copper(II) sulfate, $CuSO_4$ (Figure 2). This equation models the reaction:

$$\text{hydrated copper(II) sulfate} \rightleftharpoons \text{anhydrous copper(II) sulfate} + \text{water}$$

$$\underset{\text{(blue)}}{CuSO_4.5H_2O} \rightleftharpoons \underset{\text{(white)}}{CuSO_4} + 5H_2O$$

The $\rightleftharpoons$ symbol is used in equations for reversible reactions, instead of an arrow. It combines the **forward reaction** and **backward reaction**:

- forward reaction: $CuSO_4.5H_2O \rightarrow CuSO_4 + 5H_2O$
- backward reaction: $CuSO_4 + 5H_2O \rightarrow CuSO_4.5H_2O$

A Explain the meaning of the $\rightleftharpoons$ symbol.

What happens if you change the conditions?

You have seen how blue hydrated copper(II) sulfate decomposes, when heated, to form white anhydrous copper(II) sulfate and water. This is an **endothermic** reaction because energy is transferred from the surroundings. The backward reaction is the opposite:

- when water is added, white anhydrous copper(II) sulfate becomes blue hydrated copper(II) sulfate (Figure 3)
- energy is transferred to the surroundings (it is an exothermic reaction).

Enough energy is transferred by heating to produce steam in the reaction.

Water of crystallisation

Hydrated copper(II) sulfate, $CuSO_4.5H_2O$, is more accurately called copper(II) sulfate-5-water. The water molecules in the ionic lattice are the 'water of crystallisation'.

Figure 3 *When water is added, white anhydrous copper(II) sulfate becomes blue hydrated copper(II) sulfate.*

B Write down the full name for hydrated cobalt(II) chloride, $CoCl_2.6H_2O$.

C Ammonium chloride decomposes when heated:

$$NH_4Cl(s) \rightleftharpoons NH_3(g) + HCl(g)$$

Suggest the conditions needed for a mixture of ammonia and hydrogen chloride to form ammonium chloride.

What is a dynamic equilibrium?

In a closed system, no substances can enter or leave. A stoppered flask is a closed system, and so is a beaker of solution in which all the reacting substances remain in the solvent. When a reversible reaction happens in a closed system, the rates of the forward reaction and the backward reaction become equal. This situation is called **equilibrium**. At equilibrium:

- the forward and backward reactions still happen, so it is a **dynamic** equilibrium
- the concentrations of all the reacting substances remain constant.

Study tip

The concentrations of reacting substances are constant at equilibrium – they do not change, but they are not necessarily equal.

1 Bromine reaches equilibrium in a sealed flask (Figure 4):

$$Br_2(l) \rightleftharpoons Br_2(g)$$

Explain what happens to the concentrations of bromine in the liquid state and gas state at equilibrium. *(2 marks)*

2 Compare the rates of the forward reaction and backward reaction at equilibrium. *(1 mark)*

3 Nitrogen reacts with hydrogen to form ammonia:

$$N_2(g) + 3H_2(g) \rightleftharpoons 2NH_3(g)$$

a Explain how you know that this is a reversible reaction. *(1 mark)*

b Write down balanced equations to model the forward reaction and the backward reaction. *(2 marks)*

Figure 4 *State changes are also reversible, for example, bromine in the liquid and gas states are at equilibrium in a closed system.*

Learning outcome

After studying this lesson you should be able to:

- predict the effect of changing reaction conditions on equilibrium position.

Specification reference: C5.3c

Figure 1 *This person could fall either way.*

It is difficult to balance on a narrow railing (Figure 1). One small push and you could fall either way. Chemical reactions at equilibrium are like this. If the conditions change, their equilibrium position changes.

What is the equilibrium position?

The **equilibrium position** gives you an idea of the ratio of the equilibrium concentrations of products to reactants:

- The equilibrium position is on the left when the concentration of reactants is *greater than* the concentration of products.
- The equilibrium position is on the right when the concentration of reactants is *less than* the concentration of products.

The equilibrium position may change if there are changes in the conditions, and this affects the concentrations of the reacting substances.

What happens if you change the pressure?

If the pressure is increased, the equilibrium position moves in the direction of the fewest moles of gas, as shown in the balanced equation for the reaction.

Pressure and equilibrium position

Explain the effect on the equilibrium position of increasing the pressure in this reaction:

$$2SO_2(g) + O_2(g) \rightleftharpoons 2SO_3(g)$$

Step 1: Count the total number of moles of gas on each side of the equation.

On the left: $(2 + 1) = 3\,mol$

On the right: $2\,mol$

Step 2: Determine which side has the fewest moles of gas.

There are fewer moles of gas on the right, so the equilibrium position moves to the right if the pressure is increased.

A Explain the effect on the equilibrium position of increasing the pressure in this reaction:

$$CH_4(g) + H_2O(g) \rightleftharpoons CO(g) + 3H_2(g)$$

What happens if you change the concentration?

If the concentration of a substance is increased, the equilibrium position moves in the direction away from that substance, as shown in the balanced equation for the reaction.

Concentration and equilibrium position

Predict the effect on the equilibrium position of increasing the concentration of sulfuric acid, $H_2SO_4(aq)$, in this reaction:

$$2K_2CrO_4(aq) + H_2SO_4(aq) \rightleftharpoons K_2Cr_2O_7(aq) + H_2O(l) + K_2SO_4(aq)$$

Determine which side of the equation contains sulfuric acid.

Sulfuric acid is on the left.

The equilibrium position moves to the right if the concentration of sulfuric acid is increased (Figure 2).

Figure 2 *When sulfuric acid is added, a yellow equilibrium mixture mostly containing $K_2CrO_4(aq)$ becomes an orange equilibrium mixture mostly containing $K_2Cr_2O_7(aq)$.*

B Explain the effect on the equilibrium position of increasing the concentration of $Cl^-(aq)$ in this reaction:

$$CoCl_4^{2-}(aq) \rightleftharpoons Co^{2+}(aq) + 4Cl^-(aq)$$

What happens if you change the temperature?

If the temperature is increased, the equilibrium position moves in the direction of the endothermic change.

Temperature and equilibrium position

Explain the effect on the equilibrium position of increasing the temperature in this reaction:

$$2NO_2(g) \rightleftharpoons N_2O_4(g) \qquad (\Delta H = -58\,kJ/mol)$$

Determine which direction represents the endothermic change.

The forward reaction is exothermic because ΔH is negative, so the backward reaction is endothermic ($\Delta H = +58\,kJ/mol$).

The equilibrium position moves to the left if the temperature is increased (Figure 3).

Figure 3 *These tubes all contain equilibrium mixtures of NO_2 and N_2O_4. As the temperature increases the concentration of brown NO_2 increases and the concentration of colourless N_2O_4 decreases.*

1 Explain the effect on the equilibrium position of increasing the temperature in this reaction:

$$N_2(g) + 3H_2(g) \rightleftharpoons 2NH_3(g) \qquad (\Delta H = -92\,kJ/mol)$$

(*2 marks*)

2 This reaction reaches equilibrium in a sealed container:

$$CaCO_3(s) \rightleftharpoons CaO(s) + CO_2(g) \qquad (\Delta H = +178\,kJ/mol)$$

a Explain the effect on the equilibrium position of:
  i reducing the pressure inside the container (*2 marks*)
  ii increasing the temperature. (*2 marks*)
b Very little CaO and CO_2 are present at room temperature. Explain what this tells you about the equilibrium position at room temperature. (*2 marks*)

3 Iron(III) ions react with thiocyanate ions to form a compound ion:

$$Fe^{3+}(aq) + SCN^-(aq) \rightleftharpoons FeSCN^{2+}(aq)$$

Explain the effect on the equilibrium position of adding ammonium chloride, which reduces the concentration of $Fe^{3+}(aq)$. (*2 marks*)

An engine roars into life and a huge monster truck thunders over the top of a pile of old cars (Figure 1). Monster trucks use methanol, rather than petrol, for their fuel. How do manufacturers choose the reaction conditions needed to make chemicals such as methanol?

Figure 1 *Monster trucks are fuelled with methanol.*

How is methanol made?

Methanol is made by reacting carbon monoxide with hydrogen:

$$CO(g) + 2H_2(g) \rightleftharpoons CH_3OH(g) \qquad (\Delta H = -91 \text{ kJ/mol})$$

The **equilibrium yield** is the amount of the desired product present in a reaction at equilibrium. It depends upon the:

- pressure
- temperature.

It also depends upon the concentration of reactants. Increasing the concentrations of carbon monoxide and hydrogen in the reaction mixture moves the equilibrium position to the right, forming a higher equilibrium concentration of methanol. In other words, you need to supply plenty of reactants to make plenty of product.

Go further

Henry-Louis Le Chatelier came up with the principle that bears his name, but he is also famous for his other work, including the chemistry of cement. One of his later experiments almost killed his assistant. Find out about this experiment.

Le Chatelier's principle

Le Chatelier's principle provides us with a rule to predict what happens to reactions at equilibrium when conditions change: 'when a change is made to a reaction at equilibrium, the position of equilibrium moves to oppose the change'. So, for example, when the concentration of a substance is decreased, the equilibrium position moves to increase its concentration again.

A Explain why removing methanol from the equilibrium mixture moves the equilibrium position to the right.

How do you choose a suitable pressure?

There is 1 mol of gas on the right of the balanced equation for making methanol, but 3 mol on the left. This means that if the pressure is increased, the equilibrium position moves to the right and the equilibrium yield of methanol increases.

You might think that chemical engineers choose a very high pressure to obtain a very high equilibrium yield of methanol. However, high pressures need expensive equipment to compress the gases, a lot of energy to run, and tough reaction vessels to withstand the pressure (Figure 2). In practice, chemical engineers choose a *compromise* pressure: high enough to achieve a reasonable equilibrium yield, but not so high as to be expensive or hazardous.

Figure 2 *Industrial compressors are complex and require large amounts of energy.*

B Explain why methanol is manufactured at a pressure of 5–10 MPa (between 50 and 100 times normal atmospheric pressure).

How do you choose a suitable temperature?

The forward reaction in the manufacture of methanol is exothermic, so the backward reaction is endothermic. This means that if the temperature is increased, the equilibrium position moves to the left and the equilibrium yield of methanol decreases.

You might think that chemical engineers choose a low temperature to obtain a high equilibrium yield of methanol. However, at low temperatures the rate of reaction is too low to produce methanol quickly enough to be profitable. In practice, chemical engineers choose a *compromise* temperature: low enough to achieve a reasonable equilibrium yield, but high enough to achieve a reasonable rate of reaction (Figure 3).

Figure 3 *Chemical factories are huge and require a lot of energy for heating, lighting, and moving machinery.*

1 Explain why methanol is manufactured at a temperature of 250 °C. *(3 marks)*

2 Ethanol can be manufactured by reacting ethene with steam:
$$C_2H_4(g) + H_2O(g) \rightleftharpoons C_2H_5OH(g) \qquad (\Delta H = -45 \text{ kJ/mol})$$
a Explain the effect on the position of equilibrium of:
  i increasing the pressure *(2 marks)*
  ii increasing the temperature. *(2 marks)*
b The reaction conditions used are 6 MPa pressure at 300 °C. Use your answers to **a** to explain why these conditions represent a compromise. *(3 marks)*

3 Catalysts do not change the equilibrium position. Explain why catalysts are still used in the manufacture of substances by reversible reactions. *(2 marks)*

C5.3 Equilibria

Summary questions

1 The reaction between sulfur dioxide and oxygen to form sulfur trioxide can be modelled by this equation:

$$2SO_2(g) + O_2(g) \rightleftharpoons 2SO_3(g)$$

a Explain why this equation models a *reversible* reaction.

b Explain the meaning of the term *reversible* in the context of chemical reactions.

2 The reaction between iron(II) ions and silver ions is reversible:

$$Fe^{2+}(aq) + Ag^+(aq) \rightleftharpoons Fe^{3+}(aq) + Ag(s)$$

Iron(II) nitrate solution is pale green and iron(III) nitrate solution is orange-brown.

Describe what you would observe when iron(II) nitrate solution is mixed with silver nitrate solution.

3 Ammonium chloride decomposes when heated:

$$NH_4Cl(s) \rightarrow NH_3(g) + HCl(g)$$

This reaction is reversible.

a Write an equation, including state symbols, to model the backward reaction.

b Write an equation, including state symbols, to model the decomposition of ammonium chloride as a reversible reaction.

c Figure 1 shows an experiment to investigate this reaction.

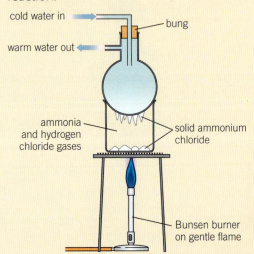

Figure 1 *Decomposition of ammonium chloride.*

i Explain how you know that the forward reaction is endothermic.

ii Explain why ammonium chloride in the solid state forms on the bottom of the round-bottomed flask.

4 Nickel metal can be obtained by heating tetracarbonyl nickel, $Ni(CO)_4$, to 200 °C: **H**

$$Ni(CO)_4(s) \rightleftharpoons Ni(s) + 4CO(g) \quad (\Delta H = +191\,kJ/mol)$$

a Describe and explain the effect on the equilibrium position of increasing the pressure in this reaction.

b **i** Explain whether the forward reaction is exothermic or endothermic.

ii Describe and explain the effect on the equilibrium position of increasing the temperature in this reaction.

5 A substance is produced in the gas state by a reversible reaction. The graph in Figure 2 shows how the equilibrium yield of this product depends on temperature and pressure.

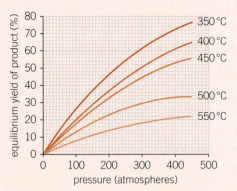

Figure 2 *Yield of product at different temperatures and pressures.*

a **i** Describe the effect on the equilibrium yield of increasing the pressure.

ii Explain what this tells you about the number of moles of gas on each side of the balanced equation that models the reaction.

b **i** Using the data at 450 atmospheres pressure, describe the effect on the equilibrium yield of increasing the temperature.

ii Explain, using your answer to **i**, whether the forward reaction or the backward reaction is endothermic.

c **i** Describe the effect on the equilibrium position of using a catalyst.

ii Explain why the manufacturers of this product might use a catalyst in the industrial process.

Revision questions

1 Which statement is true for a reversible reaction when it is in dynamic equilibrium?

  A The rate of the forward reaction is greater than the rate of the backward reaction.

  B The concentration of the reactants is decreasing.

  C The rate of the forward reaction is equal to the rate of the backward reaction.

  D The concentration of the products is increasing.

        (1 mark)

2 The equation models a reversible reaction. **H**

$$4NH_3(g) + 5O_2(g) \rightleftharpoons 4NO(g) + 6H_2O(g)$$
$$(\Delta H = -950\,kJ)$$

In a sealed container the reversible reaction forms a dynamic equilibrium.

Which one set of changes below is certain to move the position of the equilibrium to the right?

  A Increasing the pressure and increasing the temperature.

  B Increasing the pressure and decreasing the temperature.

  C Decreasing the pressure and decreasing the temperature.

  D Decreasing the pressure and increasing the temperature.

        (1 mark)

3 The reversible reaction between iron and steam makes two products, as shown in the following equation.

$$3Fe(s) + 4H_2O(g) \rightleftharpoons Fe_3O_4(s) + 4H_2(g)$$

  a Explain what is meant by the term *reversible reaction*.

        (1 mark)

  b Explain why the reversible reaction reaches dynamic equilibrium in a sealed container, and describe the features of this dynamic equilibrium. *(3 marks)*

  c Predict the effect of increasing the pressure on **H**
the position of the equilibrium.
Explain your prediction. *(2 marks)*

d A student sets up the apparatus below.

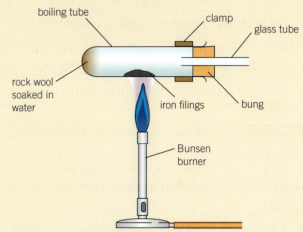

boiling tube   clamp   glass tube   rock wool soaked in water   iron filings   bung   Bunsen burner

  i Suggest how the student could use the apparatus to generate steam to react with the iron. *(1 mark)*

  ii Explain why the reversible reaction cannot reach dynamic equilibrium in this apparatus. *(1 mark)*

4* The ionic equation below models an equilibrium **H**
reaction.

The colours of the ions in solution are also shown.

$$Fe^{3+}(aq) + NCS^-(aq) \rightleftharpoons FeNCS^{2+}(aq)$$
    yellow   colourless   red

Predict the colour changes that occur if solutions containing each of the ions listed below are added separately to the solution, and explain your predictions.

    iron(III) ions, Fe^{3+}

    thiocyanate ions, NCS^-

    thiocyantoiron(III) ions, $FeNCS^{2+}$ *(6 marks)*

C5.1 Monitoring chemical reactions

- Calculate the theoretical yield of a product from a given mass of reactant.
- Calculate the percentage yield of a product.
- Define and calculate the atom economy of a reaction.
- Describe how to carry out an acid–alkali titration.

H **S**
- Explain why a particular reaction pathway is chosen to produce a specified product, using appropriate data.
- Calculate the concentration of a solution in mol/dm^3.
- Explain the relationship between concentration of solution, mass of solute, and volume of solution.
- Carry out titration calculations involving concentrations and volumes.
- Describe the relationship between molar amounts of gases and their volumes.
- Calculate the volumes of gases involved in reactions using the molar gas volume.

C5.2 Controlling reactions

- Explain what is meant by rate of reaction.
- Suggest practical methods for determining rates of reaction.
- Interpret rate of reaction graphs.
- Describe, and explain, the effect on the rate of a reaction of changing:
  - the temperature
  - the concentration of a reacting solution
  - the pressure of reacting gases
  - the size of pieces of a reacting solid.
- Describe the characteristics of catalysts and their effect on rates of reaction.
- State that enzymes act as catalysts in biological systems.
- Explain how catalysts work in terms of activation energy.

C5.3 Equilibria

S
- Describe how some reactions are reversible.
- Describe some examples of reversible reactions.
- Describe the conditions under which dynamic equilibrium occurs.

H
- Predict the effect of changing reaction conditions such as temperature, concentration, and pressure on an equilibrium position.
- Explain the appropriate conditions to make a particular product using a reversible reaction.

Dynamic equilibrium

- occurs in closed system
- rate of forward reaction ↓ over time
- rate of backward reaction ↑ over time
- rate of forward reaction = rate of backward reaction
- concentrations of all reacting substances stay constant

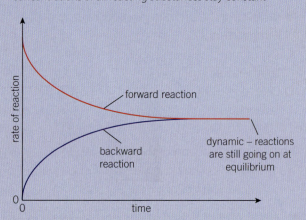

Equilibrium position **H**

- Le Chatelier's principle: when a change is made to a reaction at equilibrium, the position of equilibrium moves to oppose the change

Change in conditions	Equilibrium position moves…
↑ concentration of reacting substance	away from the reacting substance
↑ temperature	in the direction of the endothermic reaction
↑ pressure	in the direction of the fewest moles of reacting gas

Reaction calculations

$$\text{theoretical yield of product (g)} = \frac{\text{mass of limiting reactant (g)}}{\text{sum of } M_r \text{ for limiting reactant}} \times \text{sum of } M_r \text{ for product}$$

$$\text{percentage yield} = \frac{\text{actual yield}}{\text{theoretical yield}} \times 100$$

- 0 to 100%
- can be less than 100% because:
  – side reactions may make by-products
  – reaction may not go to completion
  – product lost during transfers and purification

$$\text{atom economy} = \frac{\text{sum of } M_r \text{ of the desired product}}{\text{sum of } M_r \text{ of all products}} \times 100$$

- ↑ atom economy is desirable
- 100% if only one product is made

Reaction pathway   **H**
- selected to maximise yield, atom economy, absence or usefulness of by-products, rate of reaction, and equilibrium position

C5 Monitoring and controlling chemical reactions

Calculations with volumes

Titrations   **H**
- ensure burette is vertical
- swirl flask to mix contents
- add acid drop-by-drop near end-point
- repeat to obtain two or more concordant titres

Concentration of solutions

$$\text{concentration (mol/dm}^3) = \frac{\text{amount of solute (mol)}}{\text{volume of solution (dm}^3)}$$

- $cm^3 \rightarrow dm^3$ = divide by 1000

Gas calculations

$$\text{volume of gas (dm}^3) = \text{amount (mol)} \times V_m$$

- molar volume (V_m) = 24 dm³/mol at room temperature and pressure

Reaction conditions

- for a reaction to happen, particles must collide
- particles must have activation energy or more
- ↑ frequency of successful collisions → ↑ rate of reaction

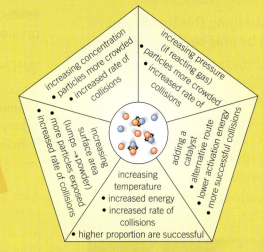

Measuring rate of reaction
- method of measurement differs depending on reaction being investigated
- rate of reaction inversely proportional to reaction time

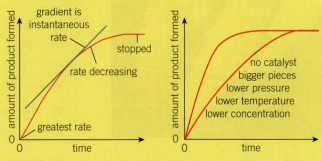

Catalysts
- increase the rate of a reaction but are unchanged at the end of the reaction
- enzymes = biological catalysts

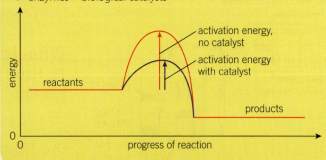

Reversible reactions
- products can react together to form original reactants
- reversible reaction symbol (⇌) combines forward reaction and backward reaction

Learning outcomes

After studying this lesson you should be able to:

- describe the importance of nitrogen, phosphorus, and potassium compounds in agriculture
- explain the importance of the Haber process in agriculture
- describe the industrial production of fertilisers.

Specification reference: C6.1g, C6.1i, C6.1j

Figure 1 *The corn cob on the left has been grown using fertiliser that provides nitrogen; the one on the right has been grown without fertiliser.*

purified nitrogen from air

compressor

iron catalyst

450 °C
200 atm

reaction vessel

condenser

purified hydrogen gas

liquid ammonia

Figure 2 *An outline of the Haber process, including the reaction conditions and catalyst used.*

Plants grow poorly and crop yields suffer if the soil only contains limited amounts of certain elements. Farmers use fertilisers to help provide the elements plants need (Figure 1).

Which elements do plants need?

Nitrogen, phosphorus, and potassium are three of the **essential elements** needed by plants. Plants do not grow well if these are in limited supply in the soil. They may show symptoms of **mineral deficiency**, some of which are described in Table 1. The quality and yield of food will also be reduced.

Table 1 *Mineral deficiencies in plants.*

Element	Typical symptoms of deficiency
nitrogen, N	poor growth, yellow leaves
phosphorus, P	poor root growth, discoloured leaves
potassium, K	poor fruit growth, discoloured leaves

Fertilisers are substances that replace the elements used by plants as they grow. Plant roots can only absorb these elements if they are in a water-soluble form. For example:

- nitrogen in nitrate ions, NO_3^-, or ammonium ions, NH_4^+
- phosphorus in phosphate ions, PO_4^{3-}
- potassium as potassium ions, K^+.

Fertilisers that provide nitrogen, phosphorus, and potassium in water-soluble compounds are called 'NPK fertilisers'.

> **A** Explain why a mixture of potassium nitrate and ammonium phosphate is an example of an NPK fertiliser.

What is the Haber process?

The **Haber process** (Figure 2) manufactures ammonia from nitrogen and hydrogen:

$$N_2(g) + 3H_2(g) \rightleftharpoons 2NH_3(g)$$

Over 150 million tonnes of ammonia are manufactured in the world each year, and more than 80% of this is used to produce fertilisers.

The **raw materials** for the Haber process are air, natural gas, and steam.

- Nitrogen is manufactured by the fractional distillation of **liquefied** air (air is 78% nitrogen).
- Hydrogen is manufactured by reacting natural gas (mostly methane) with steam.

Synoptic link

The hydrogen needed for the Haber process can also be produced by cracking crude oil. You can learn more about cracking in C6.2.6 *Cracking oil fractions*.

B Explain how you can tell that the Haber process is important for making fertilisers.

What happens in a fertiliser factory?

Several processes happen in a fertiliser factory. A variety of raw materials are needed, including sulfur for making sulfuric acid, and phosphate rock for making phosphoric acid (Figure 3). The different processes in a fertiliser factory are integrated so that a range of compounds for fertilisers can be made. These include:

● ammonium nitrate, NH_4NO_3

● ammonium sulfate, $(NH_4)_2SO_4$

● ammonium phosphate, $(NH_4)_3PO_4$

● potassium nitrate, KNO_3.

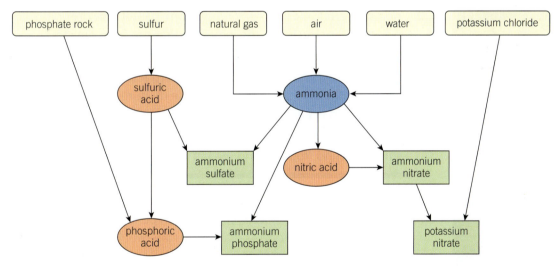

Figure 3 *Some of the raw materials and processes in fertiliser manufacture.*

C Name the raw materials needed to produce ammonium sulfate.

1 ✏ Explain why NPK fertilisers are important to farmers. *(4 marks)*

2 Write down balanced equations, including state symbols, for the formation of the following compounds from ammonia solution, $NH_3(aq)$:
   a ammonium nitrate solution using $HNO_3(aq)$ *(2 marks)*
   b ammonium phosphate solution using $H_3PO_4(aq)$. *(2 marks)*

3 ✏ Write down a balanced equation, including state symbols, to model the Haber process. Describe the reaction conditions and catalyst needed. *(4 marks)*

Learning outcomes

After studying this lesson you should be able to:

- describe how to make potassium sulfate and ammonium sulfate in the laboratory
- compare the industrial and laboratory production of fertilisers.

Specification reference: C6.1h

Figure 1 *Urea is in the solid state at room temperature, so it is easily spread as granules.*

Urea is found in urine, but can also be manufactured by reacting ammonia with carbon dioxide. It has the highest nitrogen content of all common fertilisers (Figure 1). Urea is too difficult to make in the laboratory, but there are laboratory methods to make your own fertiliser.

How can you make fertilisers in the lab?

Potassium sulfate and ammonium sulfate are two compounds found in fertilisers that are easy to make in the laboratory.

Making potassium sulfate

Potassium sulfate can be made from potassium hydroxide and sulfuric acid.

1 Put dilute potassium hydroxide solution, $KOH(aq)$, into a conical flask and add a few drops of phenolphthalein indicator. Phenolphthalein changes colour at the end-point of the titration, so adding the indicator enables you to determine when the alkali has been neutralised.

2 Add dilute sulfuric acid, $H_2SO_4(aq)$, from a burette or dropping pipette, stopping when the indicator changes from pink to colourless (Figure 2).

3 Add 'activated charcoal' (Figure 3). This attracts the phenolphthalein, and you can filter the mixture to remove the activated charcoal with the phenolphthalein attached to it.

4 Warm the filtrate to evaporate the water, leaving potassium sulfate behind. You must not heat this to dryness.

A Explain why, when carrying out the experiment *Making potassium sulfate*, the potassium sulfate would be found in the filtrate.

B Write down the balanced equation for the laboratory production of potassium sulfate solution, $K_2SO_4(aq)$, including state symbols.

Figure 2 *A white tile makes it easier to see the colour change at the end-point in a titration.*

Making ammonium sulfate

To make ammonium sulfate, you can use a similar method to the one used to make potassium sulfate.

1   Place the dilute ammonia solution, NH_3(aq), in a conical flask with methyl orange indicator. Ammonia solution releases small amounts of ammonia in the gas state, which has an irritating sharp smell, so you need to take care to avoid breathing it in.

2   Add dilute sulfuric acid, H_2SO_4(aq), from a burette or dropping pipette, stopping when the indicator changes from yellow to red.

3   When you reach the end-point, you can add a little extra ammonia solution to ensure that the reaction is complete. Any remaining ammonia will be lost during evaporation.

This experiment carries some hazards: ammonia solution and potassium hydroxide solution are alkaline, and ammonia also gives off an irritating sharp smell. Excess ammonia is given off in the gas state when the solution is warmed.

Figure 3 *Activated charcoal is a very fine carbon powder with a large surface area. It absorbs many substances, and can be found in shoe inserts for smelly feet.*

C   Write down the balanced equation for the laboratory production of ammonium sulfate solution, $(NH_4)_2SO_4$(aq), including state symbols.

How are industrial processes different?

When you make a substance in the school laboratory, you usually make a small amount at one time, using a **batch process**. On the other hand, many industrial processes are **continuous processes**. They make large amounts and go on all the time. Table 1 summarises some of the differences.

In the laboratory, you start with pure substances bought from a chemical manufacturer. Fertiliser factories start with raw materials (substances obtained from the ground, air, or sea). These must be purified before use, or the product must be purified at the end.

Table 1 *Comparison of batch and continuous processes.*

Feature	Batch process	Continuous process
rate of production	low	high
relative cost of equipment	low	high
number of workers needed	large	small
shut-down periods	frequent	rare
ease of automating the process	low	high

1   **a**   Describe two hazards in making fertiliser compounds in the laboratory.   (*2 marks*)

    b   Explain how the risks presented by these hazards may be controlled.   (*4 marks*)

2   Suggest two reasons why a fertiliser factory may require more energy to make ammonium sulfate than a laboratory does.   (*2 marks*)

3   Explain why batch processes are more suited to making small amounts of expensive 'speciality chemicals' such as medicines, and continuous processes are more suited to making large amounts of cheaper 'bulk chemicals' such as fertilisers.   (*4 marks*)

Learning outcomes

After studying this lesson you should be able to:

- explain the trade-off between rate of production and position of equilibrium in the Haber process
- interpret graphs involving yields and reaction conditions.

Specification reference: C6.1d, C6.1e

Figure 1 *The X-15 was launched from a modified bomber aircraft. It was powered by ammonia, made using the Haber process.*

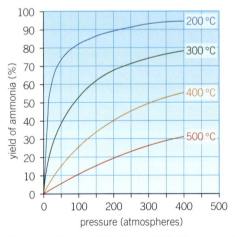

Figure 2 *The equilibrium yield of ammonia at different pressures and temperatures.*

Aircraft travel fast using jet engines, but to go really fast you need a rocket motor. NASA's X-15 manned hypersonic research programme in the 1960s set speed records that are still unbroken over half a century later. A rocket motor fuelled by liquefied ammonia powered the X-15 (Figure 1) to an incredible 2 kilometres per second, more than 4500 miles per hour.

What are the conditions chosen for the Haber process?

Remember that the Haber process manufactures ammonia from nitrogen and hydrogen:

$$N_2(g) + 3H_2(g) \rightleftharpoons 2NH_3(g) \qquad (\Delta H = -93 \, \text{kJ/mol})$$

The conditions chosen are usually:

- a pressure of 200 atmospheres (20 MPa)
- a temperature of 450 °C
- an iron catalyst.

Under these conditions, the equilibrium yield of ammonia is about 30%. This is clearly much less than 100%, so why are these conditions chosen?

What factors determine the pressure chosen?

In the balanced equation for the Haber process, there are (1 + 3) = 4 mol of gas on the left, but only 2 mol of gas on the right. If the pressure is increased, the equilibrium position moves to the right and the equilibrium yield of ammonia increases. However, it would be hazardous and expensive to choose a very high pressure. The higher equilibrium yield would not justify the additional costs, so the pressure chosen is a compromise.

Ammonia-fuelled cars

Petrol and diesel engines can be converted to run on ammonia. Some scientists and engineers believe it has a future as a 'green' fuel.

A   Write a balanced equation for the combustion of ammonia with oxygen to produce nitrogen dioxide and water.

B   Suggest an environmental benefit of using ammonia, rather than petrol or diesel, as a fuel.

C   Use the graph in Figure 2 to estimate the equilibrium yield of ammonia at 400 °C when the pressure is 200 atmospheres, and when it is 400 atmospheres.

What factors determine the temperature chosen?

The forward reaction of the Haber process is exothermic, so the backward reaction is endothermic. If the temperature is increased, the equilibrium position moves to the left and the equilibrium yield of ammonia decreases. A high equilibrium yield is favoured by a low temperature.

The temperature chosen is a compromise: low enough to achieve a reasonable equilibrium yield, but high enough to achieve a reasonable rate of reaction. In addition, the iron catalyst works more efficiently above 400 °C.

> **D** Use the graph in Figure 2 to estimate the equilibrium yield of ammonia at 200 atmospheres when the temperature is 300 °C and when it is 500 °C.

What other conditions are chosen?

The mixture of gases leaving the reaction vessel is cooled so that the ammonia is liquefied. This allows the ammonia to be removed, and unreacted nitrogen and hydrogen to be recycled. This improves the overall yield to around 97%.

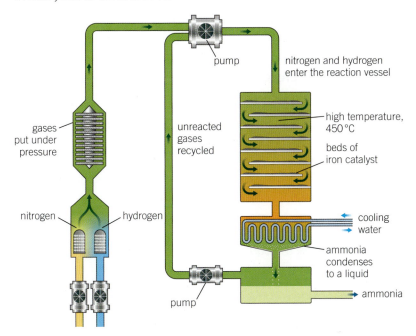

Figure 3 *The Haber process.*

Go further

Fritz Haber invented the process to make ammonia from nitrogen and hydrogen (Figure 3). Find out why he was a controversial scientist in his day.

> 1 Describe the effect on the equilibrium yield of ammonia in the Haber process of:
>   **a** increasing the pressure (*1 mark*)
>   **b** increasing the temperature (*1 mark*)
>   **c** using an iron catalyst. (*1 mark*)
> 2 Explain the effect of liquefying ammonia on the equilibrium yield of the Haber process. (*3 marks*)
> 3 Describe and explain the conditions used in the Haber process. (*6 marks*)

C6.1.4 The Contact process

Learning outcome

After studying this lesson you should be able to:

● explain the trade-off between rate of production and position of equilibrium in the Contact process.

Specification reference: C6.1d, C6.1e

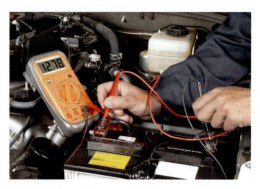

Figure 1 *The acid in a 'lead–acid' accumulator, the rechargeable battery used in cars, is sulfuric acid.*

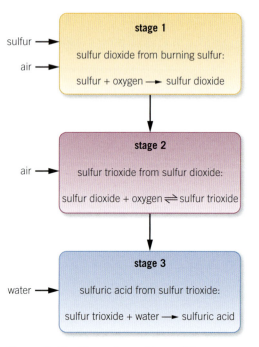

Figure 2 *The three stages for making sulfuric acid.*

Sulfuric acid is an important substance (Figure 1). Around 70% of the sulfuric acid manufactured each year is used to make fertilisers. Its other uses include oil refining, metal extraction, and making paints and polymers. Sulfuric acid is synthesised in a three step process, including the Contact process.

What happens in the Contact process?

Three raw materials are needed for making sulfuric acid:

● sulfur

● air (a source of oxygen)

● water.

These are used in three stages (Figure 2).

In stage 1, sulfur burns in air to produce sulfur dioxide (Figure 3):

$$S(s) + O_2(g) \rightarrow SO_2(g) \qquad (\Delta H = -297\,kJ/mol)$$

A Explain how you know that stage 1 involves an exothermic reaction that is not reversible.

In stage 2, the **Contact process**, sulfur dioxide and oxygen react together to produce sulfur trioxide:

$$2SO_2(g) + O_2(g) \rightleftharpoons 2SO_3(g) \qquad (\Delta H = -144\,kJ/mol)$$

The conditions chosen for this reversible reaction are usually:

● a pressure of 2 atmospheres (200 kPa)

● a temperature of 450 °C

● a vanadium(V) oxide catalyst, V_2O_5.

Under these conditions, the equilibrium yield of sulfur trioxide is about 96%.

In stage 3, sulfur trioxide is converted to sulfuric acid:

$$H_2O(l) + SO_3(g) \rightarrow H_2SO_4(aq)$$

B Write down the balanced equation for the overall reaction in which sulfuric acid is formed from sulfur, oxygen, and water.

What factors determine the pressure chosen?

In the balanced equation for the reaction in stage 2, there are $(2 + 1) = 3$ mol of gas on the left, but only 2 mol of gas on the right. If the pressure is increased, the equilibrium position moves to the right and the equilibrium yield of sulfur trioxide increases. However, in this reaction, the equilibrium position is already far to the right. There is no need for high pressures, and 2 atmospheres is just enough to push the gases through the converter.

What factors determine the temperature chosen?

The forward reaction in stage 2 is exothermic, so the backward reaction is endothermic. If the temperature is increased, the equilibrium position moves to the left and the equilibrium yield of sulfur trioxide decreases. A high equilibrium yield is favoured by a low temperature.

The temperature chosen is a compromise: low enough to achieve a reasonable equilibrium yield, but high enough to achieve a reasonable rate of reaction. In addition, the vanadium(V) oxide catalyst only works above 380 °C.

> **C** Sulfur dioxide is a common waste product formed during the processing of crude oil, natural gas, and some metal ores. Suggest one advantage and one disadvantage to a sulfuric acid manufacturer of using sulfur dioxide produced this way.

How are the hazards controlled?

The reaction between sulfur trioxide and water in stage 3 is very exothermic. It would produce a hazardous acidic mist, so stage 3 is carried out in two steps:

1 Sulfur trioxide is passed through concentrated sulfuric acid (made previously) to make a compound called oleum, $H_2S_2O_7$:

$$H_2SO_4(l) + SO_3(g) \rightarrow H_2S_2O_7(l)$$

2 The oleum is then added to water, and the reaction makes a larger volume of concentrated sulfuric acid:

$$H_2S_2O_7(l) + H_2O(l) \rightarrow 2H_2SO_4(aq)$$

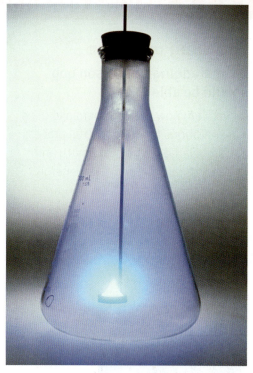

Figure 3 *Sulfur burns in air with a bright blue flame.*

> **1** Describe the effect on the equilibrium yield of sulfur trioxide of:
>   **a** increasing the pressure   (*1 mark*)
>   **b** increasing the temperature   (*1 mark*)
>   **c** using a vanadium(V) catalyst.   (*1 mark*)
> **2** Suggest how the reactions in stages 1 and 2 could help to reduce the energy costs of making sulfuric acid.   (*3 marks*)
> **3** Describe and explain the conditions used in the second stage of the Contact process. Include a balanced chemical equation in your answer.   (*6 marks*)

Learning outcome

After studying this lesson you should be able to:

- explain how the commercially used conditions for an industrial process are related to the availability and cost of raw materials and energy supplies, control of equilibrium position, and rate of reaction.

Specification reference: C6.1f

Figure 1 *Ethanol is used in antibacterial hand gels.*

Some substances can be made by more than one industrial process. For example, ethanol is the alcohol in wine and beer. It is also useful as a fuel, as a solvent for perfume and deodorants, and to kill bacteria (Figure 1). Ethanol is manufactured in two very different ways. How do manufacturers choose which way to use?

How is alcohol made from renewable raw materials?

Renewable raw materials can be replaced as they are used and, in principle, should not run out. Ethanol is made from plant sugars using **fermentation**. This process relies on single-celled fungi called yeast. Yeast cells contain enzymes that catalyse the conversion of glucose solution to carbon dioxide and ethanol:

$$\text{glucose} \rightarrow \text{carbon dioxide} + \text{ethanol}$$

$$C_6H_{12}O_6(aq) \rightarrow 2CO_2(g) + 2C_2H_5OH(aq)$$

You can carry out fermentation in a school laboratory using simple apparatus, as shown in Figure 2. Yeast cells become inactive if the temperature is too low, and their enzymes become **denatured** and stop working above about 50 °C. This means that fermentation is carried out at about 35 °C under normal atmospheric pressure. Industrial fermentation uses the same conditions but with more complex equipment.

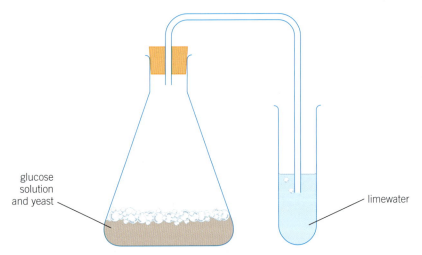

Figure 2 *Laboratory fermentation using a conical flask, delivery tube, and boiling tube.*

A Explain what will happen to the limewater during fermentation in the apparatus in Figure 2.

How is alcohol made from non-renewable raw materials?

Non-renewable raw materials are used faster than they can be replaced. They will run out one day if you keep using them. Ethene is obtained from crude oil, which is a non-renewable raw material. Ethanol can be produced by the **hydration** of ethene (Figure 3):

ethene + steam ⇌ ethanol

$C_2H_4(g) + H_2O(g) \rightleftharpoons C_2H_5OH(g)$ ($\Delta H = -45 \text{ kJ/mol}$)

You cannot do this reaction in a school laboratory. It is only suitable as an industrial process because it needs a temperature of 300 °C, and a pressure of 60 atmospheres in the presence of a phosphoric acid catalyst.

> **B** Use your knowledge of reversible reactions and equilibria to explain why a high pressure is chosen for the hydration of ethene.

What are the other differences between the processes?

Table 1 summarises some differences between the two processes for making ethanol.

Table 1 *Comparison of the two methods of making ethanol.*

Feature	Fermentation of sugars	Hydration of ethene
cost of raw materials	low	high
conditions	moderate temperature normal pressure	high temperature high pressure
energy requirements	low	high
rate of reaction	low	high
percentage yield	low – about 15%	high – about 95%
purity of product	low – needs filtering and fractional distillation	high – there are no by-products

The process chosen by a manufacturer will depend upon factors such as the availability and cost of the raw materials, and the cost of the energy needed. Different decisions will be made in different situations.

> **1** In terms of the raw materials and energy requirements, explain which process for the manufacture of ethanol is more suitable for sustainable development. *(3 marks)*
>
> **2** Other than the raw materials needed, describe two advantages of making ethanol:
>   **a** from ethene and steam *(2 marks)*
>   **b** by fermentation. *(2 marks)*
>
> **3** 🔧 An oil company wants to begin the large-scale production of ethanol in the grounds of their existing factory, which is sited close to farmland. Evaluate both methods for making ethanol in this factory and suggest which one should be used. *(6 marks)*

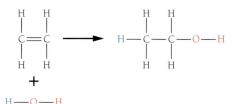

Figure 3 *The hydration of ethene, modelled using displayed formulae.*

> **Synoptic link**
>
> You can learn more about crude oil and substances obtained from it in C6.2.5 *Alkanes from crude oil*.

Learning outcomes

After studying this lesson you should be able to:

- explain how the industrial process used to extract a metal is chosen
- describe how copper may be extracted from copper oxide.

Specification reference: C6.1a

Each year, the world's mines produce metal ores containing 18 million tonnes of copper (Figure 1), 49 million tonnes of aluminium, and an astonishing 3 billion tonnes of iron. How are these metals extracted from their ores?

Figure 1 *This huge copper mine is 2 km across.*

What is an ore?

An **ore** is a rock or mineral, for example malachite (Figure 2), that contains enough metal (or metal compound) to make it economical to extract the metal – the value of the metal is more than the cost of extracting it. Different metal compounds are found in different ores. Some examples are shown in Table 1.

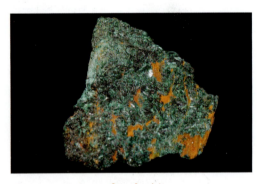

Figure 2 *A piece of malachite, a copper ore.*

Table 1 *Metal compounds found in different ores.*

Ore	Metal compounds found in the ore
malachite	copper carbonate
bauxite	aluminium oxide
haematite	iron(III) oxide

An ore must be mined, and then processed to separate the metal compound from the other substances in the ore. The metal is extracted from the pure metal compound using chemical reactions. The method chosen to extract a metal depends upon its position in the reactivity series (Figure 3).

> **A** Define the term 'ore'.

What extraction methods are there?

In principle, all metals could be extracted from their compounds using electrolysis, but electricity is expensive. If the metal is less reactive than carbon, cheaper methods are used instead. Copper and iron are less reactive than carbon, so they can be extracted by heating their compounds with carbon or with carbon monoxide.

> **B** Explain why aluminium is extracted using electrolysis.

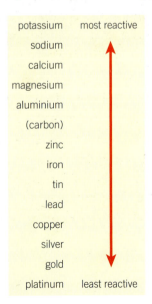

Figure 3 *A reactivity series of metals, with carbon shown for comparison.*

How is copper extracted?

Copper can be extracted from copper(II) sulfide in two stages.

Stage 1: First the copper(II) sulfide is 'roasted' in air:

copper(II) sulfide + oxygen → copper(II) oxide + sulfur dioxide

$$2CuS(s) + 3O_2(g) \rightarrow 2CuO(s) + 2SO_2(g)$$

Stage 2: The copper(II) oxide is heated with carbon:

copper(II) oxide + carbon → copper + carbon dioxide

$$2CuO(s) + C(s) \rightarrow 2Cu(s) + CO_2(g)$$

This is an example of a **redox** reaction:

- copper(II) oxide loses oxygen and is reduced
- carbon gains oxygen and is oxidised.

In this reaction, carbon is acting as a reducing agent.

Copper(II) oxide can also be reduced to copper by heating it with methane or with hydrogen (Figure 4).

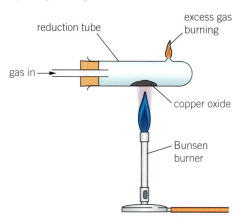

Figure 4 *Heating copper(II) oxide with methane.*

> **C** Write a balanced equation to model the reaction between copper(II) oxide and methane, CH_4, forming copper, carbon dioxide, and water.

1 Malachite contains copper(II) carbonate, $CuCO_3$. This decomposes to copper(II) oxide and carbon dioxide when heated. Write a balanced equation, including state symbols, to model this reaction. (*2 marks*)

2 **a** Write a balanced equation, including state symbols, to model the reaction in which copper(II) oxide reacts with carbon to form copper and carbon monoxide gas, CO. (*2 marks*)
   b Explain whether copper(II) oxide is oxidised or reduced in this reaction. (*2 marks*)

3 Zinc can be extracted from zinc oxide, ZnO, by reduction with carbon or with carbon monoxide. Explain, with the help of balanced equations, the role of carbon and carbon monoxide. (*4 marks*)

Extracting copper

You can reduce copper(II) oxide using charcoal, which is mostly carbon.

1 Mix the two powders in a crucible, replace the lid. The lid must be kept on the crucible during heating to stop the powders escaping and to stop air getting in (because then the carbon would burn).

2 Heat it strongly.

3 After several minutes, allow the crucible to cool.

4 When the crucible is cool, transfer its contents to a beaker of water. The copper sinks to the bottom while excess charcoal is suspended in the water.

5 Separate the copper by washing it.

Excess charcoal powder is used to ensure that all of the copper(II) oxide is reduced to copper.

Eye protection should be worn throughout this practical.

Learning outcome

After studying this lesson you should be able to:

- describe how iron may be extracted from iron oxide.

Specification reference: C6.1a

The world's first cast iron bridge was opened in 1781. Made from about 385 tonnes of iron, it spans almost 31 m across the River Severn at Ironbridge in Shropshire (Figure 1). You can still walk across it today. Modern bridges are made from steel, a form of iron, and can span over half a kilometre. How is iron produced from its ore?

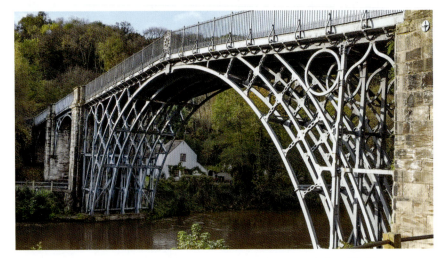

Figure 1 *The Iron Bridge in Ironbridge, Shropshire, carried traffic for over 150 years.*

What is a blast furnace?

Iron is extracted from its ore using a large reaction container called a **blast furnace** (Figure 2). Modern blast furnaces can be around 30 m high and produce about 10 000 tonnes of iron per day.

These raw materials are added to the top of the blast furnace:

- iron ore, such as **haematite**, which contains iron(III) oxide
- **coke**, which is mostly carbon, and is made by heating coal in the absence of air
- limestone, which is used to purify the iron.

In addition, hot air is forced in at the bottom of the blast furnace.

> **A** Write down the four raw materials used to make iron.

What reactions happen in a blast furnace?

Carbon can reduce iron(III) oxide to iron, but carbon monoxide is the main reducing agent in the blast furnace. Carbon monoxide is formed when coke reacts with carbon dioxide.

Stage 1: Coke burns in the hot air, making carbon dioxide:

$$C(s) + O_2(g) \rightarrow CO_2(g)$$

Stage 2: More coke reduces the carbon dioxide, making carbon monoxide:

$$C(s) + CO_2(g) \rightarrow 2CO(g)$$

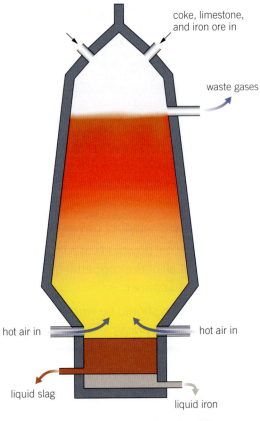

coke, limestone, and iron ore in

waste gases

hot air in

hot air in

liquid slag

liquid iron

Figure 2 *The temperature inside a blast furnace varies from 1900 °C at the bottom to 300 °C at the top.*

Stage 3: Carbon monoxide reduces iron(III) oxide to iron at around 1500 °C:

$$3CO(g) + Fe_2O_3(s) \rightarrow 3CO_2(g) + 2Fe(l)$$

B Suggest why the reaction between coke and oxygen helps to heat up the blast furnace.

The molten iron trickles downwards in the blast furnace. It contains sandy impurities from the iron ore. These are removed using the limestone, which is mostly calcium carbonate:

Stage 1: Calcium carbonate decomposes in the high temperatures:

$$CaCO_3(s) \rightarrow CaO(s) + CO_2(g)$$

Stage 2: The calcium oxide formed reacts with silica from the sandy impurities to form calcium silicate:

$$CaO(s) + SiO_2(g) \rightarrow CaSiO_3(l)$$

The molten calcium silicate is called **slag**. It floats on the molten iron, and both the iron and slag are removed separately at the bottom of the blast furnace (Figure 3).

C In the blast furnace, 1 mol of iron(III) oxide makes 2 mol of iron. Calculate the relative formula mass of Fe_2O_3 (A_r of Fe = 55.8 and A_r of O = 16.0).

D The Iron Bridge was made from 3.845×10^8 g of iron. Calculate the mass of iron(III) oxide needed to make this mass of iron, giving your answer to three significant figures.

Figure 3 *Iron leaves the blast furnace in the liquid state at over 1540 °C.*

1 a State the name and formula of the main reducing agent in the blast furnace. *(2 marks)*

b Write a balanced equation, including state symbols, to model the reaction in which iron(III) oxide is reduced to iron in the blast furnace. *(2 marks)*

2 Outline how iron is produced from iron(III) oxide in the blast furnace. In your answer, include only the redox reactions that occur. *(3 marks)*

3 Explain why the reaction between calcium oxide and silica is a neutralisation reaction. *(3 marks)*

C6.1.8 Extracting aluminium

Learning outcomes

After studying this lesson you should be able to:

- explain why electrolysis is used to extract some metals from their ores
- describe how aluminium may be extracted from aluminium oxide.

Specification reference: C6.1a, C6.1b

At 309.6 m high, The Shard is one of Europe's tallest buildings (Figure 1). It has 56 000 m² of glass held in place by 11 000 aluminium window frames. Aluminium is also useful for making aircraft, overhead electricity cables, cooking foil, and drinks cans (Figure 2). How is aluminium produced from its ore?

Figure 1 *The Shard rises 87 storeys into the London sky.*

What is aluminium like?

Aluminium exists naturally, mainly as aluminium oxide, Al_2O_3, found in an ore called **bauxite**. Aluminium is more reactive than carbon. This means that, unlike iron and copper, it must be extracted from its ore by electrolysis. However, there is a problem. Electrolysis only works if the compound is in solution or if it is molten – but aluminium oxide does not dissolve in water and its melting point is very high. There is a solution to this problem, as you are about to see.

A Explain, in terms of ions, why electrolysis does not work when aluminium oxide is in the solid state.

How is aluminium oxide electrolysed?

The melting point of aluminium oxide is over 2000 °C. It would be very expensive to heat aluminium oxide to this temperature. To get around this problem, aluminium oxide is dissolved in molten **cryolite**. Cryolite has a much lower melting point than aluminium oxide, and allows electrolysis to happen at about 950 °C.

The molten mixture of aluminium oxide and cryolite is contained in a huge electrolysis cell (Figure 3 and Figure 4). This is made from steel, lined with graphite. The graphite lining acts as the **cathode**, the negative

Figure 2 *Aluminium is used to make drinks cans.*

Synoptic link

You can learn more about electrolysis in C3.4.1 *Electrolysis of molten salts.*

electrode. A series of large graphite blocks act as the **anodes**, the positive electrodes. During electrolysis:

- aluminium is produced at the cathode
- oxygen is produced at the anodes.

The oxygen reacts with the hot graphite anodes, making carbon dioxide.

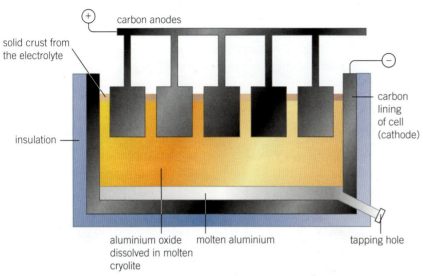

Figure 3 *A cross-section through an aluminium electrolysis cell.*

B Suggest why the anodes in an aluminium electrolysis cell may need to be replaced every few weeks.

What electrode reactions happen during electrolysis of aluminium?
These half-equations model the electrode reactions:

- at the cathode: $Al^{3+} + 3e^- \rightarrow Al$
- at the anode: $2O^{2-} \rightarrow O_2 + 4e^-$

C Explain, in terms of electrons, why aluminium is reduced and oxygen is oxidised during the electrolysis of aluminium oxide.

1 Explain why aluminium oxide is reduced during the extraction of aluminium. *(2 marks)*

2 Describe how aluminium is extracted from its oxide. Include the structure of the electrolysis cell, the function of cryolite, and the processes that occur at each electrode. *(6 marks)*

3 **a** Write down half-equations to model the electrode reactions that happen during the electrolysis of aluminium oxide. *(2 marks)*

   b Write down the balanced equation to model the overall reaction that happens in the electrolysis of aluminium oxide. *(1 mark)*

Study tip
Cryolite reduces the temperature needed for electrolysis to happen, reducing energy costs. The cost of the electricity needed is the most important factor determining the cost of extracting aluminium (Figure 4).

Figure 4 *These long rows of electrolysis cells in this aluminium smelter use huge amounts of electricity.*

Go further
Charles Hall in America and Paul Héroult in France independently worked out how to extract aluminium using electrolysis. There are some remarkable coincidences in the lives of these two scientists. Find out what they are.

Learning outcome

After studying this lesson you should be able to:

- evaluate bioleaching and phytoextraction as alternative biological methods of metal extraction.

Specification reference: C6.1c

The water in Spain's Rio Tinto river is unusual. Water draining into the river from nearby metal mines is acidic. It contains dissolved iron compounds that colour the river water red (Figure 1). The chemistry involved in acid mine drainage is also used in one of two biological methods for extracting metals.

Figure 1 *The Rio Tinto river – its name means 'red river'.*

What is acid mine drainage?

Mines often flood when they are abandoned. Metal sulfides oxidise underwater, producing sulfuric acid, which reacts with other metal ores. Soluble metal compounds form, and these leave with the water as it drains from the mine. These reactions happen naturally, but certain bacteria make them go faster. This is the basis of **bioleaching** (Figure 2).

What is bioleaching?

In respiration, glucose is oxidised and your cells use the energy transferred. In a similar way, in bioleaching, bacteria oxidise iron(II) and sulfide ions and use the energy transferred. In the process, sulfuric acid forms in the presence of water and oxygen. This sulfuric acid breaks down copper sulfide ores and other minerals, releasing copper(II) ions and other metal ions.

> **A** Name the type of reaction involved in bioleaching.

Figure 2 *Copper(II) sulfate deposits form in this bioleaching bath.*

Bioleaching is cheaper than traditional mining and processing. It allows metals to be extracted from ores that contain too little metal for traditional methods to be profitable. These are called **low-grade ores**. The bacteria occur naturally and do not need any special treatment. Bioleaching does not release harmful sulfur dioxide into the atmosphere, but it is slow. Toxic substances are sometimes produced. Care must be taken to avoid these, and sulfuric acid, escaping into water supplies and the soil.

What is phytoextraction?

Plants absorb dissolved ions through their roots. Some plants are particularly good at absorbing certain metal ions, which then accumulate in their roots, shoots, and leaves. In **phytoextraction** (Figure 3), a crop is planted in soil containing a low-grade ore or mine waste. A 'complexing agent' may be added so the plants can absorb the metal ions more easily. The plants are harvested, then burnt to produce an ash with a high concentration of the metal. The metal can then be extracted, just as if the ash was a **high-grade ore**.

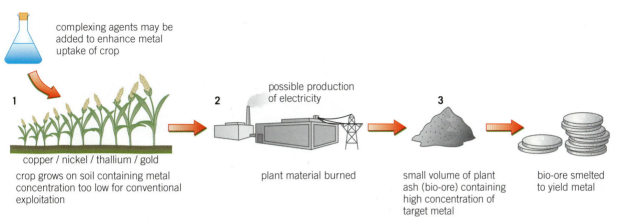

complexing agents may be added to enhance metal uptake of crop

possible production of electricity

1 copper / nickel / thallium / gold
crop grows on soil containing metal concentration too low for conventional exploitation

2 plant material burned

3 small volume of plant ash (bio-ore) containing high concentration of target metal

bio-ore smelted to yield metal

Figure 3 *An outline of phytoextraction.*

B Suggest why the second step in Figure 3 could be of advantage to the phytoextraction company.

Phytoextraction is cheaper than traditional mining and processing. It produces less waste and involves smaller energy transfers, but it is slow. Crops may need replanting and harvesting for several years before the available metal is removed from the soil. On the other hand, it is closer to being a **carbon-neutral** activity that can contribute to sustainable development. Burning plants release carbon dioxide, but they absorb carbon dioxide for photosynthesis as they grow.

C Suggest why the weather can affect phytoextraction.

'Bio' and 'phyto'

A **prefix** is a word or part of a word that goes in front of another word. The prefix 'bio' means the word is to do with living things. The prefix 'phyto' means the word is specifically to do with plants.

1 **a** Outline how bioleaching works. (*4 marks*)
  b Outline how phytoextraction works. (*4 marks*)

2 Make a table to compare the advantages and disadvantages of biological methods of metal extraction with the traditional methods you have studied in this chapter. (*6 marks*)

3 During bioleaching, iron(II) ions are oxidised to iron(III) ions. Explain, with the help of an equation, why this is oxidation. (*2 marks*)

Learning outcomes

After studying this lesson you should be able to:

- describe the composition of some important alloys
- describe how the properties of alloys relate to their uses.

Specification reference: C6.1o

Aircraft need to be lightweight yet strong. Aluminium is a metal with a low density, so objects made from it are light for their size. It can be mixed with copper to make an alloy called **duralumin** (**Figure 1**). This still has a low density but is much stronger than aluminium alone. Alloys often have useful properties that are different from those of the metals they contain.

Figure 1 *Duralumin is used to make strong yet lightweight aircraft parts.*

What are alloys?

An alloy is a mixture of two or more elements, at least one of which is a metal. Steel is an alloy in which iron is mixed with smaller amounts of other metals and carbon. Steel typically has high **tensile strength** and can also be ductile.

Table 1 shows information about steel and some other alloys.

Table 1 *Uses of different alloys.*

Name of alloy	Main metal(s)	Typical uses
steel	iron	buildings, bridges, cars
duralumin	aluminium and copper	aircraft parts
solder	tin and copper	joining electrical components and copper pipes
brass	copper and zinc	musical instruments and coins
bronze	copper and tin	bells, propellers for ships

A Name two alloys that contain copper.

What is solder like?

There are several types of solder. One type of solder is made from tin and copper. Tin melts at 232 °C and copper melts at 1085 °C, so you might think that a mixture of the two metals would melt between these temperatures. However, solder begins to melt at about 227 °C. This makes it useful for joining electrical components without damaging them. Hot solder in the liquid state flows into the gap between them, then solidifies quickly. Metals are good conductors of electricity, so the soldered joint allows an electric current to pass.

B Explain how solder may be used to join copper pipes (Figure 2).

What are brass and bronze like?

Layers of metal atoms slide over one another when metals are stretched or bent. Alloys are often stronger and harder than the individual metals they contain. Brass is like this. Its copper and zinc atoms are different sizes, and this makes it more difficult for copper atoms to slide over each other (Figure 3).

Figure 2 *Solder melts at a low enough temperature to avoid damaging copper water pipes.*

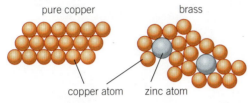

pure copper · brass · copper atom · zinc atom

Figure 3 *The larger zinc atoms distort the regular arrangement of copper atoms in brass.*

Brass is almost as good as copper at conducting electricity, but it is stronger. Brass resists **corrosion** – it does not react easily with air or water (Figure 4). These properties make brass useful for making the pins for electrical plugs.

Like brass, bronze resists corrosion, and it is stronger and harder than copper. This makes it useful for making propellers for ships. It is also used to make cymbals and bells, and to make metal artwork. Molten bronze expands slightly as it solidifies, so it fills in tiny details in an artist's mould.

C Suggest why the properties of brass and bronze make them useful for making coins.

Figure 4 *Brass has a shiny, gold-like appearance, and is useful for making musical instruments.*

1 Describe what is meant by the term 'alloy'. *(2 marks)*

2 Suggest why brass is useful for making door knobs. *(2 marks)*

3 Carbon steels are alloys of iron and carbon. Use the data in Table 2 to help you answer these questions.
  a Describe the trend in:
    i tensile strength *(2 marks)*
    ii ductility. *(2 marks)*
  b Explain why steel C is used for making bridges and buildings. *(2 marks)*
  c Explain which steel would be best for making car body panels. *(2 marks)*

Table 2 *Different carbon steels and their properties.*

Carbon steel	Percentage of carbon	Relative tensile strength	Relative ductility
A	0.10	1.00	1.00
B	0.20	1.17	0.89
C	0.27	1.23	0.82

Learning outcomes

After studying this lesson you should be able to:

- describe the process of corrosion
- describe the conditions that cause corrosion.

Specification reference: C6.1p

Have you ever left your bike out in the rain? If you have, you may have found that parts of it became rusty (Figure 1). Rusting, and other forms of metal corrosion, are a big problem.

Figure 1 *The metal in this bike chain has reacted with water and air, and has become rusty.*

What is corrosion?

Corrosion is the reaction of a metal with substances in its surroundings, such as air and water. Silver does not easily react with oxygen in the air or with water. However, it will corrode in the presence of hydrogen sulfide, H_2S, a gas produced naturally by bacteria. The hydrogen sulfide reacts with silver when oxygen and water are also present. The reaction corrodes the silver, producing a thin layer of black silver sulfide, Ag_2S. This makes objects made out of silver, such as trophies and jewellery, turn black, so they need to be cleaned (Figure 2).

Other metals corrode too. Only very unreactive metals, such as gold and platinum, do not corrode. When iron and steel corrode, it is called **rusting**.

Figure 2 *The silver cup on the left shows corrosion and the one on the right has been cleaned.*

A Explain the difference between corrosion and rusting.

What happens when objects rust?

Rusting is a **redox** reaction. Iron is oxidised to hydrated iron(III) oxide when it reacts with oxygen and water:

$$\text{iron} + \text{oxygen} + \text{water} \rightarrow \text{hydrated iron(III) oxide}$$

The familiar orange-brown rust you see is hydrated iron(III) oxide. It easily flakes off the surface of the object, exposing fresh metal underneath. Rusting can continue until an iron or steel object has completely corroded away.

B Explain, in terms of oxygen or electrons, why iron is oxidised during rusting.

Investigating rusting

You can investigate the substances needed for rusting to happen using steel nails.

1   Set up three boiling tubes as shown in Figure 3.

2   The first tube contains anhydrous calcium chloride, which absorbs water vapour and keeps the nail dry.

3   The second tube contains boiled water, which does not contain dissolved oxygen, and a bung, which stops air getting back in.

4   The nail in the third tube is exposed to air and water.

5   Observe and record the appearance of the nails a few days later.

C Explain how the results of the experiment in Figure 3 show that both oxygen *and* water are needed for rusting to occur.

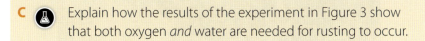
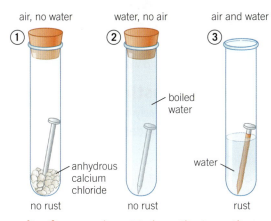

Figure 3 *The results of an experiment to investigate rusting.*

D Explain why iron railings rust faster at the seaside than they do inland.

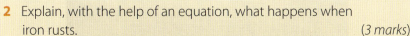

1   Describe how silver corrodes.   (*3 marks*)

2   Explain, with the help of an equation, what happens when iron rusts.   (*3 marks*)

3   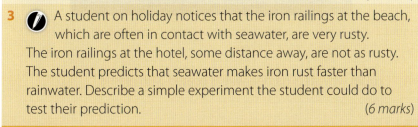 A student on holiday notices that the iron railings at the beach, which are often in contact with seawater, are very rusty.
The iron railings at the hotel, some distance away, are not as rusty. The student predicts that seawater makes iron rust faster than rainwater. Describe a simple experiment the student could do to test their prediction.   (*6 marks*)

Figure 4 *Dissolved salt is not needed for rusting to occur but it does increase the rate of rusting. This fishing trawler is becoming rusty very quickly due to the salt in seawater.*

Go further

A microscopically thin layer of aluminium oxide forms on the surface of aluminium when it reacts with oxygen. Unlike rust, this layer does not flake off. It seals the metal underneath from further contact with air and water, so aluminium does not corrode under normal circumstances. Find out what happens when a drop of mercury touches a piece of aluminium.

Stainless steel contains chromium, which forms an invisible layer of chromium oxide on the surface of the alloy. This stops oxygen and water reaching the metal below, so the iron in the steel cannot rust (Figure 1). As stainless steel is more expensive than ordinary steel, other ways to reduce corrosion are also needed.

Figure 1 *The stainless steel drum in a washing machine does not rust.*

How can you reduce rusting?

Many methods of rust prevention rely on stopping air and water reaching the surface of the metal. These include:

● painting

● coating with oil, grease, or plastic

● plating with zinc (galvanising)

● plating with tin.

If paint is damaged, rusting starts on the exposed metal. It continues underneath the paint. The paint eventually flakes off, exposing fresh metal to air and water (Figure 2). Oil, grease, and plastic coatings have similar problems.

Figure 2 *Car bodywork continues to rust once the paint is chipped, perhaps by a stone thrown up from the road.*

A Explain why steel garden wire is coated with plastic.

What is sacrificial protection?

Sacrificial protection involves a metal that is more reactive than iron, such as magnesium or zinc. As long as the iron or steel object is in contact with it, the more reactive metal corrodes first – it 'sacrifices' itself to protect the iron or steel. This method is useful where painting is difficult. Ships have zinc or magnesium blocks bolted onto their hulls below the waterline (Figure 3). These protect the hull from rusting, but they gradually corrode away and have to be replaced.

H

Figure 3 *New blocks of sacrificial metal bolted onto a ship's hull and rudder.*

How does sacrificial protection work?

During rusting, iron atoms lose electrons and are oxidised to iron(III) ions:

$$Fe \rightarrow Fe^{3+} + 3e^-$$

The more reactive the metal, the more easily it loses electrons. Sacrificial metals like zinc and magnesium lose electrons more easily than iron does. They are more readily oxidised than iron.

B Sodium is more reactive than iron. Suggest why it is not suitable as a sacrificial metal.

How does metal plating work?

A layer of metal plated onto an iron or steel metal object prevents air and water reaching the iron or steel below.

Galvanising involves dipping the metal object in molten zinc. After it has cooled and solidified, the thin layer of zinc does two things:

1 It stops air and water reaching the iron or steel below.

2 It acts as a sacrificial metal so that the object is protected, even if the zinc layer is damaged (Figure 4).

Tin plating involves electroplating the steel object with tin, or dipping it in molten tin. The inside of steel food cans is protected by tin plating. However, tin is less reactive than iron. If the tin layer is damaged, the steel acts as a sacrificial metal for the tin, and it rusts even faster than normal.

Figure 4 *Galvanised steel is used for farm gates.*

C Describe a disadvantage of tin plating for rust prevention.

1 Other than by using stainless steel, which is expensive, suggest the most appropriate way to prevent the following steel items from rusting.
   a a bridge *(2 marks)*
   b the inside of a car engine. *(2 marks)*

2 Explain why a galvanised steel farm gate does not rust. *(4 marks)*

3 Magnesium blocks are attached, underwater, by steel cables to the steel legs of a North Sea oil platform. Explain why this is done, and how it prevents corrosion of the steel legs. *(4 marks)*

Learning outcome

After studying this lesson you should be able to:

- compare quantitatively the physical properties of glass and clay ceramics, polymers, and metals.

Specification reference: C6.1r

Figure 1 *The electricity must be turned off for this electrical engineer to work on the pylon.*

Electricity is dangerous, so tall pylons carry high-voltage power lines across the country, high above the ground (Figure 1). How do engineers choose the best materials for these cables?

What are ceramics like?

Ceramics are hard, non-metallic materials. Brick, china, porcelain, and glass are all ceramic materials. Ceramics contain metals and non-metals, combined to form giant ionic lattices or giant covalent structures. These give ceramics their typical properties:

- high melting points
- hard and stiff, but brittle
- poor conductors of electricity and heat.

The compounds in ceramics are mostly oxides, and this makes them unreactive.

Glass is made by melting sand, then allowing it to cool and solidify (Figure 2). Glass has an irregular giant structure without crystals, and is usually transparent. Other ceramics are produced differently. They are made by heating clay to very high temperatures. Tiny crystals form, joined together by glass. China and porcelain are usually coated in a glaze and reheated. The glaze forms a smooth, hard, and waterproof surface.

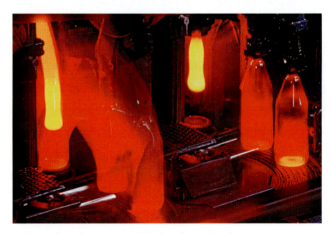

Figure 2 *Molten glass being moulded to form bottles.*

A Describe two common properties of ceramics.

How can you choose the material for an electrical cable?

The material chosen for an overhead electrical cable must be a good conductor of electricity. It must be lightweight and strong so it does not stretch or snap easily, and it should resist corrosion. Metals are good conductors of electricity. Table 1 shows properties of three metals that might be suitable.

Table 1 *Properties of three metals.*

Metal	Relative conductivity	Relative tensile strength	Density (g/cm³)
aluminium	1.0	1.0	2.7
copper	1.7	0.7	8.8
steel	0.6	2.6	7.8

Copper is the best electrical conductor of the three, but its tensile strength is low – it is not as strong when stretched as aluminium or steel. It also has the highest density, so a copper cable will be very heavy. Aluminium is chosen for overhead electricity cables. Although it is not as strong as steel, it is a better conductor and less dense than steel.

B Suggest a reason, not given in Table 1, for choosing aluminium rather than steel.

How can you choose the material for an insulator?

Insulators must prevent an electric current passing from the cable to the pylon (Figure 3). It makes sense to look at poor conductors of electricity, such as ceramics and polymers, rather than metals. Table 2 shows properties of three materials that might be suitable for making an insulator.

Table 2 *Properties of three materials.*

Material	Electrical resistance	Relative tensile strength	Relative compressive strength
glass	1.0	1.0	1.0
porcelain	0.4	0.02	7.0
poly(propene)	0.03	0.01	0.04

Poly(propene) is a tough, flexible polymer. It has the poorest electrical resistance of the three materials, and its tensile strength is the lowest, so it would snap easily when stretched. It also has the lowest **compressive strength**, so it would squash easily. Porcelain has a lower resistance than glass, but it has a higher compressive strength. From the data in Table 2, porcelain would be the most suitable for insulators where cables are fixed on top of a structure.

C Explain, using data from Table 2, why glass would be suitable for making an insulator where a cable is suspended from a structure.

1 Use data from Tables 1 and 2 to help you explain why:
  a Steel cables are used to carry a lift in a lift shaft. (*2 marks*)
  b Copper is used for household cables, but not for overhead cables. (*2 marks*)

2 Electrical insulators must not contain any bubbles or impurities. Suggest why this gives glass an advantage over porcelain. (*2 marks*)

3 Explain why poly(propene) is used to make ropes and crates, but glass and porcelain are not. (*3 marks*)

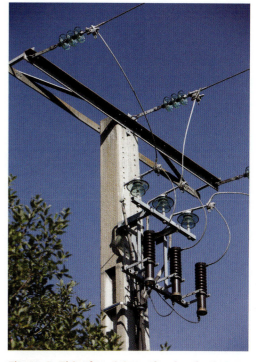

Figure 3 *This electricity pylon has both glass and porcelain insulators.*

Cob is a traditional building material made from a mixture of clay, sand, straw, and water (Figure 1). It dries to form a strong composite material with properties similar to reinforced concrete.

Figure 1 *A cob house in South Africa; cob is cheap, strong, and fireproof.*

What are composite materials?

A **composite material** is made from two or more materials combined together, each with different properties. The properties of a composite material are different from those of the materials it contains. For example, it may be stronger, less dense, or cheaper than its separate components alone.

Your clothes are likely to be made from composite materials. Cotton fabric is lightweight and comfortable to wear, but it is not very hardwearing. Polyester–cotton is a composite material made by weaving cotton thread with polyester fibre, an artificial polymer. It is still comfortable, but harder wearing than cotton, and easier to wash and dry.

A Describe the meaning of the term 'composite material'.

Many composite materials consist of fibres embedded in a polymer **resin**. The fibres have a low density and high tensile strength but are brittle, whereas the resin is hardwearing but is not strong. The composite material is lightweight, strong, and hard. Fibreglass consists of glass fibres in a resin (Figure 2). It is used for canoes, boats, and surfboards. Carbon fibre consists of carbon fibres in a resin. It is more expensive than fibreglass and is used for sports equipment, racing cars, and aircraft parts.

How are composites used in buildings?

Concrete is a composite material that consists of aggregate (small stones), sand, and cement. When water is added to this mixture, chemical reactions happen that bind the ingredients together.

Figure 2 *This technician is adding resin to fibreglass in a mould.*

Concrete has a high compressive strength – it resists being squashed, so it is useful for foundations. However, its tensile strength is low – concrete beams crack and break if you put heavy loads on them. On the other hand, steel has a high tensile strength. By embedding steel rods in concrete as it sets, you get a composite material called steel-reinforced concrete (Figure 3). This has high compressive strength *and* high tensile strength.

Figure 3 *These builders are pouring wet concrete into steel mesh to produce steel-reinforced concrete.*

B Explain why steel-reinforced concrete is better for building motorway bridges than concrete alone.

Wood is a natural material consisting of long fibres lying side by side to make a 'grain'. Wood is stronger *along* the grain than it is *across* the grain. Plywood is a composite material in which thin sheets of wood are glued together in layers. The grain in each layer is set at right angles to the layer below (Figure 4). This produces a composite material that resists bending in both directions, so it is useful for floors and walls.

C Use data from Table 1 to compare the strength of plywood to wood on its own.

1 ⚡ Explain why fibreglass is a composite material. *(4 marks)*

2 Plasterboard is a composite material used for indoor walls. It consists of a thick layer of non-flammable but brittle plaster, sandwiched between sheets of tough paper. Suggest why it is used in preference to plaster or paper alone. *(2 marks)*

3 ⚡ Compare cob with steel-reinforced concrete. In your answer, explain why both materials are strong in tension and compression. *(4 marks)*

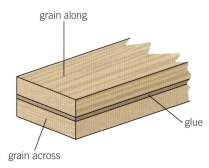

grain along

glue

grain across

Figure 4 *The structure of plywood.*

Table 1 *Strength of different types of wood.*

Material	Strength (MPa)
wood (along the grain)	40
wood (across the grain)	8
plywood (both directions)	48

A broken bone must be supported so that it can mend itself. Plaster casts are the traditional way to do this, but modern fibreglass casts are water resistant and more lightweight (Figure 1).

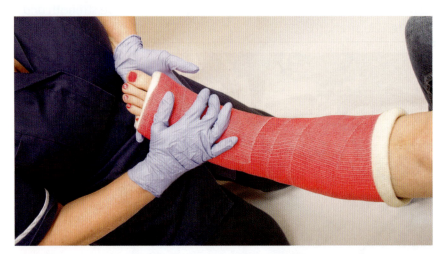

Figure 1 *This nurse is treating a broken leg with a fibreglass cast.*

How are properties and uses related?

A sodium boat would easily float because sodium is less dense than water, but it would react violently with the water and explode into flames. You usually need to consider more than one property when deciding whether to use a material for a particular purpose. For example, plastic cups may be used to hold drinks. Table 1 shows some properties of three common polymers that could be used.

Table 1 *Properties of three different polymers.*

Polymer	Cost (£ per kg)	Tensile strength (MPa)	Melting point (°C)	Maximum usable temperature (°C)
poly(ethene)	0.74	15	120	85
PET	1.20	78	254	70
poly(propene)	0.92	35	176	160

Although poly(ethene) is the cheapest, it is unsuitable. Boiling water will not melt it, but the cup will become too soft to hold safely. PET (polyethylene terephthalate) is the most expensive and, although it has the highest tensile strength and melting point, hot drinks will soften it too. Poly(propene) is the most suitable as its maximum usable temperature is above 100 °C and it has a high tensile strength.

A Explain which of the three polymers would be most suitable to make bottles for pressurised fizzy cold drinks.

What is a life-cycle assessment?

A **life-cycle assessment** or LCA is a 'cradle-to-grave' analysis of the impact of making, using, and disposing of a manufactured product (Figure 2). It should include data about:

- sustainability, including the use of raw materials and energy
- environmental impact, including waste products and pollution
- the lifespan of the product and whether any of it can be recycled
- disposal, including how easily the materials decompose (break down).

CRADLE
- **raw materials** obtained and processed to make useful materials
- **materials** used to make the product
- **energy and water** used in processing and manufacturing.

USE
- **energy** needed to **use** the product (e.g., electricity for a computer)
- **energy** needed to **maintain** the product (e.g., cleaning, mending)
- **water** and **substances** needed to maintain it.

GRAVE
- **energy** needed to **dispose** of the product
- **space** needed to dispose of it.

Figure 2 *This chart gives examples of the areas where you may need to collect data for an LCA.*

B Describe two examples of data, other than energy use, that could be included in an LCA.

C Use the chart in Figure 2 to describe three different stages in the life cycle of a product in which energy could be used.

How can you use life-cycle assessment data?

Data collected for an LCA should allow you to identify stages that could be improved, or alternative materials that might do the same job. For example, an LCA for a pair of trousers might show that 20% of the total energy used is in its production, 75% in its use, and 5% in its disposal. It would make little sense to try to save energy at the disposal stage. However, it would be worthwhile to investigate how to use less energy when washing and drying the trousers, since most energy is used at this stage.

Figure 3 *Poly(ethene) sheeting is used to make long polytunnels to protect growing crops.*

1 Use the data from Table 1 to explain why polytunnels (Figure 3) are made from poly(ethene) rather than from PET or poly(propene). *(2 marks)*

2 Describe two reasons why a life-cycle assessment may be carried out. *(2 marks)*

3 Table 2 shows energy data from an LCA comparing wooden and uPVC (unplasticised PVC) window frames. Evaluate the data. What conclusions can you make about the window frames? *(6 marks)*

Table 2 *Energy data for two types of window frame.*

	Energy used (MJ)	
	Wooden frames	uPVC frames
producing the material	4.2	12.6
manufacture	4.0	2.8
transport	5.1	3.4
maintenance	1.2	0.2
end-of-life disposal	2.5	2.5

Learning outcomes

After studying this lesson you should be able to:

- describe how materials or products are recycled for a different use
- explain why recycling may be viable
- evaluate factors that affect decisions on recycling.

Specification reference: C6.1m, C6.1n

In 1992 thousands of plastic bath toys were washed overboard from a ship in the Pacific Ocean. They floated for thousands of kilometres over the next few years (Figure 1). Some even travelled through the Arctic to the North Atlantic Ocean. Polymers and many other materials do not easily decompose, which means they must be disposed of carefully or recycled.

Figure 1 *Oceanographers were able to test their models of ocean currents by recording where the plastic ducks washed up.*

Why should you recycle materials?

Unless they can be recycled, most materials and products will end up in **landfill** sites as waste. Disposal like this is not an efficient use of resources. **Recycling** is important for many reasons, such as:

- conserving limited raw materials and energy resources
- reducing the release of harmful substances into the environment
- reducing waste.

Table 1 shows the energy saved by recycling metals rather than extracting them from ores.

A Name the two metals with the greatest energy saving when recycled.

Whether or not a particular material or product should be recycled depends upon factors such as:

- how easily the waste can be collected and sorted
- the amount and type of any by-products released by recycling
- the cost of recycling compared to disposal in landfill or by incineration (burning)
- the amount of energy involved at each stage.

Table 1 *Energy saved from recycling metals rather than extracting them from ores*

Metal	% energy saved
aluminium	95
zinc	61
iron	65
lead	58
copper	84

Waste rock

Waste rock is left behind after extracting metals from ores. Use the data in Table 2 to sort the metals into order of increasing total masses of waste rock produced per year. What assumption have you made?

Table 2 *Waste rock from extracting metals*

Metal	Mass of metal used per year (kg)	Mass of waste rock from 1 kg of metal (kg)
iron	2.2×10^{12}	2
aluminium	5.0×10^{10}	3
copper	2.0×10^{10}	150
zinc	1.3×10^{10}	11
lead	1.2×10^{10}	14

How are materials recycled?

Waste materials and products must be collected and transported to a recycling plant (Figure 2). It helps if the different materials have already been sorted, but more sorting is usually needed at the plant. It is important to ensure that, for example, glass is not contaminated by metal bottle tops. The sorted waste is then shredded or crushed into smaller pieces ready for processing.

B Suggest how iron and steel may be separated from other metals.

During processing:

- Metals are melted by heating, and the molten metal poured into moulds to produce new blocks called **ingots**.

- Paper is mixed with water, cleaned, then rolled and heated to make new paper.

- Glass is melted by heating, and moulded into new glass objects.

- Polymers like poly(ethene), PET, and poly(propene) are melted and formed into new objects (Figure 3).

1. Use your knowledge of how aluminium is extracted from its ore to suggest why the energy savings are particularly high for recycling aluminium. *(2 marks)*

2. Describe how metals and glass are recycled. *(4 marks)*

3. PET drinks bottles can be recycled to produce clothing such as fleeces. The PET is melted and then squeezed through narrow tubes to make long fibres. Suggest why it is important to separate PET from other materials before recycling. *(3 marks)*

Figure 2 *Recycling stations help people sort their waste ready for recycling.*

Figure 3 *This recycled poly(ethene) is being heated and moulded.*

Study tip

Avoid using the term 'environmentally friendly' in your answers. Be specific about the advantages of recycling.

Summary questions

1 Pyrolusite is an ore that contains manganese(IV) oxide, MnO_2.
   a Describe the meaning of the term *ore*.
   b Manganese can be produced from manganese(IV) oxide in two stages. In the first stage, manganese(IV) oxide is heated with carbon monoxide, CO. This produces manganese(II) oxide, MnO, and carbon dioxide, CO_2.
      i Write down a balanced equation to model the reaction.
      ii Explain why manganese(IV) oxide is reduced in this reaction.
      iii Explain why carbon monoxide acts as a reducing agent in this reaction.
   c In the second stage, manganese(II) oxide is heated strongly with carbon, producing manganese. Describe the role of carbon in this reaction.
   d Use your answers to b and c to suggest why manganese is not extracted using electrolysis.

2 The gold in jewellery is usually an alloy. Table 1 shows some properties of gold–copper alloys used in jewellery. The higher the hardness, the more pressure is needed to dent or scratch the metal. Copper alone has a hardness of 80.

   Table 1 *Hardness of different gold–copper alloys.*

Name of gold	% gold	Hardness
24 carat	100	45
22 carat	92	70
18 carat	75	105
14 carat	58	145
9 carat	38	85

   a Describe the meaning of the term *alloy*.
   b Calculate the percentage of copper in 18 carat gold.
   c i Describe the trend in hardness from 24 carat gold to 14 carat gold.
      ii Explain the effect of adding more copper to gold.
      iii Suggest why 9 carat gold has a much lower hardness than 14 carat gold.

3 Iron and steel objects rust unless protected.
   a Name the other substances needed for iron and steel to rust.
   b Explain why painting a steel object protects it from rusting.
   c An underground steel pipeline is connected to a piece of zinc, buried alongside it.
      i Write down the name of this type of rust protection.
      ii Explain why zinc is used rather than copper.

4 The Space Shuttle was a re-usable spacecraft designed to go into orbit and then return to Earth. It was made from aluminium alloy covered by ceramic tiles.

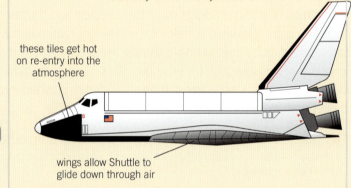

these tiles get hot on re-entry into the atmosphere

wings allow Shuttle to glide down through air

Figure 1 *The Space Shuttle.*

   a Use your knowledge and understanding of materials to explain:
      i why aluminium alloy was used to make the frame and outer walls of the Shuttle
      ii why ceramic tiles were chosen to cover the outer walls of the Shuttle where re-entry temperatures could reach 1260 °C.
   b The nose and front edges of the Shuttle's wings were heated to more than 1260 °C during re-entry. Suggest why these areas were covered with a carbon-fibre composite material.

5 Ammonia is made using the Haber process:
$$N_2(g) + 3H_2(g) \rightleftharpoons 2NH_3(g) \qquad \Delta H = -93 \, kJ/mol$$
The typical temperature and pressure used are 450 °C and 200 atmospheres. Explain, with reference to the equilibrium position, why these conditions are regarded as a compromise.

6 Outline how phytoextraction of copper works.

Revision questions

1 Look at the following diagrams.
In which of the four test tubes will the iron nail corrode? *(1 mark)* [S]

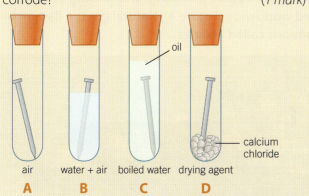

air   water + air   boiled water   drying agent

A   B   C   D

2 The list below shows part of the reactivity series of metals. Carbon is also included.

sodium
calcium
magnesium
aluminium
carbon
zinc
iron
tin
lead

Which row in the table correctly shows how the metals are extracted from their ores? *(1 mark)*

	Metals extracted by electrolysis	Metals extracted by heating with carbon
A	sodium, aluminium, zinc	magnesium, iron, tin
B	aluminium, zinc, tin	lead, iron, zinc
C	calcium, magnesium, aluminium	sodium, iron, tin
D	sodium, calcium, magnesium	lead, tin, zinc

3 Potassium sulfate is a fertiliser. [S]
A student makes some potassium sulfate at school. Look at how he carries out the experiment.

i He uses a measuring cylinder to measure out some potassium hydroxide solution, and pours it into a conical flask. He places the conical flask on the table.

ii He adds some phenolphthalein indicator.

iii He adds sulfuric acid from a burette until the indicator changes colour. He notes the reading on the burette at the end.

iv He pours the mixture into an evaporating basin, and warms it to evaporate the water.

How should the student improve his method? Explain your answer. *(4 marks)*

4 Describe how aluminium is obtained from aluminium oxide on an industrial scale, and explain why this method is chosen. *(6 marks)*

5 Ammonia is used to manufacture nitric acid. [H] [S]
In the first stage of the process, ammonia is oxidised to nitrogen monoxide, NO.

$$4NH_3(g) + 5O_2(g) \rightleftharpoons 4NO(g) + 6H_2O(g)$$
$$\Delta H = -950 \text{ kJ}$$

a Predict the effect of increasing the temperature on the yield of nitrogen monoxide at equilibrium. Explain your answer. *(2 marks)*

b Predict the effect of increasing the pressure on the yield of nitrogen monoxide at equilibrium. Explain your answer. *(2 marks)*

c Industrially, the process is carried out at 900 °C and at a pressure of 7 MPa. Suggest why. *(1 mark)*

d The reaction is carried out by passing the reactants over a platinum gauze. Suggest why. *(1 mark)*

C6.2 Organic chemistry

C6.2.1 Alkanes

Learning outcomes

After studying this lesson you should be able to:

- name the first four straight chain alkanes and draw their displayed formulae
- explain why alkanes form a homologous series
- predict the products of reactions of the alkanes.

Specification reference: C6.2a, C6.2b, C6.2c, C6.2l

Every year, the world uses an astonishing 3500 km³ of natural gas as a fuel for generating electricity, heating, and fuelling vehicles, and is transported worldwide (Figure 1). Natural gas is a flammable mixture of compounds called alkanes.

Figure 1 *This tanker carries natural gas, which is cooled and put under pressure so that it is in the liquid state.*

What are alkanes?

Hydrocarbons are compounds that contain only hydrogen atoms and carbon atoms. The **alkanes** are hydrocarbons, and they form a **homologous series** because they have features in common:

- they have the same **general formula**
- their carbon atoms are joined to each other by single covalent bonds – all alkanes are **saturated**.

Table 1 gives some information about four alkanes. Notice that their names end in 'ane'.

The number of hydrogen atoms in the formula is twice the number of carbon atoms, plus two. This gives you the general formula for alkanes:

$$C_nH_{2n+2}$$

The chemical formula for pentane, an alkane with five carbon atoms, is C_5H_{12}.

As you move down Table 1, the number of carbon atoms increases by one each time and the number of hydrogen atoms increases by two. This is another feature of a homologous series – each successive member differs by a CH_2 group.

Table 1 *The first four alkanes.*

Number of carbon atoms	Name of alkane	Chemical formula
1	methane	CH_4
2	ethane	C_2H_6
3	propane	C_3H_8
4	butane	C_4H_{10}

A Write down the chemical formula for hexane, an alkane with six carbon atoms.

How do you model alkanes?

You can model alkane molecules using displayed formulae. These show how the atoms in the molecule are joined together. You show:

- each atom by its chemical symbol, C or H
- each covalent bond by a straight line (Figure 2).

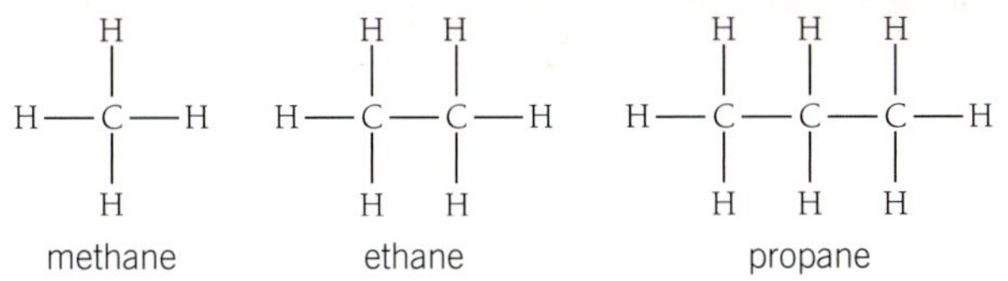

H—C—H   H—C—C—H   H—C—C—C—H

methane   ethane   propane

Figure 2 *Displayed formulae for three straight chain alkanes.*

Carbon atoms form four covalent bonds, and hydrogen atoms form one covalent bond. When you draw these formulae, it helps to draw the carbon atoms in a line first, then the correct number of bonds, and finally the hydrogen atoms.

B Draw the displayed formula for butane.

How do alkanes react?

Combustion is the main type of reaction involving alkanes. They react with oxygen in the air, and burn. If there is a plentiful supply of oxygen or air, **complete combustion** happens:

- the carbon atoms are oxidised to carbon dioxide
- the hydrogen atoms are oxidised to water.

This balanced equation models the complete combustion of propane:

propane + oxygen → carbon dioxide + water

$$C_3H_8 + 5O_2 \rightarrow 3CO_2 + 4H_2O$$

If there is a poor supply of oxygen or air, **incomplete combustion** happens instead (Figure 3). Water is still formed, but carbon is only oxidised to carbon monoxide, CO. This is a problem because carbon monoxide is a toxic gas and can cause suffocation. Some carbon atoms may not be oxidised at all, and are released as carbon particles.

C Write down a balanced equation for the complete combustion of methane.

Figure 3 *Black smoke containing carbon particles is released during incomplete combustion of an alkane.*

1 **a** State the general formula for alkanes.   *(1 mark)*
   b Write down the chemical formula for eicosane, an alkane with 20 carbon atoms in its molecules.   *(1 mark)*

2 Explain why the alkanes form a homologous series of hydrocarbons.   *(4 marks)*

3 Write down balanced equations for these reactions:
   a the complete combustion of ethane   *(2 marks)*
   b the incomplete combustion of ethane, forming water, and equal amounts of carbon monoxide and carbon.   *(2 marks)*

Learning outcomes

After studying this lesson you should be able to:

- name the first four straight chain alkenes and draw their displayed formulae
- explain why alkenes form a homologous series
- predict the products of reactions of the alkenes
- describe how functional groups determine the reactions of organic compounds.

Specification reference: C6.2a, C6.2b, C6.2c, C6.2i

Bananas ripen quickly after you have bought them, yet they have travelled thousands of kilometres from where they were grown (Figure 1). After picking, unripe bananas are transported at around 13 °C to slow down ripening. At their destination, they are exposed to ethene, an alkene, which restarts the ripening process.

Figure 1 *Green unripe bananas growing on a tree.*

Table 1 *The first three alkenes.*

Number of carbon atoms	Name of alkene	Chemical formula
2	ethene	C_2H_4
3	propene	C_3H_6
4	butene	C_4H_8

Systematic names

Compounds like alkanes and alkenes have **systematic names**. The start of each name tells you how many carbon atoms the molecules contain (Table 2).

Table 2 *The first six prefixes.*

Prefix	Number of carbon atoms
meth-	1
eth-	2
prop-	3
but-	4
pent-	5
hex-	6

What are alkenes?

The **alkenes** are hydrocarbons. Like alkanes, they form a homologous series because they have features in common. Unlike alkanes, alkene molecules contain a carbon–carbon double covalent bond. This means that alkenes are **unsaturated**.

Table 1 gives some information about three alkenes. Notice that their names end in 'ene'.

The number of hydrogen atoms in the formula is twice the number of carbon atoms. This gives you the general formula for alkenes:

$$C_nH_{2n}$$

The chemical formula for hexene, an alkene with six carbon atoms, is C_6H_{12}.

A Write down the chemical formula for pentene, an alkene with five carbon atoms.

How do you model alkenes?

You can model alkene molecules using displayed formulae. Notice that you show the double bond as '=' rather than as '–'. It helps if you draw the carbon atoms in a line first, then the correct number of bonds (remembering to include a double bond), and finally the hydrogen atoms (Figure 2).

B What do hexane and hexene have in common?

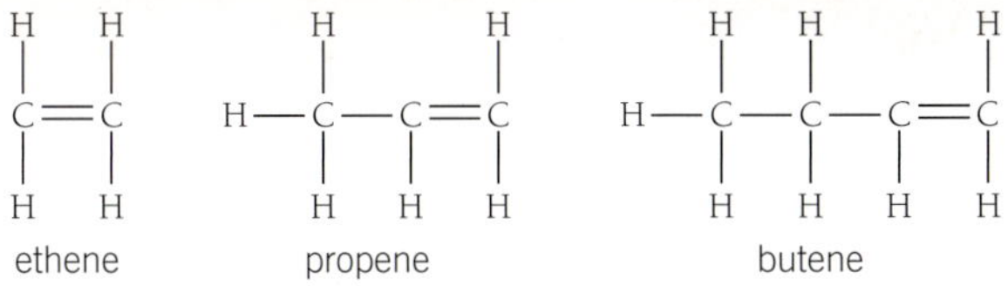

H H
| |
C=C
| |
H H
ethene

H H H
| | |
H—C—C=C
| | |
H H H
propene

H H H
| | |
H—C—C—C=C
| | | |
H H H H
butene

Figure 2 *Displayed formulae for three straight chain alkenes.*

C Draw a displayed formula for hexene.

How do alkenes react?

Just like alkanes, alkenes can burn in oxygen or air, and the products of combustion are the same. However, alkenes have a **functional group**, which allows them to take part in a wider range of reactions. A functional group is an atom, group of atoms, or type of bond in a molecule that is responsible for the characteristic reactions of the substance. In alkenes, this is the carbon–carbon double bond, C=C. It lets alkenes undergo addition reactions.

What are addition reactions?

In an **addition reaction**, an atom or group of atoms combines with a molecule to form a larger molecule, with no other product. For example, alkenes react with bromine in addition reactions to form colourless 'dibromo' compounds:

ethene + bromine → dibromoethane

$$C_2H_4 + Br_2 \rightarrow C_2H_4Br_2$$

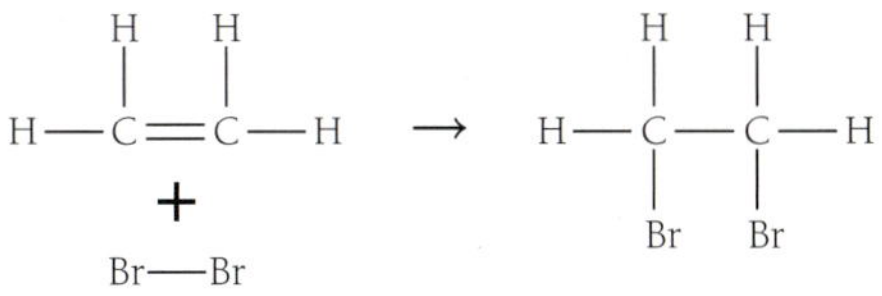

H H
| |
H—C=C—H
+
Br—Br

→

H H
| |
H—C—C—H
| |
Br Br

This reaction is the basis of a test for unsaturation. Bromine reacts with alkenes but not with alkanes. Bromine water (bromine dissolved in water) is orange–brown. It turns colourless when it is mixed with alkenes, but not when it is mixed with alkanes (Figure 3).

D Explain why ethene decolorises bromine water but ethane does not.

Alkenes undergo addition reactions with hydrogen, in the presence of a nickel catalyst, to form alkanes. For example:

ethene + hydrogen → ethane

$$C_2H_4 + H_2 \rightarrow C_2H_6$$

Figure 3 *Alkanes cannot decolorise bromine water, but alkenes can decolorise bromine water.*

1 a State the general formula for alkenes. (*1 mark*)
b Write down the chemical formula for hexadecene, an alkene with 16 carbon atoms in its molecules. (*1 mark*)

2 Explain why the alkenes form a homologous series. (*3 marks*)

3 Hexane and hexene are both in the liquid state at room temperature. Outline how you could carry out a simple laboratory test to tell them apart. Include an equation in your answer. (*5 marks*)

Learning outcomes

After studying this lesson you should be able to:

- name the first four straight chain alcohols, and draw their displayed formulae
- explain why alcohols form a homologous series
- predict the products of reactions of the alcohols.

Specification reference: C6.2a, C6.2b, C6.2c

The alcohol in wine and beer, ethanol, is just one member of the alcohol homologous series. There are many more alcohols. Methanol, for example, is toxic. It is used as a fuel (Figure 1) and in antifreeze, rather than in drinks.

Figure 1 *Methanol is very flammable.*

What are alcohols?

The **alcohols** form a homologous series. They contain a different functional group from the alkenes, so they can take part in different chemical reactions from alkenes. The functional group in alcohols is the **hydroxyl group**, –OH.

Table 1 gives some information about four alcohols. Notice that their names end in 'ol'.

You show the hydroxyl group separately in chemical formulae for alcohols, so the general formula for alcohols is:

$$C_nH_{2n+1}OH$$

The chemical formula for pentanol, an alcohol with five carbon atoms, is $C_5H_{11}OH$.

A Write down the chemical formula for hexanol, an alcohol with six carbon atoms.

Table 1 *The first four alcohols.*

Number of carbon atoms	Name of alcohol	Chemical formula
1	methanol	CH_3OH
2	ethanol	C_2H_5OH
3	propanol	C_3H_7OH
4	butanol	C_4H_9OH

Organic compounds

In everyday use, the word 'organic' is often applied to food produced without the use of artificial substances. To a chemist, however, an **organic compound** consists of molecules with carbon atoms.

How do you model alcohols?

You can model alcohol molecules using displayed formulae. You should show the hydroxyl group as –O–H in these formulae, as in Figure 2.

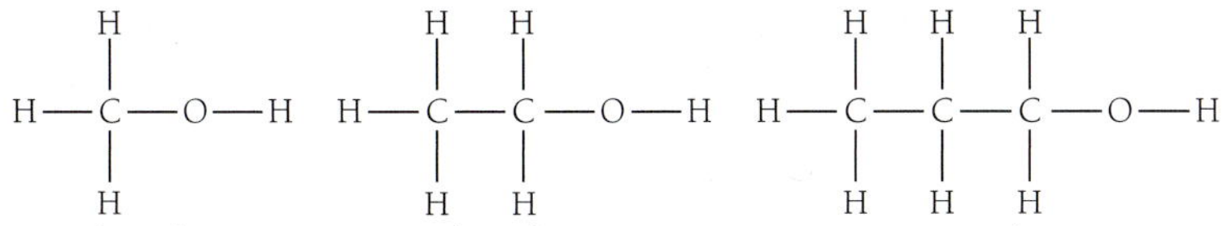

methanol   ethanol   propanol

Figure 2 *Displayed formulae for three alcohols.*

B Draw a displayed formula for butanol.

How do alcohols react?

Just like alkanes and alkenes, alcohols can burn in oxygen or air. The products of combustion are the same:

- water vapour and carbon dioxide are produced during complete combustion
- water vapour, carbon monoxide, and carbon are produced during incomplete combustion.

These equations model the complete combustion of methanol:

ethanol + oxygen → carbon dioxide + water

$$C_2H_5OH + 3O_2 \rightarrow 2CO_2 + 3H_2O$$

Make sure you include the oxygen atom in the alcohol molecule when balancing these equations.

C Write down a balanced equation for the complete combustion of methanol.

The hydroxyl functional group lets alcohols undergo reactions other than combustion. For example, they can be oxidised to form carboxylic acids. The other atoms in alcohol molecules usually do not take part in these reactions (Figure 3).

Go further

Ethene reacts with steam in the presence of a phosphoric acid catalyst, to produce ethanol. Find out about the reverse of this reaction, the dehydration of ethanol.

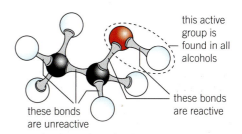

this active group is found in all alcohols

these bonds are reactive

these bonds are unreactive

Figure 3 *The hydroxyl functional group determines the majority of alcohol reactions. In this molecular model, black represents a carbon atom, white a hydrogen atom, and red an oxygen atom.*

1 **a** State the general formula for alcohols. (*1 mark*)
   b Write down the chemical formula for dodecanol, an alcohol with 12 carbon atoms in its molecule. (*1 mark*)
2 Write down three reasons that explain why the alcohols are said to form a homologous series. (*3 marks*)
3 **a** Explain why alcohols can be oxidised to carboxylic acids. (*2 marks*)
   b Explain how alcohols are oxidised to carbon dioxide and water. (*2 marks*)

C6.2.4 Carboxylic acids

Learning outcomes

After studying this lesson you should be able to:

- name the first four straight chain carboxylic acids, and draw their displayed formulae
- explain why carboxylic acids form a homologous series
- describe how carboxylic acids are made from alcohols.

Specification reference: C6.2b, C6.2c

Ant stings and bites contain methanoic acid (Figure 1), a carboxylic acid similar to the acid found in vinegar. Its old name is *formic acid*, after the Latin word for 'ant'. Chemists used to produce methanoic acid by distilling crushed ants, but fortunately there are now more convenient ways to make it.

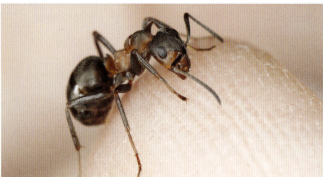

Figure 1 *Ant bites and stings contain methanoic acid.*

What are carboxylic acids?

The **carboxylic acids** form a homologous series. Its functional group is the **carboxyl group**, –COOH. It is different from the functional groups in alkenes and alcohols, so carboxylic acids can take part in different reactions from these organic compounds.

Table 1 gives some information about four carboxylic acids. Notice that their names end in 'anoic acid'.

Table 1 *The first four carboxylic acids.*

Number of carbon atoms	Name of carboxylic acid	Chemical formula
1	methanoic acid	$HCOOH$
2	ethanoic acid	CH_3COOH
3	propanoic acid	C_2H_5COOH
4	butanoic acid	C_3H_7COOH

You show the carboxyl group separately in chemical formulae for carboxylic acids, so the general formula for carboxylic acids is:

$$C_nH_{2n+1}COOH$$

Notice that there is a carbon atom in the carboxyl group. This means that the chemical formula for pentanoic acid, a carboxylic acid with five carbon atoms, is C_4H_9COOH.

A Write down the chemical formula for hexanoic acid, a carboxylic acid with six carbon atoms.

How do you model carboxylic acids?

You can model carboxylic acid molecules using displayed formulae. You should show all the bonds in the carboxyl group, as in Figure 2.

methanoic acid ethanoic acid propanoic acid

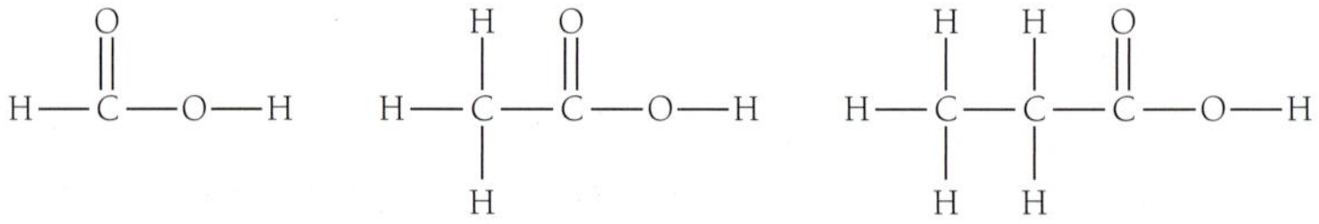

Figure 2 *Displayed formulae for three carboxylic acids.*

B Draw the displayed formula for butanoic acid.

How are carboxylic acids made?

Carboxylic acids are formed when alcohols react with oxidising agents, such as potassium manganate(VII) solution.

Ethanol can be oxidised to ethanoic acid. You can show this using a test tube of ethanol. You add some oxidising agent to the ethanol and warm it in a water bath. The oxidising agent is potassium manganate(VII) solution, $KMnO_4(aq)$, acidified with a little dilute sulfuric acid.

The colour change (which you can see in Figure 3) occurs because manganate(VII) ions are purple in solution, and are reduced to very pale pink manganese(II) ions when acidified potassium manganate(VII) oxidises ethanol to ethanoic acid.

If you carry out practical work it is important to assess all of the associated risks. You would need to wear eye protection, and to stir the reaction carefully, as the reaction can be violent. Both of the reactants carry hazard warnings: ethanol is harmful and flammable, and potassium manganate(VII) solution is harmful and will stain hands and clothing.

 C If you were carrying out the oxidation of ethanol, suggest why you should warm the mixture using a hot water bath, rather than a Bunsen burner.

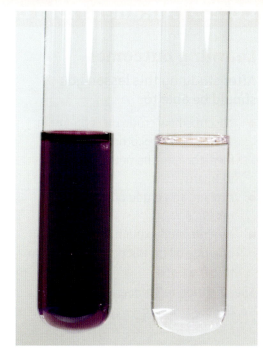

Figure 3 *Before and after the oxidation of ethanol using acidified potassium manganate(VII) solution.*

How do carboxylic acids react?

Carboxylic acids are weak acids, and their reactions are typical of acids. They react with:

- metals to produce salts and hydrogen
- alkalis and bases to produce salts and water
- carbonates to produce salts, water, and carbon dioxide (Figure 4).

For example:

$$2CH_3COOH(aq) + Na_2CO_3(aq) \rightarrow 2CH_3COONa(aq) + H_2O(l) + CO_2(g)$$

> **1 a** Write down the chemical formula for octadecanoic acid, which has 18 carbon atoms in its molecules. *(1 mark)*
> **b** Explain why fizzing is observed when magnesium ribbon is added to ethanoic acid. *(2 marks)*
>
> **2** Outline how you could use an indicator to test whether ethanoic acid forms an acidic solution. *(2 marks)*
>
> **3** The oxidation of ethanol to ethanoic acid using an oxidising agent in solution can be modelled by this equation:
>
> $$C_2H_5OH + 2[O] \rightarrow CH_3COOH + H_2O$$
>
> The oxygen atoms provided by the oxidising agent are shown as [O].
> **a** Write down an equation to model the oxidation of methanol, CH_3OH, to methanoic acid. *(1 mark)*
> **b** Describe a suitable oxidising agent for the reaction. *(3 marks)*

Figure 4 *Dilute ethanoic acid reacting with sodium carbonate.*

Learning outcomes

After studying this lesson you should be able to:

- describe how crude oil is the main source of hydrocarbons and a feedstock for the petrochemical industry
- explain why crude oil is a finite resource
- describe and explain the separation of crude oil by fractional distillation into fractions.

Specification reference: C6.2j, C6.2k, C6.2l, C6.2m, C6.2n

Have you travelled by car or bus today, or used something containing plastic? Crude oil is the main source of hydrocarbons used for fuels and as a feedstock for the production of many materials, such as plastics. It is transported worldwide (Figure 1).

Figure 1 *The Trans-Alaska Pipeline System includes a 1288 km pipeline that crosses over 500 rivers and streams in Alaska, and carries 3.4 billion litres of crude oil per day.*

What is crude oil?

Crude oil (Figure 2) is a **fossil fuel**. It was formed from the remains of marine organisms that lived millions of years ago. These became buried deep in the sea bed after they died. Chemical reactions eventually turned them into crude oil.

Finite resources like crude oil are no longer being made, or are being made extremely slowly. Crude oil is a non-renewable resource because it is being used up faster than it is being formed. Crude oil will run out one day if you keep using it.

A Explain why crude oil is a finite resource.

How do you make crude oil useful?

Crude oil is a complex mixture of hydrocarbons, mostly alkanes (Figure 3). These are separated from one another using fractional distillation. This works because the different alkanes have different boiling points. In general, the more carbon atoms there are in the alkane molecules:

- the larger the molecules
- the stronger the intermolecular forces
- the higher the boiling point.

Figure 2 *Crude oil is a viscous (not very runny) mixture of hydrocarbons.*

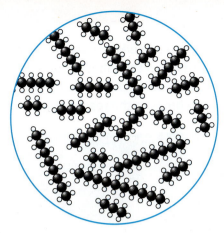

Figure 3 *Crude oil contains alkane molecules with differing sizes.*

Synoptic link

You can learn more about fractional distillation in C2.1.5 *Distillation*.

Study tip

When simple molecular substances like alkanes boil, the intermolecular forces are overcome but the strong covalent bonds between atoms in the molecules do not break.

B Explain the property of alkanes that allows them to be separated by fractional distillation.

Crude oil is heated and its vapours are piped into the bottom of a fractionating column. This has a temperature gradient – it is hot at the bottom and cold at the top. The vapours cool as they rise through the column. They condense to the liquid state if they reach a part that is cool enough.

The separated parts of the crude oil are called **fractions**. Each fraction contains many substances with similar boiling points. Bitumen has the highest boiling point. It leaves at the bottom of the column and is in the solid state at room temperature. Refinery gases (methane, ethane, propane, and butane) have the lowest boiling points. They reach the top and leave the column without cooling enough to condense. The other fractions are in the liquid state at room temperature. They fall into trays and leave from the middle of the column (Figure 4).

C Name five fractions from crude oil that are in the liquid state at room temperature, in order of increasing boiling point.

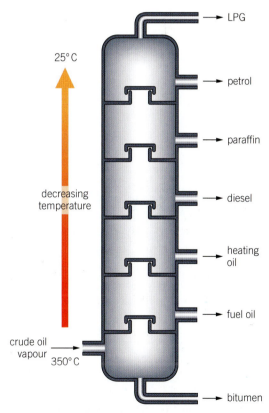

Figure 4 *The different fractions from a fractionating column. LPG (liquefied petroleum gas) is made from the refinery gases. It contains propane and butane.*

1 Write down the two main uses of crude oil. *(2 marks)*

2 Describe how crude oil is separated by fractional distillation. *(4 marks)*

3 Explain, in terms of alkane molecules and the forces between them, why crude oil can be separated by fractional distillation. *(2 marks)*

Go further

Find out some uses of each of the fractions obtained from crude oil.

Learning outcomes

After studying this lesson you should be able to:

- describe the cracking of crude oil fractions
- explain why cracking is carried out
- explain how modern life is crucially dependent upon hydrocarbons.

Specification reference: C6.2m, C6.2o

Fractional distillation of crude oil produces more of the fractions with high boiling points than the refinery can sell, and not enough of the fractions with low boiling points that its customers need. Fortunately, **cracking** comes to the rescue (Figure 1).

Figure 1 *Oil refineries are complex, with towers for fractional distillation of crude oil and cracking of fractions.*

What is cracking?

Cracking is a chemical reaction that converts large alkane molecules into smaller alkane molecules and alkene molecules. For example:

$$\text{octane} \rightarrow \text{hexane} + \text{ethene}$$

$$C_8H_{18} \rightarrow C_6H_{14} + C_2H_4$$

Cracking involves heating oil fractions to a high temperature (600 °C to 700 °C), and passing them over a hot catalyst of alumina or silica. Under these conditions, covalent bonds between atoms in the large alkane molecules break (Figure 2).

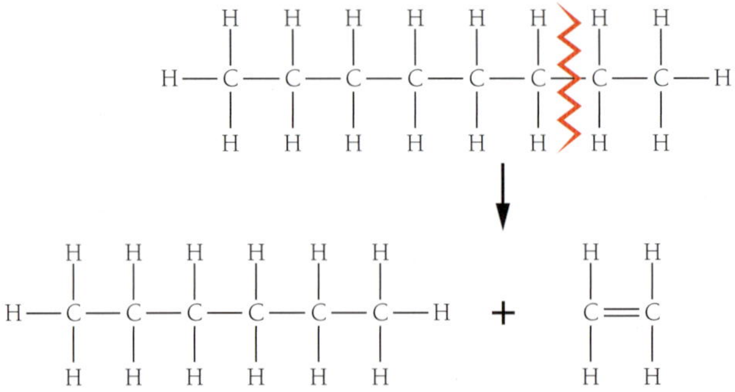

Figure 2 *Modelling the cracking of octane using displayed formulae.*

Cracking is carried out on a huge scale in oil refineries, but it can be done on a small scale in the laboratory using pieces of broken pot as the catalyst (Figure 3).

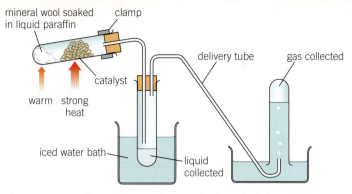

Figure 3 *Liquid paraffin can be cracked in the laboratory using this apparatus.*

A Write down a balanced equation to show the cracking of octane, C_8H_{18}, to produce an alkane and an alkene, each with the same number of carbon atoms in their molecules.

Why is cracking done?

Fractions with relatively large alkane molecules have high boiling points. They do not flow easily and they are difficult to ignite, so they are not useful as fuels. On the other hand, fractions with relatively small alkane molecules have low boiling points. They flow more easily and are easier to ignite, which makes them useful as fuels. Fractional distillation alone produces:

- more of the fractions with large molecules and high boiling points than can be sold
- not enough of the fractions with small molecules and low boiling points that are in high demand.

Cracking helps an oil refinery match its supply of useful products, such as petrol, with its customers' demand for them. Cracking also produces alkenes. These are useful for making polymers.

B From the bar chart in Figure 4, name the fractions with a higher demand than supply.

Why is crude oil important?

Crude oil is vital to our modern lifestyles (Figure 5). Around 90% of it is used to produce fuels such as LPG, petrol, diesel, and paraffin (which is used as aircraft fuel). The remainder is used as a feedstock for the chemical industry, for making products such as polymers. Although alternative fuels are being developed, crude oil cannot be easily replaced as a raw material.

C Describe what most crude oil is used for.

1 a Describe the conditions needed for cracking. *(2 marks)*
b Explain two reasons why oil refineries carry out cracking. *(2 marks)*
2 Cracking is an endothermic reaction. Suggest why it can also be described as a thermal decomposition reaction. *(2 marks)*
3 Use the bar chart in Figure 4 to help you explain how cracking could help satisfy the demand for fuels. *(3 marks)*

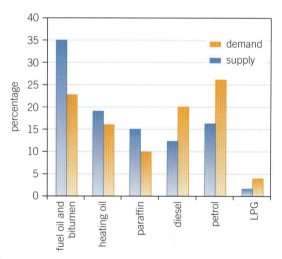

Figure 4 *An example of the supply of crude oil fractions by fractional distillation alone, and the customers' demand for them.*

Figure 5 *Some of the products made using crude oil as a raw material.*

Learning outcomes

After studying this lesson you should be able to:

- describe the basic principles of addition polymerisation
- work out the structure of an addition polymer from its monomer
- work out the monomer from the structure of an addition polymer.

Specification reference: C6.2d, C6.2g

Have you ever made a paper clip chain (Figure 1)? This models what happens when a polymer molecule is made. Lots of smaller molecules join together to make one long molecule.

Figure 1 *Modelling polymerisation; each paper clip represents a monomer molecule, and the chain represents a polymer molecule.*

What are addition polymers?

Plastics are materials made from long molecules called polymer molecules. These are made from monomer molecules, smaller molecules joined together. For example, many ethene molecules join together to form a polymer called poly(ethene). You probably know this as **polythene**. It is the polymer used to make plastic carrier bags.

Poly(ethene) is an **addition polymer** because the ethene monomers undergo addition reactions when they join together. They can do this because they contain a carbon–carbon double bond. All alkene molecules contain a carbon–carbon double bond, so all alkenes can act as monomers for addition polymers. Addition **polymerisation reactions** usually need high pressures and a catalyst.

Table 1 describes some addition polymers. Notice that the systematic name of an addition polymer comes from the name of its monomer.

Synoptic link

You can learn more about polymers in C2.2.7 *Polymer molecules*.

Table 1 *Some addition polymers.*

Monomer	Name of monomer	Systematic name of polymer	Common name of polymer
$CH_2{=}CH_2$	ethene	poly(ethene)	polythene
$CH_2{=}CHCH_3$	propene	poly(propene)	polypropylene
$CH_2{=}CHCl$	chloroethene	poly(chloroethene)	polyvinyl chloride, PVC

A Name the monomer used to make poly(tetrafluoroethene) (Figure 2).

Figure 2 *Poly(tetrafluoroethene), also called PTFE or Teflon®, forms the non-stick coating on pans.*

How do you model addition polymers?

Figure 3 models how propene molecules can join together to form poly(propene). The model is limited because real poly(propene) molecules may be made from thousands of propene molecules.

Notice that there is a repeating pattern in the section of the polymer molecule. You can model this using the idea of a repeating unit.

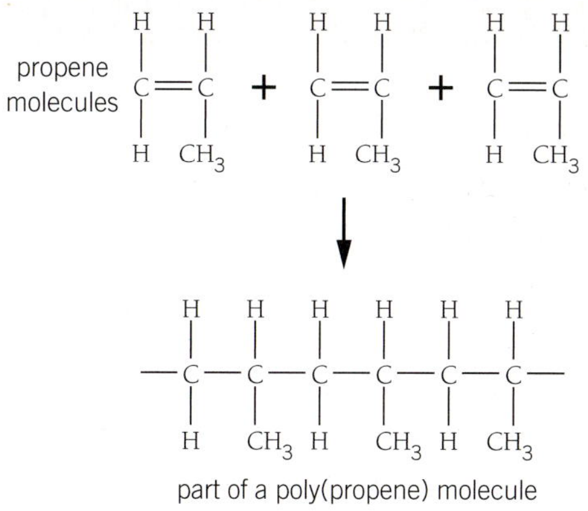

Repeating units

Draw the repeating unit for poly(propene).

Step 1: Draw the monomer with bonds at 90°, if this is not given to you (Figure 4, diagram A).

Step 2: Redraw the monomer, but:

- change the double bond to a single bond

- draw two long bonds either side of where the double bond was

- put square brackets round the diagram, passing through the outer bonds.

In Figure 4, diagram B shows the repeating unit for poly(propene).

Figure 3 Propene molecules join together to form a poly(propene) molecule.

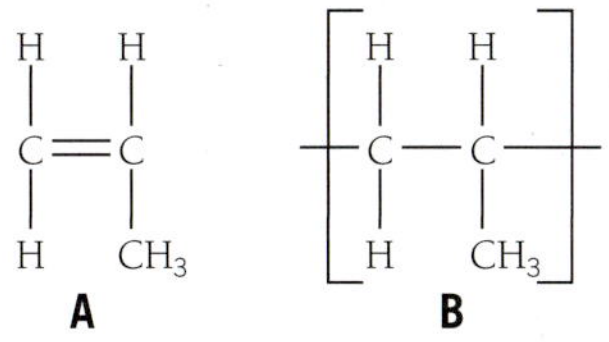

Figure 4 Propene (A) and poly(propene) (B).

B Draw the repeating unit for poly(ethene).

If you are asked to show the displayed formula for an addition polymer, draw its repeating unit but write n at the bottom of the right-hand bracket (Figure 5). The letter n stands for the number of repeating units in the polymer molecule. There is no need to write an actual number.

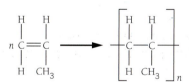

Figure 5 Modelling the reaction between n propene monomers to form a poly(propene) molecule containing n repeating units.

C Draw the displayed formula for poly(ethene).

1 Write down the feature of alkene molecules that allows them to act as monomers. *(1 mark)*

2 Describe what addition polymerisation is, and the conditions needed for it to happen. *(5 marks)*

3 **a** Draw the displayed formula of chloroethene, $CH_2=CHCl$. *(1 mark)*
   b Draw the repeating unit of poly(chloroethene). *(1 mark)*
   c Use your answer to **b** to help you to write down an equation modelling the formation of poly(chloroethene). *(1 mark)*

Learning outcome

After studying this lesson you should be able to:

- describe DNA and other naturally occurring polymers in terms of their monomers.

Specification reference: C6.2h

Figure 1 *A double spiral staircase; the structure of DNA is very similar.*

Schools often have a rule about which side of the stairs you can use. If everyone stays on the left, for example, the people going up do not bump into the people coming down. A double spiral staircase (Figure 1) avoids this problem altogether because you go up one staircase and down the other. This 'double helix' arrangement of these staircases is similar to the one found in DNA, which is a biological polymer.

What is DNA like?

DNA molecules are polymers made from monomers called **nucleotides**. Each nucleotide consists of:

- a phosphate group
- a sugar, called deoxyribose
- an organic **base**.

The structures of these parts are complex, so they are often modelled using **block diagrams**. Rather than showing the displayed formula for each part, you show its general shape (Figure 2).

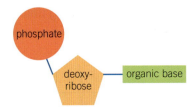

Figure 2 *A block diagram modelling the structure of a nucleotide.*

Nucleotides contain one of four different bases:

- adenine, A
- thymine, T
- cytosine, C
- guanine, G.

Nucleotides join together end to end to form a single strand of DNA. The deoxyribose of one nucleotide bonds to the phosphate group of the next nucleotide. Two strands of DNA then spiral round one another in opposite directions, just like the staircase in Figure 1. Intermolecular forces called **hydrogen bonds** form between bases on opposite strands (Figure 3). The hydrogen bonds hold the strands together, just like the supporting struts in the staircase.

> **A** Look at Figure 3, and then write down what you notice about the bases in each base pair.

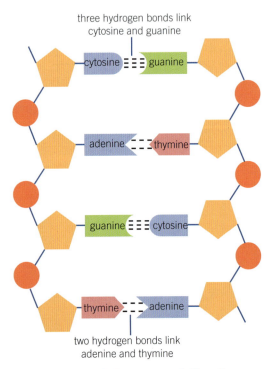

Figure 3 *A block diagram modelling the double-stranded structure of DNA.*

What are proteins like?

Protein molecules are polymers made from monomers called **amino acids**. There are about 20 naturally occurring amino acids. Each amino acid has a reactive functional group at each end, so they can join end to end to make very long protein molecules (Figure 4). Different protein molecules contain different numbers of amino acids joined together in different combinations.

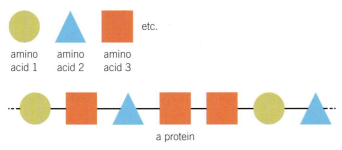

Figure 4 *A block diagram modelling the structure of a protein.*

B Name the monomers found in proteins.

What are carbohydrates like?

Carbohydrates are compounds of carbon, hydrogen, and oxygen. They include simple **sugars**, such as deoxyribose and sucrose (table sugar), and **complex carbohydrates** such as starch. Sucrose is made from two simple sugars, glucose and fructose, joined together. Starch consists of very many simple sugar molecules joined together.

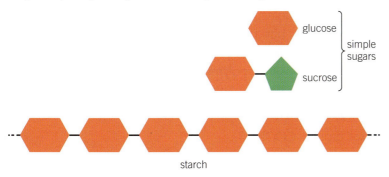

Figure 5 *Block diagrams modelling simple sugars and a section of starch.*

C Use Figure 5 to name the simple sugar that forms starch.

Study tip

If you see the word 'base', check the context used. Do not confuse bases in DNA with the bases, such as copper(II) oxide and sodium hydroxide, involved in neutralisation reactions with acids.

Synoptic link

You can learn more about DNA in B1.2.1 *DNA*.

---

1 Name the monomers found in:
  a DNA   (*1 mark*)
  b carbohydrates.   (*1 mark*)

2 Name the four bases found in DNA.   (*1 mark*)

3 Compare the structures of DNA, proteins, and carbohydrates. In your answer, include the similarities and differences between them.   (*6 marks*)

Figure 1 *A strand of nylon, made from two different monomers in solution.*

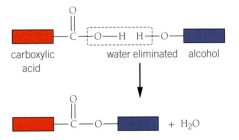

Figure 2 *A block diagram modelling the condensation reaction between a carboxylic acid and an alcohol.*

A teacher carefully pours a colourless solution into a dish containing another colourless solution. Using forceps, he reaches into the dish and pulls out a pale white strand, which he rolls onto a glass rod (Figure 1). The teacher has made nylon, a condensation polymer.

What are condensation reactions?

In a **condensation reaction**, two molecules react together to form one larger molecule and one smaller molecule. They are called condensation reactions because the smaller molecule is often water. For example, carboxylic acids react with alcohols to form organic compounds called **esters**, and water:

$$\text{ethanoic acid} + \text{ethanol} \rightarrow \text{ethyl ethanoate} + \text{water}$$
$$CH_3COOH + C_2H_5OH \rightarrow CH_3COOC_2H_5 + H_2O$$

In this reaction, the carboxyl group –COOH reacts with the hydroxyl group –OH. You can model the reaction using block diagrams (Figure 2). The remaining parts of the carboxylic acid and the alcohol are joined by an ester group –COO–.

> **A** Explain why the formation of an ester is a condensation reaction.

What are condensation polymers?

Condensation polymers are polymers formed by condensation reactions. A monomer molecule for an addition polymer has one functional group, a C=C bond. However, a monomer molecule for a condensation polymer needs *two* different functional groups, one at each end. Proteins are condensation polymers formed from amino acids. Each amino acid has two different functional groups:

- an amino group, $-NH_2$
- a carboxyl group, –COOH.

The amino group of one molecule reacts with the carboxyl group of another molecule. The organic product formed still has an amino group and a carboxyl group, so it can join with two more amino acids. This continues, forming a long polymer molecule. The remaining parts of each monomer are joined by an amide group, –CONH– (Figure 3).

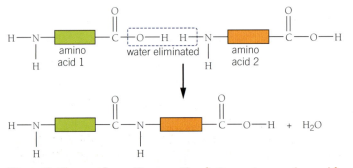

Figure 3 *The condensation reaction between two amino acids.*

B Explain why the monomers for making proteins need two different functional groups.

What are polyesters and polyamides like?

There are two main types of artificial condensation polymer. **Polyesters** are made from a carboxylic acid with two carboxyl groups, and an alcohol with two hydroxyl groups. They are called polyesters because they contain many ester groups, –COO– (Figure 4). PET is a polyester used to make clothing and fizzy drinks bottles.

Figure 4 *A section of a polyester.*

Polyamides are made from a carboxylic acid with two carboxyl groups, and an 'amine' with two amino groups. They are called polyamides because they contain many amide groups, –CONH– (Figure 5). Nylon is a polyamide used to make clothing and carpets.

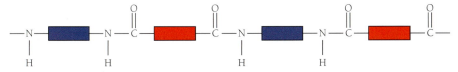

Figure 5 *A section of a polyamide.*

C Explain the difference between the block diagrams for a polyester and a polyamide.

Addition polymerisation happens with a catalyst at high temperatures and pressures. However, condensation polymerisation can happen without a catalyst at room temperature and pressure. This is why the teacher could make nylon in a school laboratory.

1 Write down the minimum number of functional groups that a monomer must have to form a condensation polymer. (*1 mark*)

2 Explain why PET and nylon are condensation polymers. (*2 marks*)

3 Nylon can be made from hexanedioyl dichloride, $ClOC(CH_2)_4COCl$, and hexane-1,6-diamine, $NH_2(CH_2)_6NH_2$. Both are corrosive and a required solvent is highly flammable.

  a Explain three precautions needed to control the hazards in this experiment. (*3 marks*)

  b Explain why hydrogen chloride is produced instead of water, but it is still a condensation reaction. (*3 marks*)

Learning outcomes

After studying this lesson you should be able to:

- describe how a chemical cell produces a potential difference until the reactants are used up
- describe how a hydrogen–oxygen fuel cell works
- evaluate the advantages and disadvantages of fuel cells for given uses.

Specification reference: C6.2p, C6.2q

For more than a century, most vehicles have been powered by petrol or diesel. This is now changing, and electric cars and buses are becoming more common (Figure 1). In some vehicles the electricity is provided by batteries, while in others it is provided by fuel cells.

How does a chemical cell work?

The batteries you use in portable devices such as mobile phones and torches, are **chemical cells**. Exothermic reactions in the cell develop a **potential difference** or voltage between its two ends. When the cell is connected to an electrical circuit, a current flows through the cell and the components of the circuit. This continues until one of the reactants is used up. The cell 'goes flat' and can no longer provide a potential difference.

Figure 1 *Electric cars at a charging station.*

Making a chemical cell

The chemist and physicist John Daniell invented one of the first chemical cells in 1836. You can make a Daniell cell using two metals and two electrolytes.

1 Pour some zinc sulfate solution into a beaker and dip a zinc strip into it.

2 Pour some copper(II) sulfate solution into another beaker and dip a copper strip into it (Figure 2). Care must be taken as the reactant solutions can be harmful at some concentrations.

3 Connect the two solutions using filter paper soaked with potassium nitrate solution.

4 Connect a voltmeter to the metal strips to measure the potential difference.

The two electrolytes used in the Daniell cell are zinc sulfate and copper(II) sulfate.

Eye protection should be worn throughout this practical.

Figure 2 *A simple Daniell cell connected to a voltmeter.*

A Explain a precaution you should take when carrying out a practical like *Making a chemical cell*.

B Most modern chemical cells are 'dry' cells. Suggest why the Daniell cell is called a 'wet' cell.

How does a hydrogen–oxygen fuel cell work?

A **fuel cell** produces electricity through a chemical reaction between a fuel and oxygen, without combustion happening (Figure 3).

Hydrogen reacts with oxygen in an exothermic reaction to produce water vapour:

$$2H_2(g) + O_2(g) \rightarrow 2H_2O(g)$$

The same overall reaction happens in a hydrogen–oxygen fuel cell, but it is separated into two reactions, one in each side of the cell (Figure 3).

1   Hydrogen molecules lose electrons and become hydrogen ions:

$$2H_2(g) \rightarrow 4H^+(aq) + 4e^-$$

Hydrogen ions pass through a 'proton exchange membrane' to the other side of the fuel cell, and electrons travel through the external circuit to the other side of the fuel cell.

2   Hydrogen ions combine with oxygen and electrons at the other side to form water vapour:

$$4H^+(aq) + O_2(g) + 4e^- \rightarrow 2H_2O(g)$$

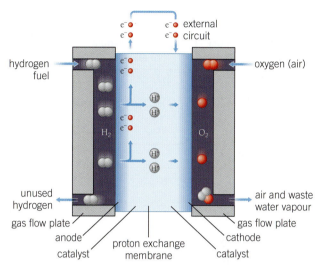

Figure 3 *A cross-section through a typical hydrogen–oxygen fuel cell.*

C   Methanol, CH_3OH, can be used in some fuel cells instead of hydrogen. Write down a balanced equation for the overall reaction with oxygen in these cells.

D   Table 1 shows some example information about a 250 km journey using a car with a battery, or a car with a fuel cell using hydrogen made from natural gas. Use this information to evaluate the advantages and disadvantages of using a fuel cell rather than a battery for this journey.

Table 1 *Comparing a battery and a fuel cell.*

Feature	Battery	Fuel cell
refuelling time (min)	450	5
cost of refuelling (£)	4.00	5.25
CO_2 emitted (kg)	60	45
mass of car (kg)	1550	1200

1   Name the waste product of a hydrogen–oxygen fuel cell.   (*1 mark*)

2   Write down equations to model the reactions that happen at each side in a hydrogen–oxygen fuel cell.   (*2 marks*)

3   Explain, in terms of electrons, whether the reactions at each side of a hydrogen–oxygen fuel cell are oxidation or reduction reactions.   (*4 marks*)   **H**

Global Challenges

C6.2 Organic chemistry

GCSE CHEMISTRY ONLY

Summary questions

1 a Write down the general formula for alkanes.
b Copy and complete Table 1.

Table 1 *Formulae of alkanes.*

Name of alkane	Formula of alkane	Displayed formula of alkane
methane	CH_4	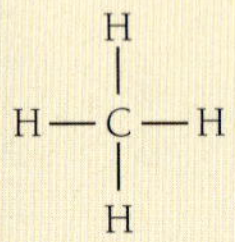 H–C–H (with H above and below)
	C_2H_6	
		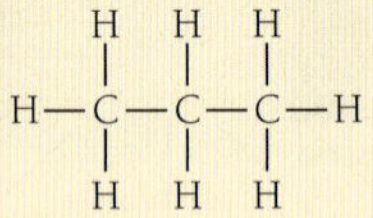 H–C–C–C–H (with H's)
butane		

2 a Write down the general formula for alkenes.
b Write down the structure of the functional group that determines the chemical properties of alkenes.
c Copy and complete Table 2.

Table 2 *Formulae of alkenes.*

Name of alkene	Formula of alkene	Displayed formula of alkene
ethene		
		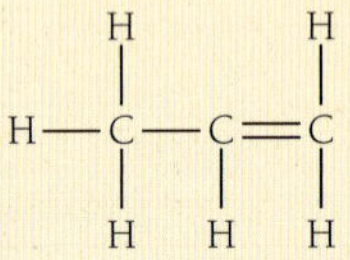 H–C–C=C (with H's)

d Describe a simple laboratory test to distinguish between hexane and hexene. In your answer, include any essential substance and the observations you expect.
e Figure 1 shows two different structures for butene:

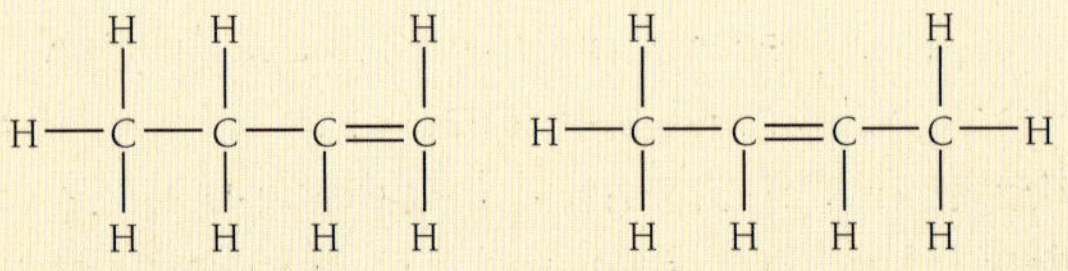

Figure 1 *Two displayed structures of butene.*

Suggest what the numbers 1 and 2 tell you about the structure of the molecule.

3 Alkanes, alkenes, alcohols, and carboxylic acids form four different homologous series. Write down four features of a homologous series.

4 Alkanes and alkenes react with oxygen in the air when they burn. During complete combustion, carbon dioxide and water are produced. Write balanced equations to show the complete combustion of:
a pentane, C_5H_{12}
b hexene, C_6H_{12}.

5 a Write down the structure and name of the functional group that determines the chemical properties of alcohols.
b Draw the displayed formula for ethanol, C_2H_5OH.
c Ethanol reacts with acidified potassium manganate(VII) to form ethanoic acid, CH_3COOH.
i Name the homologous series to which ethanoic acid belongs.
ii Write down the name of a suitable dilute acid to acidify the potassium manganate(VII).
iii Explain the role of manganate(VII) ions in the reaction.
iv Describe what you expect to observe as the reaction proceeds.

6 a Explain why alkenes can act as addition monomers.
b Figure 2 shows the displayed formula for ethyl benzene. Ethyl benzene can polymerise to form poly(ethyl benzene), better known as polystyrene.

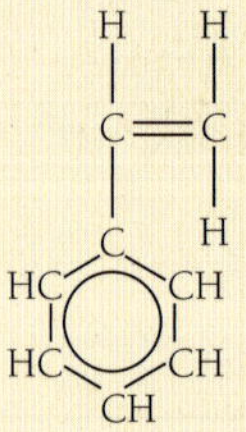

Figure 2 *Ethyl benzene.*

i Draw the displayed formula for the repeating unit of polystyrene.
ii Use your answer to **i** to help you to write down the equation to model the addition polymerisation reaction to form polystyrene, using appropriate displayed formulae.
c Nylon is a condensation polymer. Describe the differences between addition polymerisation and condensation polymerisation. **H**

250

Revision questions

1 Which displayed formula shows an alkene? *(1 mark)* **S**

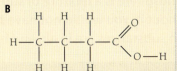

2 What is the name of the compound represented by the displayed formula below?

A ethanol
B ethanoic acid
C methanol
D methanoic acid *(1 mark)*

3 Some cars run on LPG (liquefied petroleum gas). LPG is a mixture of propane and butane.

 a Write down the name of the homologous series that includes propane and butane. *(1 mark)* **S**

 b Draw the structural formula of butane. *(1 mark)*

 c Propane and butane burn in a car engine. Construct the balanced symbol equation for the complete combustion of propane, C_3H_8. *(2 marks)*

 d LPG is one of the fractions obtained by the fractional distillation of crude oil.

 i Explain, in terms of intermolecular forces and the temperatures of different parts of the fractionating column, why LPG is removed from the top of the fractionating column. *(2 marks)*

 ii The following table shows the boiling points of alkanes present in different fractions. Each alkane is represented by a letter.

Alkane	Boiling point (°C)
W	−1
X	125
Y	317
Z	490

Predict the letter of the alkane that is in the bitumen fraction. Justify your decision. *(2 marks)*

4 The diagram shows a new way of generating electricity in fuel cells. A big data storage company uses electricity generated in one of these fuel cells to power its computers, 24 hours a day. **S**

DIGESTOR
Bacteria produces methane from sewage

FUEL CELL
1. Chemical reactions produce hydrogen from methane and water.
2. Reactions in the cell generate an electric current and a waste product, water.

Evaluate the advantages and disadvantages of generating electricity in this way compared to generating electricity in coal-fired power stations. *(6 marks)*

5 The diagrams represent two condensation polymers. **H S**

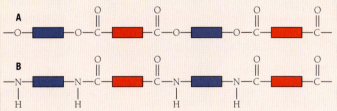

 a Write down the letter of the polyester and explain how you know it is a polyester. *(1 mark)*
 b Write down the names of the two functional groups that react together when a polyester forms from its monomers. *(1 mark)*
 c Explain why each monomer molecule has two functional groups, one at each end. *(1 mark)*
 d Write down the name of the other substance formed when a polyester forms from its monomers. *(1 mark)*

Learning outcomes

After studying this lesson you should be able to:

● explain how the atmosphere is thought to have originally formed

● describe how it is thought an oxygen-rich atmosphere developed over time.

Specification reference: C6.3a, C6.3b

The Earth's atmosphere is a mixture of substances in the gas state (Figure 1). Without it, you would not have the air you need to breathe. Which substances are in air, and how did they get there?

Figure 1 *The atmosphere is a relatively thin layer above the Earth's surface. It extends 100 km into space, which is barely 1.6% of the planet's radius.*

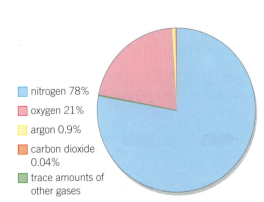

■ nitrogen 78%
■ oxygen 21%
■ argon 0.9%
■ carbon dioxide 0.04%
■ trace amounts of other gases

Figure 2 *This pie chart shows the percentages of the main gases in air today.*

What is the atmosphere like today?

Just three substances form 99.9% of the Earth's atmosphere today (Figure 2). These are nitrogen, oxygen, and argon. There are smaller percentages of many other gases, including carbon dioxide and water vapour.

A Write down the chemical formulae of the three most common substances in the Earth's atmosphere. What do you notice?

What was the atmosphere like in the past?

It is likely that Earth's early atmosphere came from substances released by volcanoes. However, no-one was around to record events as they happened, so it is difficult to be certain that this theory is correct.

The Earth is 4.54 billion years old. There was a great deal of **volcanic activity** during its early years. Volcanoes release huge volumes of water vapour and carbon dioxide. As the Earth cooled, the water vapour condensed to form oceans, leaving an atmosphere of mostly carbon dioxide. It probably contained small amounts of other gases such as ammonia and methane, but little or no oxygen (Figure 3).

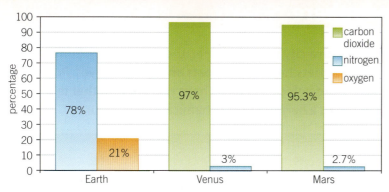

Figure 3 *The Earth's early atmosphere was probably like the atmospheres of Venus and Mars today. The Earth's atmosphere today is shown with the atmospheres of Venus and Mars for comparison.*

> **B** Describe the main differences between the Earth's early atmosphere and its atmosphere today.

Why has the atmosphere changed over time?

Theories to explain the changes to the Earth's atmosphere must explain where the oxygen came from, and where the carbon dioxide went. Plants and **algae** make their own food by **photosynthesis**. They absorb carbon dioxide from the atmosphere and release oxygen:

$$\text{carbon dioxide} + \text{water} \rightarrow \text{glucose} + \text{oxygen}$$

When plants and algae appeared, photosynthesis reduced the amount of carbon dioxide in the atmosphere. It also released oxygen. At first, the oxygen reacted with metals in rocks to produce metal oxides (Figure 4). Later, as most of the metals became oxidised, free oxygen began to accumulate in the atmosphere. Eventually the oxygen-rich atmosphere that you have today developed.

> **C** Explain how the presence of ancient layers of iron oxide provides evidence for the production of oxygen.

Figure 4 *This banded iron formation shows ancient layers of iron oxide, formed about 3 billion years ago.*

Go further

The percentage of oxygen in the atmosphere has varied over the last 600 million years. Find out when it reached its maximum, and what percentage it is thought to have reached.

1 Table 1 shows an example of the gases released by a volcano. Explain how it provides evidence for the composition of the early atmosphere. *(4 marks)*

2 Explain the change in the percentage of oxygen in the atmosphere over time. *(6 marks)*

3 a Levels of ammonia, NH_3, in the early atmosphere would have decreased because sunlight decomposes ammonia to form its elements. Write a balanced equation for this reaction. *(2 marks)*

b Levels of methane, CH_4, and hydrogen, H_2, would have decreased when they reacted with oxygen. Write balanced equations for these reactions. *(4 marks)*

Table 1 *Gases released by a volcano.*

Substance	% by volume
carbon dioxide	20.5
hydrogen	0.9
sulfur dioxide	6.9
water vapour	70.5
other gases	1.2

C6.3.2 Pollution and the atmosphere

Looking down on a city in the summer, the view is hazy. Some people on the streets have sore throats and stinging eyes. Pollutants from industry and vehicle exhausts have reacted together in the sunlight, forming a harmful photochemical smog (Figure 1).

Figure 1 *Photochemical smog contains many harmful atmospheric pollutants.*

What are pollutants?

Pollutants are substances released into the environment that may harm living things. Atmospheric pollutants are released into the air. Many of them are released as a result of burning fossil fuels. They include carbon monoxide, particulates, oxides of nitrogen, and sulfur dioxide.

A Explain what atmospheric pollutants are.

Why does carbon monoxide cause problems?

Carbon monoxide, CO, is produced during incomplete combustion of fuels that contain carbon. This can happen when coal, wood, or natural gas are burned in a poor supply of air. It also happens in vehicle engines. Carbon monoxide is a toxic gas. It is colourless, and has no taste or smell. When breathed in, carbon monoxide attaches to the haemoglobin protein in red blood cells. This reduces the amount of oxygen that the bloodstream can carry. Carbon monoxide poisoning causes drowsiness, difficulty breathing, and even death. For this reason many homes have carbon monoxide detectors like the one shown in Figure 2.

B Describe two symptoms of carbon monoxide poisoning.

Figure 2 *Household carbon monoxide detectors sound an alarm if faulty gas boilers or cookers produce carbon monoxide.*

Why do particulates cause problems?

Particulates are small particles. They are produced in industrial processes such as metal extraction (Figure 3). Like carbon monoxide, they are also produced during incomplete combustion and in vehicle engines. The smallest particulates settle deep in the lungs when they are breathed in. This causes diseases such as bronchitis and other breathing problems, and increases the chance of heart disease.

PM₁₀ particles

PM_{10} particles are 10 μm in diameter or less. There is a link between their concentration in air and health problems. Write down the maximum diameter of a PM_{10} particle in metres in standard form, giving your answer to two significant figures.

Figure 3 *Particulates are released in industrial smoke and from poorly maintained diesel engines.*

Why do acidic oxides cause problems?

Nitrogen and oxygen, the main gases in air, do not normally react together. However, they do react at the high temperatures in vehicle engines, forming nitrogen monoxide, NO. This is then oxidised in air to form nitrogen dioxide, NO_2. These oxides of nitrogen are jointly called 'NO_x'. Nitrogen dioxide dissolves in the moisture in clouds, forming an acidic solution, which eventually falls as **acid rain**. Acid rain erodes stonework and corrodes metals. It can kill trees (Figure 4), and living things in rivers and lakes.

Fossil fuels (coal, crude oil, and natural gas) naturally contain small amounts of sulfur compounds. These impurities form sulfur dioxide when the fuel is burnt. Sulfur dioxide also causes acid rain, and it can cause breathing difficulties.

C Name two substances that cause acid rain.

Figure 4 *These trees have been badly damaged by acid rain.*

1 **a** Outline how acid rain forms. *(6 marks)*

  b Describe two problems caused by acid rain. *(2 marks)*

2 Carbon monoxide may form when methane, CH_4, is used as a fuel in an industrial furnace. Explain, with the help of a balanced equation, how this happens. *(4 marks)*

3 Make a table to outline the main sources of atmospheric pollutants, and the problems these pollutants can cause. *(8 marks)*

Study tip

Remember that oxides of metals are basic and oxides of non-metals are acidic. Rainwater is naturally weakly acidic because it contains dissolved carbon dioxide from the atmosphere.

C6.3.3 Climate change

Figure 1 *The Thames Barrier when closed to prevent flooding.*

The Thames Barrier (Figure 1) is designed to protect London from flooding due to tidal surges. When it was officially opened in 1984, engineers expected it to be used about three times a year. It is now being used much more often.

What is the greenhouse effect?

Greenhouse gases such as carbon dioxide and methane absorb **infrared radiation** radiated by the Earth's surface, then emit it in all directions. This **greenhouse effect** keeps the Earth and its atmosphere warm enough for living things to exist (Figure 2).

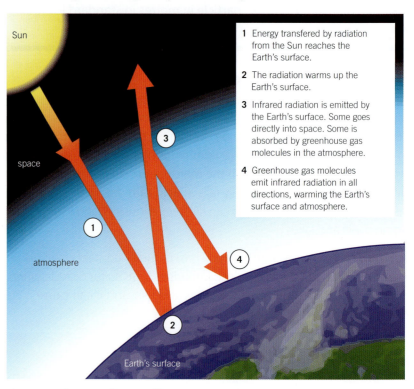

1 Energy transferred by radiation from the Sun reaches the Earth's surface.

2 The radiation warms up the Earth's surface.

3 Infrared radiation is emitted by the Earth's surface. Some goes directly into space. Some is absorbed by greenhouse gas molecules in the atmosphere.

4 Greenhouse gas molecules emit infrared radiation in all directions, warming the Earth's surface and atmosphere.

Figure 2 *The greenhouse effect.*

A Write down the type of radiation absorbed and emitted by greenhouse gases.

What is the enhanced greenhouse effect?

Carbon dioxide is released into the atmosphere by the combustion of fossil fuels. Methane is released into the atmosphere from rice paddy fields, cattle, landfill waste sites, and the use of natural gas. The release of additional greenhouse gases by human (**anthropogenic**) activities has the potential to cause an enhanced greenhouse effect, increasing the temperature of the Earth's atmosphere (Figure 3). This **global warming** leads to melting ice caps and rising sea levels, and to **climate change**. Climate change brings altered weather patterns, causing flooding, and problems with farming and disease control.

B Suggest why climate change affects farming.

How can greenhouse gas emissions be reduced?

Emissions of greenhouse gases into the atmosphere can be reduced by steps such as:

- reducing the consumption of fossil fuels, for example by using **biofuels**
- using renewable energy resources such as wind and solar energy to generate electricity
- stopping carbon dioxide escaping when fuels are used by using **carbon capture**.

Such steps are expensive, but so are steps to protect against the effects of global warming. These include flood barriers, planting different crops, and designing buildings to withstand high winds.

C Describe two ways to reduce greenhouse gas emissions.

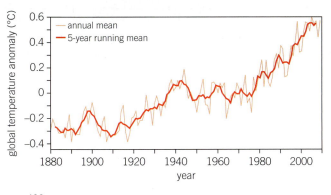

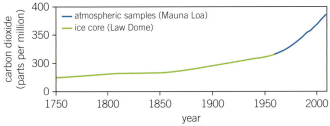

Figure 3 *Data comparing the change in global temperatures with the change in carbon dioxide levels.*

1 Describe the greenhouse effect. (*4 marks*)

2 Table 1 shows the relative ability of two greenhouse gases to warm the atmosphere.
   a Water vapour is a greenhouse gas. Suggest two reasons why emissions of water vapour are not seen as a problem. (*2 marks*)
   b Methane becomes fully oxidised in air. Explain why this causes a further problem. (*2 marks*)

3 Evaluate the information shown in the two graphs in Figure 3. To what extent do they show that carbon dioxide levels are linked to global temperature? (*6 marks*)

A scientific consensus

The Intergovernmental Panel on Climate Change (IPCC) has considered the scientific evidence for global warming and climate change. In one of their conclusions, they write that it is extremely likely that human influence was the main cause of global warming between 1951 and 2010. The IPCC says 'extremely likely' not because they are unsure, but because the Earth and its atmosphere is a very large, complex system. Scientists cannot carry out a controlled fair test in the normal way. They can make predictions based on observations, measurements, and computer models, but the actual outcome will only be revealed in the future.

Some people believe that the human race is carrying out a huge, potentially dangerous experiment on the planet by allowing carbon dioxide levels to rise. To what extent would you agree with this view?

Table 1 *Relative warming effects of two greenhouse gases.*

Greenhouse gas	Relative warming effect
carbon dioxide	1
methane	24

How much water do you use in a day? The average person in the UK uses about 150 litres per day. Very little of this is for cooking and drinking, but it is important that all tap water is potable – safe to drink (Figure 1). How is this achieved?

Figure 1 *This chemist is collecting a sample of water at a water treatment works for chemical analysis.*

Where does drinking water come from?

Tap water will originally have come from water stored in lakes, reservoirs, or **aquifers**. It may also have come from rivers or waste water. Water from all these sources contains microorganisms and many different substances:

- insoluble materials such as leaves, and particles from rocks and soil
- soluble substances, including salts, and pollutants such as pesticides and fertilisers.

Most of these must be removed at a treatment works to make the water safe to drink.

A Suggest how pesticides and fertilisers get into river water.

How is fresh water treated?

Figure 2 describes the main steps in water treatment, from reservoir to **potable water**.

> **B** Explain why the water is filtered through fine sand *after* it has passed through a screen.

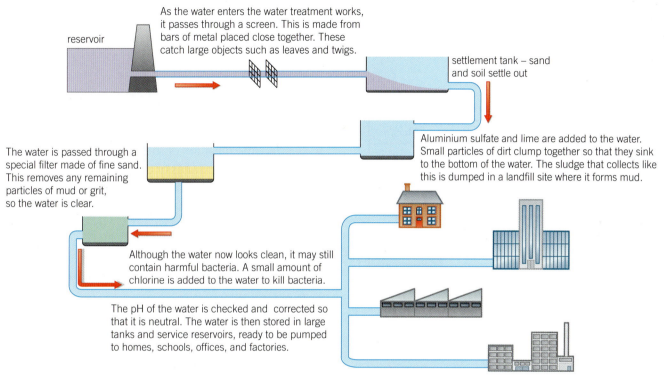

reservoir

As the water enters the water treatment works, it passes through a screen. This is made from bars of metal placed close together. These catch large objects such as leaves and twigs.

settlement tank – sand and soil settle out

Aluminium sulfate and lime are added to the water. Small particles of dirt clump together so that they sink to the bottom of the water. The sludge that collects like this is dumped in a landfill site where it forms mud.

The water is passed through a special filter made of fine sand. This removes any remaining particles of mud or grit, so the water is clear.

Although the water now looks clean, it may still contain harmful bacteria. A small amount of chlorine is added to the water to kill bacteria.

The pH of the water is checked and corrected so that it is neutral. The water is then stored in large tanks and service reservoirs, ready to be pumped to homes, schools, offices, and factories.

Figure 2 *Stages in water treatment.*

How is salt water treated?

Seawater contains high concentrations of dissolved salts. To make this water potable, these must be removed, in a process called **desalination**. For small scale desalination, 'reverse osmosis' using special 'ultrafilters' is used to filter out the salts. For large scale desalination, simple distillation is used (Figure 3).

Desalination would not be worthwhile in countries like the UK, but it is worthwhile where fresh water supplies are limited, or where the cost of energy resources is low.

> **C** Suggest why using simple distillation to provide drinking water on a large scale is not worthwhile in the UK.

> **1** Outline the stages, in order, used in treating fresh water to make it safe to drink. *(6 marks)*
>
> **2** Explain why chlorine can keep water safe to drink during its journey from the water treatment works to peoples' homes. *(2 marks)*
>
> **3** Describe the advantages and disadvantages of distilling large volumes of seawater to provide drinking water. *(6 marks)*

Study tip

Chlorine does not clean the water but it does kill bacteria in the water.

Figure 3 *A desalination plant on the coast in the Middle East for producing fresh water from seawater.*

C6.3 Interpreting and interacting with Earth Systems

Summary questions

1 Write down simple definitions of the following terms.
   a atmosphere
   b greenhouse gas
   c pollutant.

2 Table 1 shows the composition of the atmospheres of Venus, Earth, and Mars today.

Table 1 *The composition of three atmospheres.*

Formula of gas	Percentage of gas in the atmosphere		
	Venus	Earth	Mars
Ar	0.007	0.9	1.8
CO_2	97	0.04	95.3
CH_4	0	0.002	trace
H_2O	0.002	1 (varies)	0
N_2	3	78	2.7
O_2	0	21	0.14

   a Describe three significant differences between the Earth's atmosphere today and the atmospheres of Venus and Mars.
   b The two gases that are in the greatest abundance in the atmospheres of Venus and Mars today are thought to have been two of the main gases in the early atmosphere of Earth. Name these two gases.
   c The Earth's early atmosphere is thought to have contained more water vapour than today. Explain why the amount of water vapour in the atmosphere decreased over time.
   d Explain the role of plants and algae in changing the composition of the Earth's atmosphere over time.

3 Oxides of nitrogen are a cause of acid rain. They are produced when fossil fuels burn in vehicle engines.
   a Write down a balanced equation to show how nitrogen monoxide, NO, is produced.
   b Nitrogen monoxide is oxidised to nitrogen dioxide, NO_2, in air. This gas reacts with water droplets in the clouds to produce equal amounts of nitrous acid, HNO_2, and nitric acid, HNO_3. Write down a balanced equation for this reaction.

4 Sulfur dioxide is also a cause of acid rain.
   a Explain why sulfur dioxides are produced when fossil fuels are used.
   b Name one health problem linked to exposure to sulfur dioxide.
   c Describe two environmental problems caused by acid rain.

5 Vehicle catalytic converters reduce the emissions of harmful gases such as carbon monoxide, CO, and nitrogen monoxide, NO.
   a Describe the meaning of the term catalyst.
   b It takes about 5 minutes of driving for the catalyst in a converter to become hot enough to work effectively. Suggest why this is a problem in built-up areas such as towns and cities.

6 Carbon dioxide is a greenhouse gas.
   a Outline how the atmospheric greenhouse effect works.
   b The graph in Figure 1 shows the percentage of carbon dioxide in the atmosphere, measured over several decades from the top of a mountain in Hawaii.
      i Suggest why the measurements are made from a mountain top rather than from a city.
      ii Describe the trend in carbon dioxide levels.
      iii Suggest a reason for the annual zig-zag pattern seen in the readings.

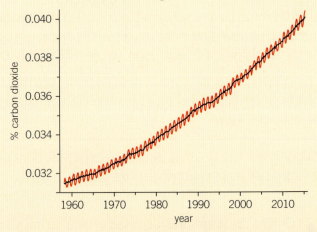

Figure 1 *Change in percentage of carbon dioxide over time.*

Revision questions

1 What is the major source of carbon monoxide in the atmosphere?

  A The combustion of impurities in coal.

  B High temperature reactions in car engines.

  C The combustion of impurities in natural gas.

  D Incomplete combustion of fossil fuels and wood.

                                            (1 mark)

2 Which row in the table lists three oxides that dissolve in rain to make it acidic?

A	carbon monoxide, carbon dioxide, sulfur dioxide
B	carbon dioxide, sulfur dioxide, nitrogen dioxide
C	carbon monoxide, carbon dioxide, nitrogen dioxide
D	carbon monoxide, sodium oxide, nitrogen dioxide

                                            (1 mark)

3 The following diagram shows how the greenhouse effect is caused.

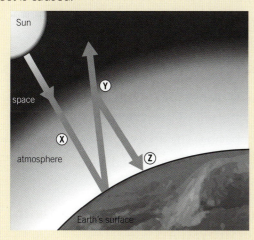

Copy and complete the following sentences.

Arrow **X** shows energy being transferred from the Sun to the Earth by _____. This warms the surface of the Earth.

Arrow **Y** represents _____ radiation emitted by the Earth's surface. Some of this radiation goes directly into space. Some is absorbed by greenhouse gas molecules in the _____.

Greenhouse gas molecules emit infrared radiation in all directions. Arrow **Z** represents some of this radiation, which _____ the Earth's surface. Radiation emitted by greenhouse gas molecules also warms the _____.

                                            (5 marks)

4 In 2015 leaders of many countries met in Paris to discuss how to reduce emissions of greenhouse gases.

  a Name two greenhouse gases.   *(2 marks)*

  b Explain why it is important to reduce greenhouse gas emissions, and describe how emissions of these gases can be reduced.   *(6 marks)*

5 The table shows annual total UK sulfur dioxide emissions. **H**

Year	Total UK sulfur dioxide emissions (million tonnes)
1995	2.4
2000	1.2
2005	0.7
2010	0.4
2014	0.3

  a **i** Finish plotting the data on a copy of the axes below. Show each piece of data with a cross. Do not draw any lines.   *(2 marks)*

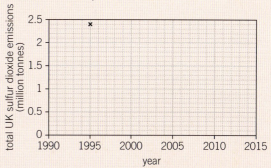

    ii Describe the overall trend shown on your graph.   *(1 mark)*

    iii Give the 5-year period with the biggest change in sulfur dioxide emissions.   *(1 mark)*

    iv Describe and explain two possible benefits resulting from the overall change in sulfur dioxide emissions.   *(3 marks)*

  b The following graph shows the total UK sulfur dioxide emissions from 1978 to 1983.

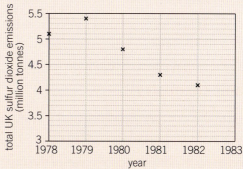

    Describe the change in emissions between 1978 and 1979 and suggest a possible reason for it.   *(2 marks)*

C6.1 Improving processes and products

- Describe the importance of nitrogen, phosphorus, and potassium compounds in agriculture, and explain the importance of the Haber process. **S**

- Describe and compare the industrial and laboratory production of fertilisers.

- Explain the trade-off between rate of production and position of equilibrium in the Haber process and the Contact process. **H**

- Interpret graphs involving yields and reaction conditions.

- Explain how the conditions for an industrial process are chosen.

- Explain how the method of metal extraction depends upon the reactivity of the metal.

- Describe how copper, iron, and aluminium are extracted from their oxides.

- Evaluate bioleaching and phytoextraction. **H**

- Describe the composition of some alloys, and how their properties relate to their uses. **S**

- Describe the process of corrosion and the conditions that cause it.

- Explain how corrosion is reduced by physical barriers and by sacrificial protection.

- Compare the physical properties of glass and clay ceramics, polymers, and metals.

- Explain the physical properties of composite materials.

- Explain how the choice of material for a given use is related to its properties.

- Describe the basic principles of life-cycle assessments and interpret data from them.

- Describe how materials or products are recycled, and explain why this is viable.

- Evaluate factors that affect decisions on recycling.

C6.2 Organic chemistry

- Recognise functional groups and identify members of the same homologous series. **S**

- Name and draw the displayed formulae of the first four straight chain alkanes, alkenes, alcohols, and carboxylic acids.

- Predict the formulae and structures of the products of reactions of alkanes, alkenes, and alcohols.

- Explain how modern life is crucially dependent upon hydrocarbons.

- Explain why crude oil is a finite resource, and recall that crude oil is the main source of hydrocarbons and a feedstock for the petrochemical industry.

- Describe and explain the fractional distillation of crude oil into fractions.

- Describe the cracking of crude oil fractions and explain why it is carried out.

- Describe the basic principles of addition polymerisation. **S**

- Work out the structure of an addition polymer from its monomer, and *vice versa*.

- Describe DNA and other naturally occurring polymers in terms of their monomers.

- Explain the basic principles of condensation polymerisation. **H**

- Describe practical techniques to make polymers.

- Recall that chemical cells produce a potential difference until the reactants are used up.

- Describe how a hydrogen–oxygen fuel cell works.

- Evaluate the advantages and disadvantages of fuel cells for given uses.

C6.3 Interpreting and interacting with Earth Systems

- Explain how the atmosphere is thought to have originally formed.

- Describe how an oxygen-rich atmosphere might have developed over time.

- Describe the major sources of atmospheric pollutants and the problems they cause.

- Describe the greenhouse effect, the problems caused by an enhanced greenhouse effect, and how they may be reduced.

- Evaluate evidence for causes of climate change.

- Describe how different sources of water are treated to make them safe to drink.

Metals

Extracting metals
- reactivity series used to determine method of extraction
- biological extraction = bioleaching, phytoextraction **H**

Corrosion
- iron and steel rust in the presence of oxygen and water
- prevent rust by:
  – stopping air and water reaching metal
  – sacrificial protection (zinc)
  – galvanising does both
- aluminium and stainless steel resist corrosion

Alloys
- mixture of two or more elements
- many and varied uses

Life-cycle assessment
- cradle-to-grave analysis
- consider properties and uses of glass, ceramics, polymers, and composite materials

C6 Global challenges

Fertiliser manufacture
- fertilisers replace essential elements (N, P, K) used in plant growth
- Haber process → ammonia **H**
- Contact process → sulfuric acid

Making electricity

Chemical cells
- 'batteries'
- p.d. until reactants used up

Fuel cells
- hydrogen fuel
- p.d. while fuel is supplied

Ethanol manufacture
- fermentation of plant sugars (renewable)
- hydration of ethane (non-renewable)

Water for drinking
- ground or waste water: sieving → sedimentation → filtration → chlorination → potable water
- seawater → desalination → potable water

Pollution and the atomosphere

Pollutant	Major source	Problems caused
carbon monoxide	hydrocarbon fuels	toxic gas
particulates	hydrocarbon fuels, industrial processes	lung and heart disease, blackened buildings
sulfur dioxide	sulfur impurities in fuels	acid rain (damages buildings and living things)
oxides of nitrogen	engines	

- greenhouse effect
- climate change

Atmosphere timeline

4.5 billion years ago

Earth forms
early atmosphere from volcanic activity

Earth cools
oceans form as water vapour condenses
CO_2 dissolves in water

plants and algae evolve, photosynthesis begins
metals in rocks react with O_2

carbon dioxide + water ↓ glucose + oxygen

mostly CO_2 and water vapour

CO_2 levels decrease

O_2 levels increase

today

Homologous series
- same general formula
- same functional group
- members differ by CH_2
- trends in physical properties
- similar chemical reactions

Alkanes	Alkenes	Alcohols	Carboxylic acids
C_nH_{2n+2}	C_nH_{2n}	$C_nH_{2n+1}OH$	$C_nH_{2n+1}COOH$
saturated	unsaturated	hydroxyl group	carboxyl group
complete combustion to form CO_2 and H_2O	addition reactions (e.g., with H_2 or Br_2)	oxidation to carboxylic acids	react to produce salts

MAIN COMPONENT IN CRUDE OIL, SEPARATED BY...

LARGER ALKANES SPLIT INTO SMALLER, MORE USEFUL ALKANES AND ALKENES BY CRACKING

Fractional distillation
- fractions separated by heating and condensation because larger alkane molecules have:
  – stronger intermolecular forces
  – higher boiling points

Polymers
- alkenes act as monomers for addition polymers
- naturally-occurring polymers = nucleotides → DNA; amino acids → proteins; glucose → starch
- condensation polymers = polyester, polyamide **H**

Learning outcomes

After studying this lesson you should be able to:

- safely and carefully handle gases, liquids, and solids
- mix reagents under controlled conditions
- use appropriate apparatus to explore chemical changes and products.

Specification reference: PAG C1

The reactivity of a metal is how readily it takes part in chemical reactions (Figure 1). You cannot measure the reactivity of a single metal directly, but you can compare the reactions of different metals to put them in a reactivity series.

Figure 1 *This scientist is preparing to test a metal part to see how resistant it is to corrosion by concentrated acids.*

How can you use reactions with dilute acid to put metals in a reactivity series?

If a metal reacts with a dilute acid, you see bubbles of hydrogen leaving the surface of the metal. The faster the bubbling, the more reactive the metal is.

It is not safe to react very reactive metals (such as calcium and lithium) with dilute acid. You can use small pieces of metals such as magnesium, aluminium, zinc, iron, and copper (Figure 2).

The dilute acid you should use is typically 1.0 mol/dm³ hydrochloric acid or 0.5 mol/dm³ sulfuric acid. You could test each metal in acid at room temperature and record your observations. Then, for any metals that show no reaction or a very slow reaction, you could warm the mixture and record any further observations.

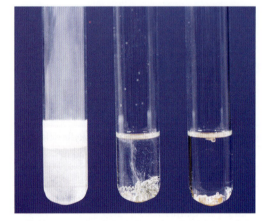

Figure 2 *Dilute hydrochloric acid reacting with magnesium (left), zinc (middle), and copper turnings (right).*

Synoptic link

Determining a reactivity series for elements in Group 7 (IUPAC Group 17) is described in C4.1.3 *Halogen displacement reactions.*

Warming safely

Make sure you wear eye protection. Do not overfill test tubes – they should be no more than a quarter full. You can warm the reaction mixture by placing the test tube in a beaker of hot water. If you use a Bunsen burner:

- half-open the air hole
- hold the test tube securely using a test-tube holder
- tilt the test tube so that it points away from you and other people
- hold the test tube a few centimetres above the flame, with the liquid's surface over the flame
- move the test tube in and out of the flame, and take it away before boiling begins.

Have a test tube rack ready to put the test tube into after warming.

Collecting and testing hydrogen

Hydrogen is less dense than air, so you can collect it by upward delivery. Hold an empty test tube upside down over the reaction mixture. The hydrogen will rise and fill this test tube. Allow time to collect enough gas, then hold a lighted splint to the mouth of the test tube and observe whether there is a pop.

Designing your practical

1   List the apparatus you need to safely carry out an investigation of the reaction of metals with dilute acid.

2   Write a detailed, step-by-step method suitable for this investigation.

3   Write a risk assessment for this investigation.

4   Draw an appropriate results table, including headings.

Analysing your results

5   Use the results of your experiments to put the metals in order of reactivity, starting with the most reactive.

Evaluating your practical

6   Compare your order of reactivity with an accepted one. Suggest reasons for any differences.

7   Evaluate the method:

   a   Describe what you did to obtain accurate results.

   b   Describe any difficulties you had.

   c   Suggest how you could use the reaction of metals with water to check the validity of your reactivity series.

Magnesium is more reactive than copper, so it displaces copper from copper(II) sulfate solution. The reaction is exothermic and the temperature rise can be measured using a thermometer. Two students decided to investigate the reactivity of different metals using displacement reactions. Table 1 shows their results.

1   Write down two variables the students should keep the same to ensure they obtain valid results.   (2 marks)

2   Suggest two reasons, other than error, for the results obtained with metal Y.   (2 marks)

3   Use the results to list the metals W, X, Y, and Z in order of decreasing reactivity.   (1 mark)

Table 1   *The students' results.*

Metal added to copper(II) sulfate solution	Temperature rise (°C)
W	8.4
X	3.9
Y	0.0
Z	12.1

C2 Electrolysis

Learning outcomes

After studying this lesson you should be able to:

- use appropriate apparatus and techniques to draw, set up, and use electrochemical cells to separate compounds and produce elements
- use appropriate methods to identify unknown products in the gas state.

Specification reference: PAG C2

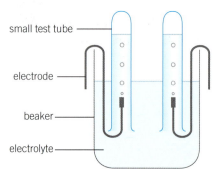

Figure 2 *This electrolysis cell uses two small test tubes to collect gases. The gases can be analysed later.*

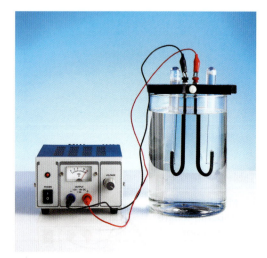

Figure 3 *An electrolysis experiment using a power pack, leads, and electrodes suspended from a bar.*

The fuel for hydrogen-powered vehicles can be produced by the electrolysis of water using renewable energy resources (Figure 1). You can investigate the electrolysis of solutions in the laboratory, collecting and analysing the products formed.

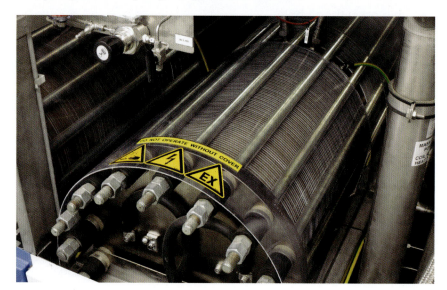

Figure 1 *This industrial electrolyser at a hydrogen refuelling station produces hydrogen from water.*

How do you set up an electrolysis cell?

You need a container for the solution (the electrolyte), and two electrodes connected to a battery or power pack. There are various designs of electrode. Figure 2 shows a common type of graphite electrode that you can hook over the side of a beaker.

The electrode you connect to the negative terminal of the battery or power pack becomes the cathode, and the electrode you connect to the positive terminal becomes the anode (Figure 3). You can collect products in the gas state in small test tubes so that you can identify them later. Remember that electrolytes and their products may be harmful, so wear eye protection.

Electrical connections

If you do not see any changes in your electrolysis cell, carry out these checks.

- Make sure that the end of each electrode is under the surface of the electrolyte.
- Make sure that the electrical connections are made securely – try cleaning the crocodile clips, for example.

If you still do not see any changes, replace each piece of apparatus in turn.

How can you identify the products of electrolysis?

Products in the solid state, such as copper, usually appear as a coating on an electrode. You can identify products in their gas states using different tests.

- A lighted splint ignites hydrogen with a pop.
- A glowing splint relights in oxygen.
- Damp blue litmus paper turns red, then white, in chlorine. Take care: chlorine is a toxic gas. You need to be especially careful not to breathe in any chlorine gas if you are asthmatic.

Designing your practical

1 List the apparatus you need to safely carry out the electrolysis of a solution, and to collect any products in the gas state.

2 Draw a labelled diagram to model the apparatus you will use.

3 Write a detailed, step-by-step method suitable for investigating the electrolysis of solutions.

4 Copper(II) sulfate solution is irritating to the eyes and skin. Explain two precautions you need to take to control the risk of harm when using this substance.

5 Draw an appropriate results table, including headings, for this practical activity.

6 Predict the product formed at each electrode (cathode and anode) during the electrolysis of:
   a copper(II) sulfate solution
   b sodium chloride solution.

Analysing your results

7 Use the results of your gas tests to identify each product in the gas state.

Evaluating your practical

8 Compare your results with your predictions. Suggest reasons for any differences.

9 Evaluate the method:
   a Describe what you did to obtain accurate results.
   b Describe any difficulties you had.
   c Suggest how you could use alternative apparatus or methods to obtain results that are more accurate.

When it is dissolved in water, potassium chloride can be decomposed by electrolysis.

1 Name a suitable metal element for the inert electrodes. *(1 mark)*

2 Some of the solution formed after electrolysis was tested using universal indicator solution. The indicator turned purple. Explain this result. *(2 marks)*

3 If the concentration of potassium chloride is very low, oxygen is produced at the anode instead of chlorine. Suggest a reason for this observation. *(2 marks)*

Collecting gases by displacement

Fill each small test tube with the electrolyte, then quickly turn them upside down in a beaker of electrolyte. The aim is to keep the open ends under the surface. This way, the test tubes remain filled with electrolyte until you are ready to connect the battery.

Synoptic link

Tests for gases are described in detail in C4.2.1 *Detecting gases*.

Learning outcomes

After studying this lesson you should be able to:

- safely use a range of equipment to purify and separate chemical mixtures, including chromatography.

Specification reference: PAG C3

Biochemists are scientists who study the chemistry of living things and the substances they produce. They often use chromatography, for example, to analyse the coloured pigments in plant cells, such as chlorophyll and carotene (Figure 1).

Figure 1 *These biochemists are preparing plant extracts, before analysing them using chromatography.*

How can you prepare leaf pigments for chromatography?

To extract the pigments from the plant cells, you need to break up the leaves in the presence of a suitable solvent. First, cut up some leaves finely. Then grind them for several minutes in a mortar with a pinch of sand and a little propanone. You could apply this plant extract directly to a strip of chromatography paper, or you may prefer to purify it first using filtration.

How do you carry out chromatography?

Mark a baseline on a strip of chromatography paper, then apply a small sample of the leaf extract. Place the paper in a stoppered boiling tube containing propanone as the mobile phase. The propanone level must be below the baseline. Remember to remove the chromatography paper before the solvent reaches the top, mark the level of the solvent, then let it dry. For each coloured spot, record its shape, colour, and R_f value.

Applying a sample to chromatography paper

Apply the sample on the baseline of the chromatography paper. Your results will be more accurate if you concentrate your sample into a small spot or line. If it is more spread out, the different substances present in the sample may tend to merge and streak. You could use a short length of capillary tube (Figure 2) to apply your sample.

- Dip one end of the capillary tube in the leaf extract.
- Put your finger over the other end, then briefly tap the tube onto the baseline.
- Allow the spot to dry, then add another spot on top.

Repeat this process several times to obtain a concentrated spot of the leaf extract.

Figure 2 *Doctors and nurses may use a capillary tube to collect a small sample of blood for analysis.*

Designing your practical

1   List the apparatus you need to safely carry out an investigation of the pigments contained in leaf cells.

2   Write a detailed, step-by-step method suitable for this investigation.

3   Propanone vaporises easily. It is highly flammable and it can cause severe eye damage. Explain two precautions you need to take to control the risk of harm when using this substance.

4   Draw an appropriate results table, including headings.

Analysing your results

5   Measure the distance travelled by the solvent from the baseline, and the distance travelled by the centre of each coloured spot from the baseline. Use these quantities to calculate the R_f value for each spot.

$$R_f = \frac{\text{distance travelled by spot}}{\text{distance travelled by solvent}}$$

Evaluating your practical

6   Compare your chromatogram with ones obtained by other students, or with the one shown in Figure 3. Suggest reasons for any differences.

7   Evaluate the method:

   a   Describe what you did to obtain accurate results.

   b   Describe any difficulties you had.

   c   Suggest how you could improve your method or use a different method to separate plant pigments.

Figure 3 *A chromatogram obtained from grass leaves.*

You can extract and analyse the yellow pigment in daffodil petals. You crush some petals using a mortar and pestle, add them to some ethanol in a boiling tube, then heat, with care, to 60 °C to produce a yellow solution.

1   Suggest why ethanol is used to extract the yellow pigment rather than water.   (*1 mark*)

2   Ethanol is highly flammable. Describe how you could heat it safely.   (*2 marks*)

3   Name a suitable method to separate the yellow solution from the petals, and explain why it is suitable for this mixture.   (*2 marks*)

4   Suggest how you could remove some ethanol from the yellow solution to make it more concentrated.   (*1 mark*)

Learning outcomes

After studying this lesson you should be able to:

- safely use a range of equipment to purify and separate chemical mixtures, including distillation
- safely use appropriate heating devices and techniques, including a Bunsen burner, water bath, or electric heater
- use appropriate apparatus to make and record a range of measurements accurately, including temperature.

Specification reference: PAG C4

Figure 1 *Tomatoes need fresh water to grow.*

In a seawater greenhouse, sea water and energy from the Sun are used to produce cool humid air and fresh water. Water vapour evaporates from the seawater at one end of the greenhouse, and cools and condenses, providing fresh water at the other end (Figure 1). You can use a similar process to obtain pure water from solutions such as ink, orange juice, or sodium chloride solution.

How can you produce water by distillation?

You need to heat the solution so that water vapour (or steam, if it is boiling) evaporates from it. The water vapour is then cooled and condensed. Figure 2 shows distillation apparatus that includes a water-cooled condenser. This is effective but can be complex to set up, so you may use simpler apparatus instead.

Figure 3 shows simple apparatus for distillation. Instead of using a condenser, a delivery tube carries steam to a boiling tube. The boiling tube is in a beaker of iced water to cool it.

If you use this sort of apparatus, take care to:

- avoid contact with steam, as it can cause burns
- clamp the apparatus securely so that it cannot fall over.

You will need to adjust the Bunsen burner flame to avoid an excessive rate of boiling. You may put insoluble anti-bumping granules or pieces of broken pot in the conical flask. These help the solution to boil more evenly and safely, but you still need to observe carefully and adjust the flame.

Figure 2 *Distilling ink in the laboratory using a condenser. Cooling water enters the condenser through the lower tubing at the right and leaves through the upper tubing on the left.*

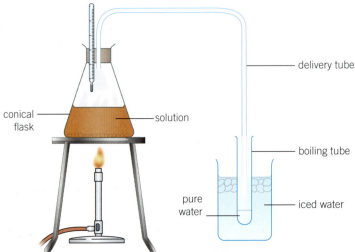

Figure 3 *Simple distillation apparatus.*

- conical flask
- solution
- delivery tube
- boiling tube
- pure water
- iced water

How do you test the purity of your product?

Pure water boils at 100 °C at normal atmospheric pressure. You can use a thermometer or a temperature probe connected to a data logger to measure the boiling point of your distillation product. Water that contains dissolved substances has a boiling point higher than 100 °C.

Using a thermometer

Ethanol-filled thermometers used in schools usually measure in the range −10 °C to 110 °C, and mercury-filled thermometers measure in the range −10 °C to 360 °C. To measure the boiling point of water, choose a thermometer with a maximum that is close to 100 °C. Many thermometers have a resolution of ±1 °C, but some measure to ±0.5 °C or better. If you move the thermometer, allow time for the temperature to settle before taking a reading.

Designing your practical

1 List the apparatus you need to produce pure water safely from a solution.

2 Write a detailed, step-by-step method suitable for this investigation.

3 Write a risk assessment for this investigation.

4 Write down a list of the observations you should make.

Analysing your results

5 Calculate the percentage difference between the boiling point of your purified water and the accepted value.

$$\text{percentage difference} = 100 \times \frac{\text{(your value in °C)} - 100\,°C}{100\,°C}$$

Evaluating your practical

6 If your water is pure, the percentage difference calculated above should be 0%. Suggest reasons for any difference observed.

7 Evaluate the method:

   a Describe what you did to obtain accurate results.

   b Describe any difficulties you had.

   c Suggest how you could improve your method.

A student used fractional distillation to separate ethanol from a mixture of ethanol and water.

1 Suggest why they used an electric heater to heat the mixture rather than a Bunsen burner. *(1 mark)*

2 Write down the physical property that fractional distillation depends on. *(1 mark)*

3 Describe a physical test the student could use to determine whether the water remaining after the fractional distillation is pure. *(2 marks)*

C5 Identification of species

Figure 1 *Waste water from a chemical factory.*

Synoptic link

Tests for cations are described in detail in C4.2.2 *Detecting cations*.

Table 2 *Results of halide ion tests.*

Halide ion	Silver halide precipitate colour
chloride, Cl⁻	white
bromide, Br⁻	cream
iodide, I⁻	yellow

What dissolved substances are there in waste water from a factory (Figure 1)? Environmental chemists analyse samples to make sure that any harmful substances present don't exceed permitted limits. You can use simple laboratory tests to analyse unknown samples by identifying cations and anions.

How do you identify cations?

You can use flame tests to identify some metal ions in solution and in ionic compounds in the solid state. Some metal ions in solution form coloured hydroxide precipitates when you add dilute sodium hydroxide solution. Table 1 summarises the results of common tests for cations.

Table 1 *Results of tests for cations.*

Metal ion	Flame test colour	Hydroxide precipitate colour
lithium, Li⁺	red	*no precipitate*
sodium, Na⁺	yellow	*no precipitate*
potassium, K⁺	lilac	*no precipitate*
calcium, Ca²⁺	orange-red	white
copper(II), Cu²⁺	green-blue	blue
iron(II), Fe²⁺	*no coloured flame*	green
iron(III), Fe³⁺	*no coloured flame*	orange-brown
zinc, Zn²⁺	*no coloured flame*	white

How do you identify anions?

You can identify carbonate ions, CO_3^{2-}, using a dilute acid such as dilute hydrochloric acid.

- The surface of a solid carbonate fizzes as carbon dioxide gas is produced.
- There is a brief fizzing when you add the acid to a solution of the carbonate.

You can identify non-metal ions in solution using precipitate tests.

- Sulfate ions, SO_4^{2-}, form a white precipitate with barium chloride solution acidified with dilute hydrochloric acid. Barium chloride is toxic.
- Halide ions form precipitates with silver nitrate solution acidified with dilute nitric acid (Table 2).

Analysing compounds

First plan your sequence of tests. If you are given unknown solutions to analyse, carry out precipitate tests for cations and anions first, as these are easier to do than flame tests. If you are given unknown solids to analyse, you could carry out flame tests and carbonate tests first, then make solutions for precipitate tests.

It is vital that you work carefully and cleanly. You will obtain false-positive results if samples and test solutions become muddled, or if you use the same dropping pipette for each solution without thoroughly rinsing it with water each time. Use small volumes of the unknown substances in test tubes for precipitate tests.

Take care when performing this practical, as some chemicals will be harmful or toxic, others may also be corrosive. You should wear appropriate eye protection.

Study tip

Ammonium carbonate and the carbonates of Group 1 elements are soluble in water, but most carbonates are insoluble.

Synoptic link

Tests for anions are described in detail in C4.2.3 *Detecting anions*.

Designing your practical

1 List the apparatus you need to carry out analytical tests on unknown solutions.

2 Write a detailed, step-by-step method suitable for analysing unknown solutions.

3 Write a risk assessment for this investigation.

4 Draw an appropriate results table, including headings. Include a column for your analysis of each result.

Analysing your results

5 For each test you carried out, decide whether it shows the presence or absence of an ion, then write this into your table.

Evaluating your practical

6 Evaluate your results. Have you correctly identified the ions present in each unknown substance?

7 Evaluate the method:

  a Describe what you did to obtain accurate results.

  b Describe any difficulties you had.

  c Suggest how you could improve your method.

The labels have fallen off four different containers of white powder in a school chemical store. They show:

- sodium chloride, $NaCl$
- sodium sulfate, Na_2SO_4
- calcium chloride, $CaCl_2$
- calcium bromide, $CaBr_2$.

1 Describe how you would test these substances to identify each one. Describe what you would do and how you would interpret the results. *(5 marks)*

Learning outcomes

After studying this lesson you should be able to:

- use appropriate apparatus, reagents, and techniques to measure pH in different situations
- use appropriate reagents and techniques to determine the concentrations of strong acids and strong alkalis
- use appropriate apparatus to make and record a range of measurements accurately, including volumes of liquids.

Specification reference: PAG C6

Some people take folic acid tablets to prevent folic acid deficiency. The manufacturers use titration to check the concentration of folic acid in their tablets (Figure 1). You can use the same technique to find the concentration of laboratory acids such as hydrochloric acid.

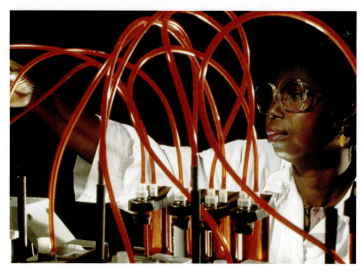

Figure 1 *This chemist is using an automated titration machine as part of the quality control in manufacturing folic acid tablets.*

How do you use titration to determine the concentration of an acid?

When carrying out a titration you must wear eye protection. You need to carry out these steps.

- Fill a burette with the acid.
- Use a volumetric pipette to transfer 25.0 cm³ of sodium hydroxide solution to a conical flask.
- Add a few drops of a single indicator to the flask.
- Gradually add acid from the burette to the flask until the indicator just changes colour.
- Record the start and end readings on the burette so you can calculate the titre.
- Repeat the method until you obtain concordant titres (similar readings, within 0.10 cm³ of each other).

If you know the concentration of the sodium hydroxide solution, you can calculate the concentration of the acid.

Using indicators

Universal indicator solution is a mixture of indicators that has a range of colours. This is useful when you want to estimate the pH of a solution, because you can compare the colour with a pH colour chart. However, having a range of colours is not useful for titration, as the colour changes too gradually for you to identify the end-point confidently. If you use a single indicator instead, you will see an obvious colour change, giving a sharp end-point (Table 1).

Indicators are weak acids or weak alkalis, so add only a few drops to avoid affecting your results. Take care: some indicators are flammable.

Table 1 *The colours of three single indicators.*

Indicator solution	Colour in acidic solution	Colour in alkaline solution
litmus	red	blue
phenolphthalein	colourless	pink
methyl orange	red	yellow

Designing your practical

1 List the apparatus you need to carry out a titration.

2 Write a detailed, step-by-step method suitable for this investigation.

3 Write a risk assessment for this investigation.

4 Draw an appropriate results table, including headings.

Analysing your results

5 Identify the concordant titres, or the ones closest to each other if none are concordant. Calculate a mean using these titres.

6 Use the mean titre, and the volume and concentration of alkali, to calculate the concentration of the acid. **H**

Evaluating your practical

7 Evaluate your results. If you are given an accepted concentration for the acid, how close to this is your result?

8 Evaluate the method:

  a Describe what you did to obtain accurate results.

  b Describe any difficulties you had.

  c Suggest how you could improve your method.

Using a burette

Make sure you clamp the burette securely, and that it is vertical looking from both the front and the side.

Practise using the tap.

- The tap is closed when it is horizontal, and open when it is vertical.

- If you are right-handed, use your left hand to control the tap, and your other hand to swirl the flask. If you are left-handed, use your right hand for the tap.

With practice, you should be able to add acid drop by drop near the end-point.

Carefully add about 10 cm³ of acid to the burette, using a plastic funnel, then drain the acid into a beaker. Now fill the burette with acid and remove the funnel. Let out enough acid to reach the zero mark or just below. Check that there are no bubbles in the apparatus before carrying out a titration.

Synoptic link

Titration methods are described in detail in C5.1.5 *Titrations*, and calculations in C5.1.6 *Titration calculations*.

A student carries out a titration using this method:

Step 1: Add 25.0 cm³ of dilute sodium hydroxide solution to a conical flask.

Step 2: Add a few drops of indicator.

Step 3: Add dilute hydrochloric acid from a burette until the indicator just changes colour.

1 Name a suitable indicator for this titration, and describe the colour change seen at the end-point. *(2 marks)*

2 Explain why a volumetric pipette is not suitable for Step 3. *(2 marks)*

3 Describe a further step the student should carry out to obtain valid results. *(2 marks)*

4 Describe a precaution, other than using eye protection, that the student should take to work safely. *(1 mark)*

Learning outcomes

After studying this lesson you should be able to:

- safely use a range of equipment to purify and separate chemical mixtures including evaporation, filtration, and crystallisation
- use appropriate apparatus to make and record a range of measurements accurately, including mass and volumes of liquids
- safely use and handle gases, liquids, and solids, including careful mixing of reagents under controlled conditions and using appropriate apparatus, to explore chemical changes and products.

Specification reference: PAG C7

Figure 2 *Lead nitrate solution and sodium iodide solution react together to form a yellow precipitate of lead iodide.*

Synoptic link

Filtration is described in detail in C2.1.4 *Filtration and crystallisation.*

Table salt is sodium chloride, but this is not the only useful salt. For example, copper(II) sulfate is useful to fruit farmers as a fungicide (Figure 1). You can make these salts, and others, in the laboratory.

Figure 1 *This machine is spraying grapevines in a vineyard to control harmful fungal diseases.*

How do you make an insoluble salt?

If your salt is insoluble in water, you can make it using a precipitation reaction between two suitable solutions. Table 1 summarises some rules about solubility.

Table 1 *Solubilities of different salts.*

Soluble	Insoluble
all sodium, potassium, and ammonium salts	
all nitrates	
most chlorides, bromides, and iodides	chlorides, bromides, and iodides of silver and lead
most sulfates	sulfates of lead, barium, and calcium
sodium, potassium, and ammonium carbonates	most carbonates

In general, to make a salt *XY*, mix *X* nitrate solution with sodium *Y* solution. For example, to make insoluble lead iodide, you would mix lead nitrate solution with sodium iodide solution (Figure 2). The insoluble salt forms as a precipitate that you can separate from the solution by filtering. You then wash it with water and dry it in a warm oven. This is not a salt that you will make in the school laboratory.

How do you make a soluble salt?

You can make ammonium salts by titrating a suitable acid with ammonia solution. You then produce the dry salt by crystallisation:

- hydrochloric acid makes chlorides
- sulfuric acid makes sulfates
- nitric acid makes nitrates.

To make a soluble salt containing metal ions, the method you choose depends on the metal involved. Table 2 summarises the methods available.

Table 2 *Methods for making soluble salts.*

Metal in soluble salt	Suitable method to use with dilute acid			
	Titrate acid with alkali	Add excess metal to acid, then filter out remaining metal	Add excess insoluble base to acid, then filter out remaining base	Add excess insoluble carbonate to acid, then filter out remaining carbonate
potassium	✓			
sodium	✓			
calcium	✓			✓
magnesium	✓	✓	✓	✓
aluminium		✓	✓	*no carbonate*
zinc		✓	✓	✓
iron		✓	✓	✓ iron(II) only
copper			✓	✓
silver			✓	✓

Using excess insoluble solids

Wearing eye protection, add the metal, base, or carbonate to the appropriate acid, then:

- stir the mixture, warming if necessary
- continue adding the solid until you see unreacted solid in the mixture, and any fizzing has stopped
- filter the mixture to remove the excess solid
- transfer the filtrate to an evaporating basin
- evaporate some water from the salt solution to form crystals
- cool, filter, and dry the crystals in a warm oven.

Take care when heating chemicals: do not heat them to dryness.

Synoptic link

Crystallisation is described in detail in C2.1.4 *Filtration and crystallisation*, and making ammonium salts is described in detail in C6.1.2 *Making fertilisers.*

Synoptic link

Making a salt by titration is described in detail in C6.1.2 *Making fertilisers.*

Designing your practical

1 List the apparatus you need to make a given salt.
2 Write a detailed, step-by-step method suitable for making this salt.
3 Write a risk assessment for making the salt using your chosen method.
4 Write down a list of the observations you should make.

Analysing your results

5 Draw a scale diagram to illustrate one or more of your crystals.

Evaluating your practical

6 Evaluate your results. How do your crystals compare with those produced by other students?
7 Evaluate the method:
  a Describe what you did to obtain accurate results.
  b Describe any difficulties you had.
  c Suggest how you could improve your method.

A student made copper(II) sulfate solution by reacting sulfuric acid with excess copper(II) oxide powder, followed by filtration.

1 Write down why an excess of copper(II) oxide was used. *(1 mark)*
2 Suggest why copper was not used to make copper(II) sulfate. *(1 mark)*
3 Describe how a pure, dry sample of copper(II) sulfate can be produced from the solution. *(4 marks)*

C8 Measuring rates of reaction

Learning outcomes

After studying this lesson you should be able to:

- use appropriate apparatus to make and record a range of measurements accurately, including mass, time, and volumes of liquids and gases
- make and record appropriate observations during chemical reactions, including measuring rates of reaction by a variety of methods such as production of gas and colour change.

Specification reference: PAG C8

Normal high-speed cameras can capture an image of a speeding bullet (Figure 1), but Japanese scientists have built one that takes 4.4 trillion images per second. This is so fast it can show particles reacting together. Fortunately, there are simpler ways to investigate reactions.

Figure 1 *High-speed photography shows a bullet leaving a gun. It is propelled by rapidly expanding gases from a fast chemical reaction.*

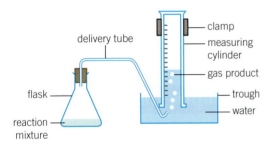

Figure 2 *Measuring the rate of reaction using a measuring cylinder.*

Synoptic link

Different ways to measure rates of reaction are described in detail in C5.2 *Controlling reactions*.

How can you measure the rate of a reaction?

There are many ways to measure the rate of a reaction. The method you choose will depend upon the reactant or product you want to measure, and the apparatus available to you. For example, for a product in the gas state you can measure the change in:

- mass, if the gas has a relatively large relative formula mass, such as carbon dioxide ($M_r = 44.0$)
- volume, unless the gas is very soluble, such as sulfur dioxide.

If you decide to measure the volume of the gas, you could measure it using:

- a gas syringe
- an upturned measuring cylinder of water (Figure 2).

What variables can you investigate?

The rates of reactions depend on the:

- temperature of the reaction mixture
- concentrations of reacting solutions
- surface area (particle size) of reactants in the solid state
- pressure of reactants in the gas state
- presence of a suitable catalyst.

It is difficult to investigate the effect of pressure in the school laboratory, but you can easily investigate the other four variables. When you investigate the effect of one of these, you must control the other variables because these may also affect the rate of reaction. You do not need to try to control pressure because atmospheric pressure changes very little.

Designing your practical

1 List the apparatus you need to investigate the effect of your independent variable on the rate of your chosen reaction.

2 Write a detailed, step-by-step method suitable for carrying out this investigation.

3 Write a risk assessment for carrying out this investigation.

4 Draw an appropriate results table, including headings.

Analysing your results

5 Plot a suitable graph of your results.

Evaluating your practical

6 Evaluate your results:

  a Identify any outliers.

  b Comment on the precision of your measurements (consider the spread of any repeat measurements).

7 Evaluate the method:

  a Describe what you did to obtain accurate measurements.

  b Describe what you did to obtain precise results.

  c Suggest how you could improve your method to obtain results that are more accurate or precise.

Timing a reaction

If you are going to time a reaction, take particular care at the start. This is when the reactant concentrations, and so the rate of reaction, are highest. Have everything ready so that you can start the stopwatch as soon as you mix the reactants. Record times in seconds, s, not in minutes and seconds.

When carrying out chemical reactions you should take care, and wear eye protection.

Marble chips (calcium carbonate) react with dilute hydrochloric acid, producing carbon dioxide. Some students investigated the effect of changing the concentration of acid on the rate of this reaction. They diluted 1.0 mol/dm^3 hydrochloric acid with different volumes of water, and timed how long the reaction took to fill a 100 cm^3 measuring cylinder with carbon dioxide.

1 The students used the same mass of marble chips each time. Write down one other property of these chips that they should control. *(1 mark)*

2 Table 1 shows the students' results. Plot a suitable graph, including a straight line of best fit. *(3 marks)*

3 One of the results is an outlier. Circle this on your graph and suggest one reason why it may have happened. *(2 marks)*

Table 1 *The students' results.*

Concentration of HCl(aq) (mol/dm³)	Time to collect 100 cm³ of CO₂ (s)	Rate of reaction (cm³/s)
1.0	17.5	5.7
0.8	21.8	4.6
0.6	26.3	3.8
0.4	43.8	2.3
0.2	87.5	1.1

Revision questions for C1–C3

1 Which of these changes is a chemical change?

   A dissolving

   B melting

   C evaporating

   D burning   (1 mark)

2 Which technique is best for separating copper sulfate crystals from a solution of copper sulfate in water?

   A chromatography

   B distillation

   C evaporation

   D filtration   (1 mark)

3 Which of the electron arrangements below is for the atom of a metal?

   A 2.8.2

   B 2.8.6

   C 2.8.7

   D 2.8.8   (1 mark)

4 The table shows the melting points and boiling points of four substances.

   Which substance is in the gas state at 20 °C?   (1 mark)

	Melting point (°C)	Boiling point (°C)
A	−101	15
B	−7	59
C	−39	357
D	98	890

5 An atom has a mass number of 52 and an atomic number of 24. Which row of the table correctly gives its numbers of protons, neutrons, and electrons?

	Number of protons	Number of neutrons	Number of electrons
A	24	28	24
B	28	24	28
C	24	28	52
D	52	24	52

                                                                     (1 mark)

6 The formula of paracetamol is $C_8H_9NO_2$. What is the relative formula mass, M_r, of paracetamol? The relative atomic mass of C is 12.0, of H is 1.0, of N is 14.0 and of O is 16.0.

   A 51

   B 67

   C 151

   D 178   (1 mark)

7 The formula of an aluminium ion is Al^{3+} and the formula of a sulfate ion is SO_4^{2-}.
What is the formula of aluminium sulfate?

A $Al(SO_4)_3$
B $Al(SO_4)_2$
C $Al_2(SO_4)_3$
D $Al_3(SO_4)_2$

(1 mark)

8 A teacher passes an electric current through liquid lead bromide, $PbBr_2$. Graphite electrodes pass electricity to and from the lead bromide. Which row of the table correctly describes which ions are attracted to each electrode? *(1 mark)*

	Ions attracted to the cathode	Ions attracted to the anode
A	lead cations	bromide anions
B	lead anions	bromide cations
C	bromide cations	lead anions
D	bromide anions	lead cations

H

9 Propane burns in oxygen to produce carbon dioxide and water vapour.

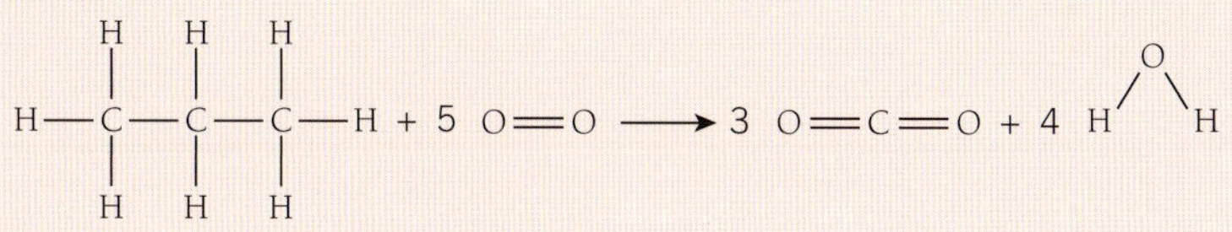

The following table shows the bond energies for the bonds that are broken and made in the reaction.

Bond	Bond energy (kJ/mol)
C–H	413
C–C	348
O–H	464
O=O	498
C=O	805

Overall, how much energy is transferred for the reaction shown in the equation?

A −892
B −2052
C −2748
D −15 032

(1 mark)

10 A student measures the pH of four different acids, each with a concentration of $1.0\,mol/dm^3$. Which is the strongest acid?

Acid	pH
A	0.5
B	1.0
C	1.5
D	2.0

(1 mark)

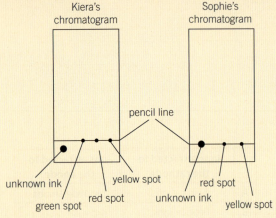

Kiera's chromatogram

Sophie's chromatogram

pencil line

unknown ink
green spot
red spot
yellow spot
red spot
unknown ink
yellow spot

11 Some students use chromatography to identify the mixture of dyes in an unknown ink.

The teacher tells them that the unknown ink was made by mixing two or three dyes.

First, the students prepare their chromatography paper by adding spots of ink.

a i Suggest one way in which Kiera could improve her experiment.

(1 mark)

ii Explain one way in which Kiera's prepared chromatography paper is better than Sophie's. *(1 mark)*

A third student, Sarah, obtains the chromatogram below.

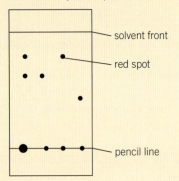

solvent front

red spot

pencil line

b Calculate the R_f value for the red spot. Use a ruler to help you.

(2 marks)

12 A student makes copper sulfate crystals by reacting copper oxide with dilute sulfuric acid.

Look at the outline of her method.

1 Add copper oxide powder to dilute sulfuric acid until the copper oxide is present in excess.

2 Separate the excess copper oxide from the copper sulfate solution.

3 Remove water from the copper sulfate solution.

a Write down the two missing state symbols from the equation for the following reaction.

$CuO(s) + H_2SO_4(\_\_) \rightarrow CuSO_4(aq) + H_2O(\_\_)$ *(2 marks)*

b Describe how to carry out each step of the experiment, giving the names of the pieces of apparatus required. Include the names of the separation techniques and explain how each one works. *(6 marks)*

13 Air is a mixture of substances, including nitrogen, oxygen, and argon.

a Nitrogen and oxygen exist as molecules, N_2 and O_2.

Explain why nitrogen and oxygen have low boiling points. *(1 mark)*

b Nitrogen, oxygen and argon are useful individually.

Industrially, they are separated from each other by cooling air until it liquefies.

Fractional distillation is then used to separate nitrogen, oxygen and argon from each other.

   i Write down the name of the change of state that occurs when
a gas becomes liquid. *(1 mark)*

   ii Explain the process of fractional distillation. *(4 marks)*

   iii The table shows the melting and boiling points of the three
substances. For each substance in the table, write down its
state at −188 °C. *(1 mark)*

   iv Use data from the table in **iii** to suggest why argon and
oxygen are difficult to separate from each other. *(1 mark)*

 c Oxygen has three stable isotopes.

   i Explain what is meant by the term *isotope*. *(1 mark)*

   ii Calculate the number of neutrons in an $^{18}_{8}O$ isotope. *(1 mark)*

Substance	Melting point (°C)	Boiling point (°C)
argon	−189	−186
oxygen	−218	−183
nitrogen	−210	−196

14 Two students, Barney and Edward, are investigating the electrolysis
of copper sulfate solution using copper electrodes. They want to find
out about the factors that affect the mass of copper deposited. The
following diagram shows their apparatus.

 a Explain why the mass of the copper cathode will increase. *(2 marks)*

 b Barney obtains the results in this table.

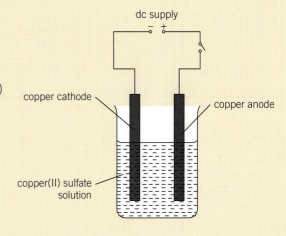

Current (A)	Mass of copper deposited (g)
0.4	0.08
0.6	0.12
0.8	0.16
1.0	0.20
1.2	0.24

   i Suggest the question Barney is investigating. *(1 mark)*

   ii Write a conclusion for Barney's investigation. *(1 mark)*

 c Edward investigates how the time the current flows affects the
mass of copper deposited. He assumes that the decrease in mass of
the anode is equal to the increase in mass of the cathode.
Look at Edward's method.

   1 Use a balance to find the starting mass of the anode.

   2 Place the anode in the circuit, close the switch, and start
the timer.

   3 After the required time, open the switch.

   4 Take out the anode and use the balance to find its new mass.

   5 Do the experiment again, but for a different length of time.

   i Suggest one variable that Edward should control in order to carry
out a fair test. *(1 mark)*

   ii Suggest why Edward measures the decrease in mass of the
anode, not the increase in mass of the cathode. *(1 mark)*

   iii Suggest and explain how Edward can change his experiment in
order to allow him to assess the precision of his results. *(2 marks)*

Time (min)	Mass of anode (g)	Change in mass (g)
5.0	99.80	0.20
10.0	99.60	0.40
15.0	99.40	
20.0	99.20	0.80
25.0	99.00	1.00

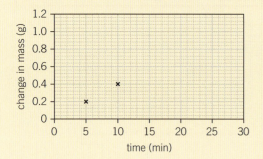

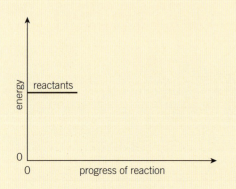

d Look at Edward's results.

Mass of anode at start = 100.00 g

i Calculate the value that is missing from the table. *(1 mark)*

ii Plot the missing results on a copy of the axes shown, and draw a line of best fit. *(2 marks)*

iii Use your graph to estimate the mass of the anode after 2.00 minutes. *(2 marks)*

15 a Magnesium is an important metal.

i Magnesium alloys are used to make parts for cars and aeroplanes. Write down what is meant by the word *alloy*. *(1 mark)*

ii In some fireworks, tiny pieces of magnesium burn to make magnesium oxide. Write a balanced chemical equation for the reaction. *(2 marks)*

b Magnesium can be obtained from a naturally-occurring compound, magnesium carbonate, $MgCO_3$.

i First, magnesium carbonate is heated strongly. The products are magnesium oxide and carbon dioxide:

$$MgCO_3(s) \rightarrow MgO(s) + CO_2(g)$$

Draw simple particle diagrams to show how the particles are arranged in the two products. Represent each particle with a circle.

Under each diagram, write down the name of the product it shows. *(4 marks)*

ii Next, one of the products of the reaction in **ii** is heated with silicon:

$$2MgO(s) + Si(s) \rightarrow SiO_2(s) + 2Mg(g)$$

Explain which substance is oxidised in this reaction. *(1 mark)*

iii The reaction in **ii** is endothermic.

Draw a line on a copy of the axes shown, to show the energy of the products compared to the energy of the reactants. *(1 mark)*

iv Use the equations given in **i** and **ii**, as well as relative atomic mass data from the Periodic Table, to calculate the maximum mass of magnesium that can be obtained from 100 kg of magnesium carbonate. *(4 marks)* **H**

c Magnesium can also be extracted from seawater. First, magnesium chloride is produced in a series of chemical reactions.

The magnesium chloride is then passed into an electrolytic cell.

i The formula of magnesium chloride is $MgCl_2$. Write down the formulae of its two ions. *(2 marks)*

ii Write a half equation for the reaction that occurs at the cathode. *(1 mark)*

Revision questions for C4–C6

1 What is potable water?

   A Water that is safe to drink.

   B Water that has been filtered.

   C Water with a pH of 7.0.

   D Water with added chlorine.   (*1 mark*)

2 Which statement is correct for each of the first four Group 7 (IUPAC Group 17) elements?

   A It reacts with water to form hydrogen.

   B It is in the gas state at room temperature.

   C It has a high boiling point.

   D It reacts with a Group 1 element to form a salt.   (*1 mark*)

3 A student has two soluble compounds, X and Y.

She does some tests on the compounds. Her results are presented in the following table.

Compound	Flame test colour	Colour of precipitate with silver nitrate solution
X	yellow	yellow
Y	lilac	white

What are the names of compounds X and Y?

   A X is sodium iodide and Y is calcium chloride.

   B X is lithium chloride and Y is sodium bromide.

   C X is sodium bromide and Y is calcium iodide.

   D X is sodium iodide and Y is potassium chloride.   (*1 mark*)

4 Ethanol can be produced by fermenting glucose.

$$C_6H_{12}O_6 \rightarrow 2C_2H_5OH + 2CO_2$$

What is the atom economy of the process?

   A 81.8%

   B 22.5%

   C 51.1%

   D 55.0%   (*1 mark*)

5 A student investigates the reaction between magnesium and dilute hydrochloric acid.

He measures the reaction time at four different temperatures.

He does the experiments using the same concentration and volume of acid, and the same length of magnesium ribbon.

Look at his results.

Which reaction took place at the **lowest** temperature?   (*1 mark*)

Temperature	Reaction time (s)
A	156
B	78
C	37
D	17

6 Which formula represents an alkane?

   A $C_{15}H_{32}$

   B $C_{16}H_{32}$

   C $C_{17}H_{35}$

   D $C_{18}H_{42}$ (*1 mark*)

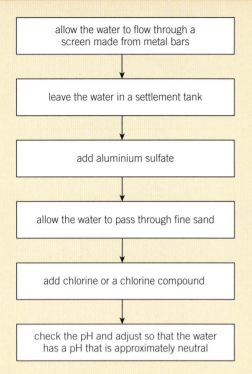

allow the water to flow through a screen made from metal bars

↓

leave the water in a settlement tank

↓

add aluminium sulfate

↓

allow the water to pass through fine sand

↓

add chlorine or a chlorine compound

↓

check the pH and adjust so that the water has a pH that is approximately neutral

7 Look at the flow chart.
It shows how water can be made safe for drinking.

  a Name the process that occurs when muddy water passes through sand. *(1 mark)*

  b Explain why chlorine is added to the water. *(1 mark)*

  c At a water treatment works, a scientist finds that the water pH is 9. Which substance in the list below should be added to the water? Explain your answer.

    ethanoic acid
    sodium carbonate
    sodium hydroxide *(2 marks)*

8 **a** Draw straight lines to match each structural formula to its name. **S**

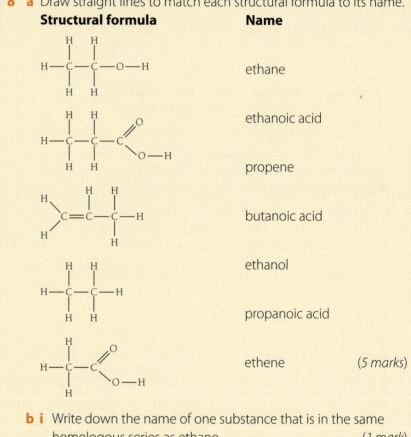

Structural formula	Name
	ethane
	ethanoic acid
	propene
	butanoic acid
	ethanol
	propanoic acid
	ethene *(5 marks)*

  b **i** Write down the name of one substance that is in the same homologous series as ethane. *(1 mark)*

    ii Draw the structural formula of the substance you named in part **i**. *(1 mark)*

  c Name the four bases present in DNA. *(4 marks)*

9 Group 0 elements are known as the noble gases.
Look at the table of data about noble gases. Some data is missing.

Element	Atomic symbol	Boiling point (°C)	Atomic radius (pm)
Helium	He	−269	50
Neon	Ne	−246	65
Argon		−186	95
Krypton	Kr		110
Xenon	Xe	−108	130

a Use the Periodic Table to write down the chemical symbol of argon.
(1 mark)

b **i** Plot a bar chart showing the boiling point data on a copy of the axes shown. *(2 marks)*
 ii Use your bar chart to predict the boiling point of krypton. *(1 mark)*
 iii Describe the trend in boiling points for the noble gases. *(1 mark)*
 iv Explain the trend you described in **iii**. *(1 mark)*

c The electronic structure of neon is 2.8
 i Write down the electronic structures of helium and argon. *(2 marks)*
 ii Use ideas about electronic structure to suggest an explanation for the trend in atomic radius shown in the table. *(2 marks)*

d The unit for atomic radius in the table is the picometre (pm).
One picometre is 1×10^{-12} m.
Calculate the atomic radius of krypton in nanometres (nm). *(2 marks)*

e Use ideas about electronic structure, and your knowledge of the chemical properties of oxygen and fluorine, to suggest an explanation for the statement below.
Helium forms no compounds. Xenon forms compounds with oxygen and fluorine, but not with other elements. *(4 marks)*

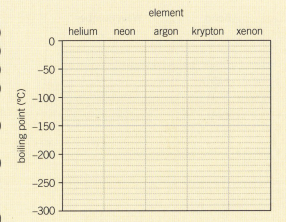

10 A student has some sodium thiosulfate solution and dilute hydrochloric acid.
Describe how he could use these solutions to investigate the effect of temperature on the rate of a reaction.
Include in your answer:
- the apparatus needed
- what the student needs to do
- how the student can make the investigation fair. *(6 marks)*

11 The table shows the properties of two materials, polypropene and sisal.
Both materials can be used to make ropes for life buoys near rivers.
Polypropene is made from substances obtained from crude oil.
Sisal is obtained from sisal plants.

Material	Does it rot?	Relative strength when pulled	Flexibility
polypropene	No	1.4	very flexible
sisal	Yes	1.8	very flexible

a Use data from the table to explain why ropes for life buoys are made from polypropene, not sisal. *(2 marks)*
b Suggest one advantage of making rope from sisal compared to making rope from polypropene. *(1 mark)*

c A Life Cycle Assessment (LCA) for a product involves collecting and analysing data about its production, use, and disposal.
Suggest and explain which parts of the LCA are the same for ropes made from both materials, and which parts of the LCA are different for the two types of rope. *(3 marks)*

Bond	Wavenumber
C–H	2850–3300
C–C	750–1100
C–O	1000–1300
N–H	3360–3500
O–H	3230–3550

12 Infrared spectrometry is an instrumental method of analysis.
a Describe two advantages of instrumental analysis. *(2 marks)*
b An infrared spectrometer detects the bonds present in simple molecular substances.
The table shows the wavenumbers for some different bonds.
A university student produced infrared spectra for two compounds, A and B.
i Both spectra show peaks in the range 900–1050.
Write down the bond they contain. *(1 mark)*
ii The molecules of bond compounds have C–H bonds.
Write down the wavenumber for the peaks these bonds will produce in spectra. *(1 mark)*
iii Compound A produced a peak at 3400 and compound B produced a peak at 3350.
Write down the bond that must be present in:
Compound A *(1 mark)*
Compound B *(1 mark)*
iv Use your answers to **iii** to suggest the identity of a bond that is present in compound A but not present in compound B. *(1 mark)*

13 A student has some sodium hydroxide solution.
She does a titration to find out its concentration. She follows the steps below.
1 She pours hydrochloric acid of known concentration into a burette.
2 She uses a measuring cylinder to measure out 25.00 cm³ of the sodium hydroxide solution.
3 She pours the sodium hydroxide solution into a conical flask and adds some indicator.
4 She adds hydrochloric acid solution from the burette to the conical flask until the indicator changes colour.
5 She repeats step D until she obtains at least two concordant titres (titres that are within 0.10 cm³ of each other.)

	Run 1 (rough)	Run 2	Run 3	Run 4	Run 5
Final burette reading (cm³)	26.00	0.10	25.00	1.05	30.00
Initial burette reading (cm³)	0.30	25.00	49.70	26.05	4.90
Titre (cm³)	25.70	24.90	24.70	25.00	

a Suggest and explain an improvement to step B. (*2 marks*) [S]

b Describe two things the student must do to obtain concordant titres in step E. (*2 marks*)

c The table shows the student's results.
  i Calculate the titre for Run 5. (*2 marks*)
  ii Write down the three concordant titres. (*1 mark*)
  iii Use your answer to **ii** to calculate the mean titre. (*2 marks*)

d The concentration of hydrochloric acid is $1.00 \, mol/dm^3$. Use your answer to **c iii** to calculate the number of moles of acid in the mean titre. (*2 marks*)

e Look at the equation for the reaction of hydrochloric acid with sodium hydroxide solution. [H]

$$HCl(aq) + NaOH(aq) \rightarrow NaCl(aq) + H_2O(l)$$

  i Use your answer to **d** to write down the number of moles of sodium hydroxide in $25.00 \, cm^3$ of solution. Explain your answer. (*2 marks*) [S]

  ii Calculate the concentration of sodium hydroxide solution in mol/dm^3. (*2 marks*)

  iii Calculate the concentration of sodium hydroxide solution in g/dm^3. [S]
   Use relative atomic mass values given in the Periodic Table. (*2 marks*)

14 Look at the equation below.
$$PCl_5(g) \rightleftharpoons PCl_3(g) + Cl_2(g) \qquad \Delta H = +124 \, kJ$$
  a Calculate the relative formula mass of PCl_5.
   Use data from the Periodic Table. (*2 marks*)
  b Write down the meaning of the $\rightleftharpoons$ symbol. (*1 mark*)

  c In a sealed container, the substances shown in the equation above reach equilibrium. [H]
   i Explain why removing phosphorus trichloride from the equilibrium mixture moves the equilibrium position to the right. (*1 mark*)
   ii Predict and explain the effect on the equilibrium position of increasing the pressure. (*2 marks*)
   iii Predict and explain the effect on the equilibrium position of increasing the temperature. (*2 marks*)

Maths for GCSE Chemistry

Figure 1 *A silicon microsphere is about 6.1×10^{-6} across.*

Figure 2 *The Sun is 1.5×10^{11} m from the Earth.*

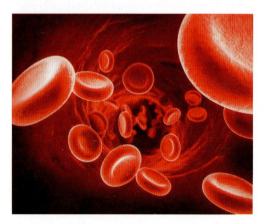

Figure 3 *These red blood cells have a diameter of approximately 7.0×10^{-6} m.*

How big are silicon microspheres in suncream, how far away is the Sun, and what is the size of a red blood cell (Figures 1–3)?

Scientists use maths all the time – when collecting data, looking for patterns, and making conclusions. This chapter includes all the maths for your GCSE chemistry course. The rest of the book gives you many opportunities to practise using maths in chemistry.

1 Decimal form

There will always be a whole number of people in a school, and a whole number of chairs in each classroom.

When you make measurements in science the quantities may *not* be whole numbers, but numbers *in between* whole numbers. They will be in **decimal form**, for example 3.2 cm, or 4.5 g.

2 Standard form

Some quantities in science are very large, like the distance from the Earth to the Sun. Other quantities are very small, such as the size of an atom.

In **standard form** (also called scientific notation) a number has two parts.

- You write a decimal number, with one digit (not zero) in front of the decimal place, for example 3.7.
- You multiply the number by the appropriate power of ten, for example 10^3. The power of ten can be positive or negative.
- For a quantity such as length, you then add the unit, for example 'm' for metres.
- This gives you a quantity in standard form, for example, 3.7×10^3 m. This is the length of one of the runways at Heathrow airport.

Table 1 explains how you convert numbers to standard form.

Table 1 *Very large and very small numbers can be written in standard form.*

The number	The number in standard form	What you did to get to the decimal number	...so the power of ten is...	What the *sign* of the power of ten tells you
1000	1.0×10^3	You moved the decimal point 3 places to the *left* to get the decimal number.	+3	The positive power shows the number is *greater* than one.
0.01	1.0×10^{-2}	You moved the decimal point 2 places to the *right* to get the decimal number.	−2	The negative power shows the number is *less* than one.

Here are two more examples. Check that you understand the power of ten, and the sign of the power, in each example.

- $20\,000\,Hz = 2.0 \times 10^4\,Hz$
- $0.0005\,kg = 5.0 \times 10^{-4}\,kg$.

Note that $1.0 \times 10^3\,m$ is the same as $10^3\,m$.

It is much easier to write some of the very big or very small quantities that you find in real life using standard form. For example:

- the distance from the Earth to the Sun is $150\,000\,000\,000\,m = 1.5 \times 10^{11}\,m$
- the diameter of an atom is $0.000\,000\,000\,1\,m = 1.0 \times 10^{-10}\,m$
- the diameter of a red blood cell is around $0.000\,007\,m = 7 \times 10^{-6}\,m$
- the speed of light is $300\,000\,000\,m/s = 3 \times 10^8\,m/s$.

You need to use a special button on a scientific calculator when you are calculating with numbers in standard form, which should have many of the buttons shown on the calculator in Figure 4. You should work out which button you need to use on your own calculator (it could be **EE**, **EXP**, **10ˣ**, or **×10ˣ**).

2.1 Multiplying numbers in standard form

When you multiply two numbers in standard form you *add* their powers of ten. When you divide two numbers in standard form you *subtract* the power of ten in the denominator (the number below the line) from the power of ten in the numerator (the number above the line). For example:

- $10^2 \times 10^3 = 10^5$ or $100\,000$ (because $2 + 3 = +5$)
- $10^2 \div 10^4 = 10^{-2}$ or 0.01 (because $2 - 4 = -2$)

How many pages?

A library, like the one shown in Figure 5, contains $200\,000$ books. Each book has 400 pages. The total length of the shelves is 4800 m.

1 Calculate the total number of pages.

Step 1: Convert the numbers to standard form.

$200\,000 = 2 \times 10^5$, $400 = 4 \times 10^2$, and $4800 = 4.8 \times 10^3$

Step 2: Calculate the total number of pages within all the books in the library.

Total number of pages = number of books × pages per book

$= (2 \times 10^5\,\text{books}) \times (4 \times 10^2\,\text{pages per book})$

$= (2 \times 4) \times (10^5 \times 10^2)\,\text{pages}$

$= 8 \times 10^7\,\text{pages}$

Figure 4 *You need a scientific calculator to do calculations involving standard form.*

Figure 5 *You can use standard form to help you work out how many pages are in this library.*

2 The total length of the shelves in the library is 4800 m. Calculate the thickness of a page.

Divide the total width of all the books by the total number of pages.

$$\text{Width of a single page} = \frac{\text{width of all the books}}{\text{number of pages in all the books}}$$

$$= \frac{4.8 \times 10^3 \, \text{m}}{8 \times 10^7 \, \text{pages}}$$

$$= \frac{4.8}{8} \times \frac{10^3}{10^7} \, \text{m/page}$$

$$= 0.6 \times 10^{-4} \, \text{m}$$

$$= 6 \times 10^{-5} \, \text{m in standard form}$$

3 Ratios, fractions, and percentages
3.1 Ratios

A **ratio** compares two quantities. For example, a ratio of 15 : 30 of Bunsen burners to beakers means that, for every 15 Bunsen burners, there are 30 beakers.

You may need to calculate a ratio in which one of the numbers is 1. This is useful if you need to compare two ratios. There are two ways to do this:

1 Divide both numbers by the *first* number (15 : 30 becomes 1 : 2 when you divide by 15). So for every Bunsen burner there are two beakers.

2 Divide both numbers by the *second* number (15 : 30 becomes 0.5 : 1 when you divide by 30). So for every half a Bunsen burner there is one beaker.

Notice that three ways are given above to describe the number of Bunsen burners compared to the number of beakers, but they mean the same thing.

Take care if you are asked to use ratios to find a fraction. The ratio of 15 : 30 in the example above does not mean that 15/30 (or half) of the objects are Bunsen burners.

Fractions from ratios

The ratio of nitrogen atoms to oxygen atoms is 2 : 4. What fraction are nitrogen atoms?

Step 1: Add the numbers in the ratio together.

$$2 + 4 = 6$$

Step 2: Divide the proportion of nitrogen atoms in the ratio by the total of the numbers in the ratio (which you found in Step 1).

$$\frac{\text{proportion of nitrogen atoms}}{\text{total atoms}} = \frac{2}{6} = \frac{1}{3}$$

One third of the atoms are nitrogen atoms.

You can simplify a ratio so that both numbers are the lowest whole numbers possible.

Simplifying ratios

A student mixed 15 cm³ of acid with 90 cm³ of water. Calculate the simplest ratio of the volume of acid to the volume of water.

Step 1: The ratio of *acid : water* is 15 : 90

Step 2: Both 15 and 90 have a common factor, 5. You can divide both numbers by 5.

$$acid : water = \frac{15}{5} : \frac{90}{5} = 3 : 18$$

Step 3: Both 3 and 18 have a common factor, 3. You can divide both numbers by 3.

$$acid : water = \frac{3}{3} : \frac{18}{3} = 1 : 6$$

Notice that to get the simplest form of the ratio, you have now divided both 15 and 90 by 15 (i.e. 3 × 5), which is the highest common factor of 15 and 90.

3.2 Fractions

The horizontal line in a fraction means 'divide', so $\frac{1}{3}$ means $1 \div 3$. This is useful to know if you have to convert a fraction into a decimal (Figure 6).

See that $\frac{1}{3} = 1 \div 3 = 0.\dot{3}$ (the dot above the number 3 shows that the number 3 recurs, or repeats over and over again).

Calculating the fraction of a quantity

A student has a 25 g sample of sodium chloride. Calculate the mass of $\frac{2}{5}$ of this sample.

Step 1: Divide the total mass of the sample by the denominator (the number on the bottom) in the fraction.

25 g ÷ 5 = 5 g

Step 2: Multiply the answer to Step 1 by the numerator (the number on the top) in the fraction.

5 g × 2 = 10 g

Figure 6 *This slice represents $\frac{1}{8}$ (or 0.125) of a cake.*

The **reciprocal** of a number is 1 ÷ (number). For example, the reciprocal of 4 is 0.25 (which is 1 ÷ 4). The reciprocal of a fraction is the fraction written the other way up. So the reciprocal of $\frac{2}{3}$ is $\frac{3}{2}$.

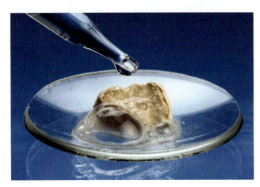

Figure 7 *Hydrochloric acid reacts with calcium carbonate in limestone but may not react with impurities in it.*

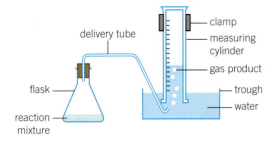

Figure 8 *Substances in the gas state can be measured using an upside down measuring cylinder.*

3.3 Percentages

A **percentage** is a number expressed as a fraction of 100.

Calculating a percentage

A student found that a 7.5 g sample of limestone contained 7.2 g of calcium carbonate (Figure 7). Calculate the percentage by mass of calcium carbonate in the sample.

Step 1: Write the two numbers as a fraction.

$$\text{fraction of limestone that is calcium carbonate} = \frac{\text{mass of calcium carbonate}}{\text{mass of limestone}}$$

$$= \frac{7.2}{7.5}$$

Step 2: Convert the fraction to a decimal.

$$\frac{7.2}{7.5} = 0.96$$

Step 3: Multiply the answer to Step 2 by 100%.

$$0.96 \times 100\% = 96\%$$

You can use this method to convert a fraction to a percentage, but you would only need to start at Step 2.

You may need to calculate a percentage of a quantity.

Using a percentage to calculate a quantity

A chemical reaction produces 80 cm³ of gas (Figure 8). Calculate 25% of this volume.

Step 1: Convert the percentage to a decimal.

$$25\% = \frac{25}{100} = 0.25$$

Step 2: Multiply the answer to Step 1 by the total volume.

$$0.25 \times 80\,\text{cm}^3 = 20\,\text{cm}^3$$

You may need to calculate a percentage increase or decrease in a quantity from its original value.

A percentage change

A student heats a 4.75 g sample of copper carbonate and finds its mass decreases to 3.04 g. Calculate the percentage change in mass.

Step 1: Calculate the actual change in mass.

$$4.75\,\text{g} - 3.04\,\text{g} = 1.71\,\text{g}$$

Step 2: Divide the actual change in mass by the original mass.

$$\frac{1.71}{4.75} = 0.36$$

Step 3: Multiply the answer to Step 2 by 100%.

$$0.36 \times 100\% = 36\%$$

Remember that this is a percentage *decrease*.

4 Estimating the result of a calculation

When you use your calculator to work out the answer to a calculation you can sometimes press the wrong button and get the wrong answer. The best way to make sure that your answer is correct is to estimate it in your head first.

Estimating an answer

You want to calculate the volume of foam made and you need to find $34 \, cm^3/s \times 8 \, s$ (Figure 9). Estimate the answer and then calculate it.

Step 1: Round each number up or down to get a whole number multiple of 10.

$34 \, cm^3/s$ is about $30 \, cm^3/s$

$8 \, s$ is about $10 \, s$

Step 2: Multiply the numbers in your head.

$30 \, cm^3/s \times 10 \, s = 300 \, cm^3$

Step 3: Do the calculation and check it is close to your estimate.

Volume $= 34 \, cm^3/s \times 8 \, s$

$= 272 \, cm^3$

This is quite close to 300 so it is probably correct.

Notice that you could do other things with the numbers:

$34 + 8 = 42$

$\dfrac{34}{8} = 4.3$

$34 - 8 = 26$

Not one of these numbers is close to 300. If you got any of these numbers you would know that you needed to repeat the calculation.

Figure 9 *Foamy soap bubbles containing oxygen are produced from hydrogen peroxide in the 'elephant's toothpaste' reaction.*

Sometimes the calculations involve more complicated equations, or standard form.

5 Significant figures

These lengths each have three **significant figures (sig. fig.)**. The significant figures are underlined in each case.

153 m, 0.153 m, 0.00153 m

If you write these lengths using standard form, you can see that they all have three significant figures (Table 2).

Table 2 *Using standard form can help you to see how many significant figures there are in a number.*

Length	153 m	0.153 m	0.00153 m
Length written in standard form	1.53×10^2 m,	1.53×10^{-1} m	1.53×10^{-3} m

Table 3 *Some measurements given in significant figures.*

Quantity	Number of sig. fig.
2358 mm	4
7 m/s	1
5.1 nm	2
0.05 s	1

Table 3 shows some more examples of measurements given to different numbers of significant figures.

How do you know how many significant figures to give when you answer a question? In general, you should round your answer to the lowest number of significant figures as you were given in the question.

If you are multiplying and dividing two numbers, and each has a different number of significant figures, work out which number has fewer significant figures. When you do your calculation, give your answer to this number of significant figures.

Significant figures are not the same as decimal places. If you are adding and subtracting decimal numbers, work out which number is given to fewer decimal places. When you do your addition or subtraction, give your answer to this number of decimal places.

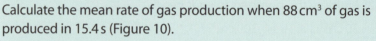

Significant figures

Calculate the mean rate of gas production when 88 cm^3 of gas is produced in 15.4 s (Figure 10).

Step 1: Write down what you know.

volume = 88 cm^3 (this number has 2 significant figures)

time = 15.4 s (this number has 3 significant figures)

Step 2: Write down the equation that links the quantities you know and the quantity you want to find.

$$\text{rate (cm}^3\text{/s)} = \frac{\text{volume (cm}^3)}{\text{time (s)}}$$

Step 3: Substitute values into the equation and do the calculation.

$$\text{rate} = \frac{88\,cm^3}{15.4\,s}$$

= 5.714 285 714 cm^3/s (you should not leave your final answer like this as there are too many sig. fig.)

= 5.7 cm^3/s to 2 sig. fig. (the values in the question have a minimum of 2 sig. fig., so the answer should be to 2 sig. fig.)

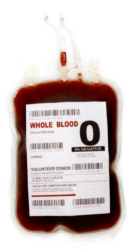

Figure 10 *A gas syringe is used to measure volumes of gas produced in a reaction.*

6 Frequency tables and bar charts
6.1 Frequency tables

The word **data** describes observations and measurements that are made during experiments. Data can be:

- qualitative (descriptive, but with no numerical measurements)
- quantitative (including numerical measurements).

Qualitative data include **categoric variables,** such as elements, oil fractions, and blood groups (Figure 11). The values of categoric variables are names not numbers. You can make a **frequency table** if you count the number of times that each categoric variable (for example, in each blood group) appears in a sample.

Figure 11 *Blood group O is the most common.*

Table 4 shows two variables:

- The **independent variable** (in this case, blood group). This is the characteristic or quantity that you decide to change.
- The **dependent variable** (in this case, frequency). This is the value you measure as you change the independent variable.

In a table, you should always show:

- the independent variable in the first column
- the dependent variable in the second column.

Quantitative data include:

- **continuous variables** – characteristics or quantities that can take any value within certain upper and lower limits, such as length and time
- **discrete variables** – characteristics or quantities that can only take particular values, such as number of paper clips, or shoe size.

Sometimes data are grouped into classes (as in Table 5). If you need to group data into classes:

- make sure the values in each class do not overlap
- aim for a sensible number of classes – usually no more than six.

6.2 Bar charts

You can use a bar chart to display the number or frequency when the independent variable is categoric (for example, blood groups, Figure 12) or has discrete values (for example, shoe size or pulse rate, Figure 13).

In a **bar chart**, you plot:

- number or frequency on the vertical axis
- the independent variable, or class, on the horizontal axis.

You should always leave equal gaps between the bars.

You can also use bar charts to compare two or more independent variables (Figure 14).

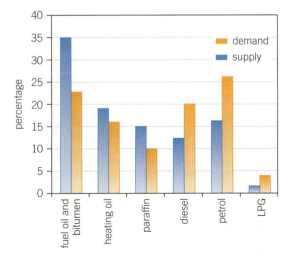

Figure 14 *Crude oil contains different fractions. Some of these are in greater supply than the demand for them. The pairs of bars are separated here to make the data easier to compare.*

Table 4 *Independent variable (first column) and dependent variable (second column).*

Blood group	Frequency
A	10
B	3
AB	1
O	11

Table 5 *Frequency table for resting pulse rate in a class of students.*

Pulse rate, r (beats per minute)	Frequency
$60 \leq r < 64$	1
$65 \leq r < 69$	4
$70 \leq r < 74$	12
$75 \leq r < 79$	8
$80 \leq r < 84$	5
$85 \leq r < 90$	1

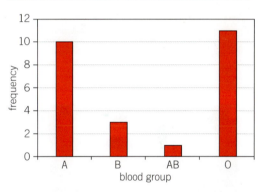

Figure 12 *A bar chart to represent the blood group data. The bars are of equal width and separated from one another.*

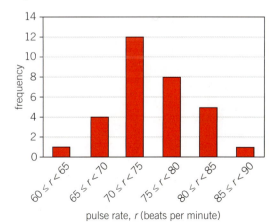

Figure 13 *A bar chart to represent the pulse rate data in Table 5.*

7 Mean values

To calculate the mean of a series of values:

Step 1: add together all the values in the series to get a total

Step 2: divide the total by the number of values in the data series.

Calculating a mean

A student repeated a titration several times. Her concordant titres (those close enough to be valid) were:

$$24.50 \, cm^3, \ 24.45 \, cm^3, \ 24.55 \, cm^3$$

Calculate the mean titre.

Step 1: Add together the recorded titres.

$$24.50 \, cm^3 + 24.45 \, cm^3 + 24.55 \, cm^3 = 73.50 \, cm^3$$

Step 2: Divide by the number of recorded values (in this case, there are three titres).

$$\frac{73.50 \, cm^3}{3} = 24.50 \, cm^3 \text{ (to 2 decimal places)}$$

The mean titre for the student's concordant results was $24.50 \, cm^3$.

8 Estimates and order of magnitude

Being able to make a rough estimate is helpful. It can help you to check that a calculation is correct by knowing roughly what you expect the answer to be. A simple estimate is an **order of magnitude** estimate, which is an estimate to the nearest power of 10. For example, to the nearest power of 10, you are probably 1 m tall and can run 10 m/s.

So, you, your desk, and your chair are all of the order of 1 m tall (Figure 15). The diameter of a molecule is of the order of 1×10^{-9} m, or 1 nanometre (nm).

9 Mathematical symbols

You have used lots of different symbols in maths, such as +, −, x, ÷. Table 6 shows other symbols that you might meet in science.

Table 6 *Maths symbols used in science.*

Symbol	Meaning	Example
=	equal to	2 m/s × 2 s = 4 m
<	is less than	the mean height of a child in a family < the mean height of an adult in a family
<<	is very much less than	the diameter of an atom << the diameter of an apple
>>	is very much bigger than	the diameter of the Earth >> the diameter of a pea
>	is greater than	the pH of an alkali > the pH of an acid
∝	is proportional to	*F* (force) ∝ *x* (extension) for a spring
~	is approximately	272 m ~ 300 m (see 8 Estimates and order of magnitude)

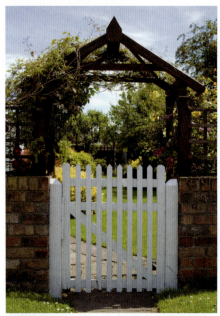

Figure 15 *These are all about 1 m high.*

10 Changing the subject of an equation

An equation shows the relationship between two or more variables. You can change an equation to make *any* of the variables become the subject of the equation.

To change the subject of an equation, you can do an opposite (inverse) operation to both sides of the equation to get the variable that you want on its own.

This means that:

- subtracting is the opposite of adding (and adding is the opposite of subtracting)
- dividing is the opposite of multiplying (and multiplying is the opposite of dividing)
- taking the square root is the opposite of squaring (and squaring is the opposite of taking the square root).

Kinetic energy

Change the equation kinetic energy $(KE) = \frac{1}{2}mv^2$ to make v the subject (Figure 16).

Step 1: Multiply by 2 and divide by m to get the v^2 on its own. Do the same on both sides of the equation.

$$2 \times KE = 2 \times \frac{1}{2}mv^2$$

$$2 \times KE = mv^2$$

$$\frac{2 \times KE}{m} = \frac{mv^2}{m}$$

$$\frac{2 \times KE}{m} = v^2, \quad \text{so} \quad v^2 = \frac{2 \times KE}{m}$$

Step 2: Take the square root of both sides.

$$v = \sqrt{\frac{2 \times KE}{m}}$$

Figure 16 *If you know the kinetic energy and mass of this roller coaster you can work out the speed.*

11 Quantities and units

11.1 International system of units

When you make a measurement in science you need to include a number *and* a unit. This is one of the differences between numbers in maths (which don't need units), and measurements in science (which do).

When you do a calculation your answer should also include both a number *and* a unit. There are some special cases where the units cancel, but usually they do not.

Everyone doing science, including you, needs to use the **international system of units (SI units)**. There are seven base units, but only six are commonly used (Figure 17). All other units are derived (worked out) from these base units.

Here are some of the quantities that you will use, along with their units.

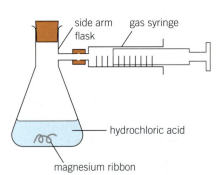

side arm
flask
gas syringe

hydrochloric acid

magnesium ribbon

Figure 17 *Time is measured in s and volume in dm³ (but cm³ is often more convenient).*

Table 7 *Base units and derived units.*

Quantity	Base unit
distance	metre, m
mass	kilogram, kg
time	second, s
current	ampere, A
temperature	kelvin, K
amount of substance	mole, mol

Quantity	Derived unit
charge	coulomb, C
concentration	mole per cubic decimetre, mol/dm³
density	kilogram per cubic metre, kg/m³
energy	joule, J
potential difference	volt, V
pressure	pascal, Pa
volume	cubic decimetre, dm³

For example, 1.5 J is a *measurement*. The number 1.5 is not a measurement because it does not have a unit.

Some quantities that you *calculate* do not have a unit because they are a ratio – for example, the relative atomic mass.

11.2 Using units in equations

When you put quantities into an equation it is best to write the number *and* the unit. This helps you to work out the unit of the quantity that you are calculating.

Speed = distance/time

A sprinter can run 100 m in 10 s. Calculate the speed of the sprinter.

Step 1: Write down what you know.

$$distance = 100 \, m$$

$$time = 10 \, s$$

Step 2: Write down the equation you need.

$$speed \; s \; (m/s) = \frac{distance \; d \; (m)}{time \; t \; (s)}$$

Step 3: Do the calculation and include the units.

$$speed = \frac{100 \, m}{10 \, s}$$

$$= 10 \, m/s$$

m/s are the units of speed.

12 Metric prefixes

You can use **metric prefixes** to show large or small multiples of a particular unit. Adding a prefix to a unit means putting a letter in front of the unit, for example km. It shows you that you should multiply your value by a particular power of 10 for it to be shown in an SI unit.

For example, 3 millimetres = 3 mm = 3×10^{-3} m. Most of the prefixes that you will use in science involve multiples of 10^3:

Table 8 *Prefixes.*

Prefix	tera	giga	mega	kilo		deci	centi	milli	micro	nano
Symbol	T	G	M	k		d	c	m	μ	n
Multiplying factor	10^{12}	10^9	10^6	10^3		10^{-1}	10^{-2}	10^{-3}	10^{-6}	10^{-9}

12.1 Converting between units

It is helpful to use standard form when you are converting between units. To do this, it's best to consider how many of the 'smaller' units are contained within one of the 'bigger' units. For example:

- There are 1000 mm in 1 m. So 1 mm = $\frac{1}{1000}$ m = 10^{-3} m.

- There are 1000 m in 1 km. So 1 km = 1000 m = 10^3 m.

13 Data and graphs

During your GCSE course you will collect data in different types of experiment or investigation (Figure 18). The data will be

- from an experiment where you have changed *one* independent variable (or allowed time to change) and measured the effect on a dependent variable (Figure 19)

- from an investigation where you have collected data about *two* independent variables to see if they are related.

13.1 Collecting data by changing a variable

In many practical experiments you change one variable (the independent variable) and measure the effect on another variable (the dependent variable).

You plot the data on a graph, called a line graph. If the gradient of the line of best fit is:

- **positive** then as the independent variable gets *bigger* the dependent variable gets *bigger*

- **negative** then as the independent variable gets *bigger* the dependent variable gets *smaller*

- **zero** then changing the independent variable has no effect on the dependent variable.

You say that there is a positive relationship, negative relationship, or no relationship between the variables.

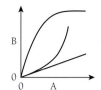

in all these graphs, if A increases then B increases

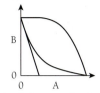

in all these graphs, if A increases then B decreases

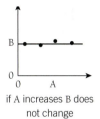

if A increases B does not change

Figure 18 *Different relationships between variables.*

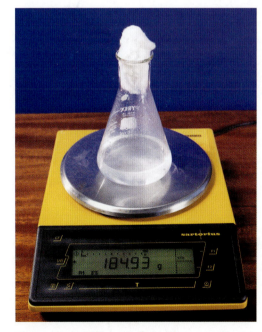

Figure 19 *In this rate of reaction experiment, time is the independent variable and mass is the dependent variable.*

14 Graphs and equations

If you are changing one variable and measuring another you are trying to find out about the relationship between them. A straight line graph tells you about the mathematical relationship between variables, but there are other things that you can calculate from a graph.

14.1 Straight line graphs

The general equation for a straight line is $y = mx + c$, where m is the **gradient** and c is the point on the y-axis where the graph intercepts, called the y-intercept.

Straight line graphs that go through the origin (0, 0) are special. For these graphs, y is directly proportional to x, and $y = mx$ (Figure 20).

When people say 'plot a graph' they usually mean plot the points then draw a line of best fit.

When you describe the relationship between two *physical* quantities, you should think about the reason why the graph might (or might not) go through (0, 0).

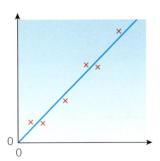

Figure 20 *A line of best fit that passes through the origin.*

Reactions and gases

Magnesium powder reacts with dilute hydrochloric acid, releasing hydrogen gas.

A line of best fit for a graph of volume of gas (y-axis) against mass of magnesium in g (x-axis) is a straight line through (0, 0). Explain what the gradient shows in this context, and state why the graph goes through (0, 0).

Step 1: Match the equation to $y = mx$ to work out what the gradient means.

> The gradient gives you the volume of hydrogen gas produced per gram of magnesium.

Step 2: Think about what happens to x when the y quantity is zero.

> The line goes through (0, 0) because when there is no magnesium to react with the acid, no hydrogen is produced.

14.2 Calculations using straight line graphs

When you draw a graph you choose a scale for each axis.

- The scale on the x-axis should be *the same* all the way along the x-axis, but it can be *different* to the scale on the y-axis.

- Similarly, the scale on the y-axis should be *the same* all the way along the y-axis, but it can be *different* to the scale on the x-axis.

- Each axis should have a label and a unit, such as 'time (s)'.

You often need to calculate a gradient from a graph. This may represent important physical quantities. Its units can give you a clue as to what it represents.

Calculate a gradient

Calculate the gradient of the graph in Figure 21 between 0 s and 10 s.

Step 1: Select two points on the *y*-axis and subtract one from the other:

8 m/s − 0 m/s = 8 m/s

Step 2: Select two points on the *x*-axis and subtract one from the other:

10 s − 0 s = 10 s

Step 3: Find the gradient between 0 s and 10 s.

$$\text{Gradient} = \frac{\text{change in } y}{\text{change in } x}$$

$$\text{Gradient} = \frac{(8\,\text{m/s} - 0\,\text{m/s})}{(10\,\text{s} - 0\,\text{s})} = 0.8\,\text{m/s}^2$$

The unit m/s² is the unit of acceleration. Therefore the gradient of a speed–time graph shows acceleration.

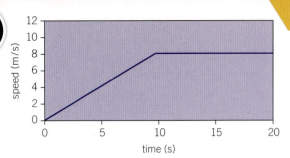

Figure 21 *A bike ride.*

You can also find the **y-intercept** of a graph. This is the value of the quantity on the *y*-axis when the value of the quantity on the *x*-axis = 0.

The meaning of the *y*-intercept depends on the quantities that you plot on your graph.

14.3 Graphs with time on the *x*-axis

For all graphs where the quantity on the *x*-axis is time, the gradient will tell you the **rate of change** of the quantity on the *y*-axis with time. For example, in chemistry the rate of change of the volume of a gas (*y*-axis) with time (*x*-axis) can tell you the **reaction rate**.

14.4 Graphs that are *not* a straight line

When you plot a graph of the relationship between certain variables, you may not get a straight line.

However, you may still need to find the gradient of the graph.

To find the **gradient** at a point, such as P in Figure 22, you need to draw a **tangent** to the curve at that point:

- Put a ruler against point P on the curve.
- Adjust the angle of the ruler so that near P it is an equal distance from the curve either side of P.
- Draw a line using the ruler.
- Make a right-angled triangle with your line as the hypotenuse.

The triangle can be as big as you like, but should be large enough for you to calculate sensible changes in values. A minimum of 8 cm works well.

The gradient is the *change in y* (the vertical side) divided by *the change in x* (the horizontal side). The Greek letter delta, or Δ, is used to mean change, so you can write this as Δy/Δx.

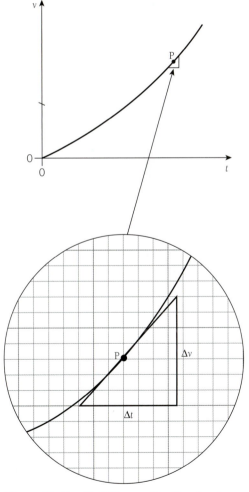

Figure 22 *You find the gradient by drawing a tangent.*

If the horizontal axis is time, the gradient will be equal to the rate of change of the variable on the vertical axis. For a chemical reaction, a graph of mass or volume against time will give you the reaction rate.

15 Areas and volumes
15.1 Surface areas

Use these expressions to calculate the surface area of regular 2D objects:

- area of a rectangle = length × width (this also works for a square)
- area of a triangle = $\frac{1}{2}$ × base × height (this works for any triangle, Figure 23)
- area of a circle = π × radius^2 (the radius is half the diameter)

You can estimate the surface area of irregular shapes by counting squares on graph paper. This method is useful for estimating the area of a leaf, for example, if you trace the leaf onto graph paper.

The surface area of a three-dimensional (3D) object is equal to the total surface area of all its faces. In a **cuboid**, the areas of any two opposite faces are equal. This allows you to calculate the surface area of the cuboid without having to draw a net.

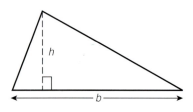

Area = $\frac{1}{2}$ hb

Figure 23 *Area of a triangle = $\frac{1}{2}$ × b × h.*

Calculating the surface area of a nanoparticle

A nanoparticle in the shape of a cuboid measures 15 nm × 20 nm × 80 nm. Calculate its surface area.

Step 1: Calculate the area of each face.

area of face 1 = 15 nm × 20 nm = 300 nm^2

area of face 2 = 15 nm × 80 nm = 1200 nm^2

area of face 3 = 20 nm × 80 nm = 1600 nm^2

Step 2: Calculate the total area of the three different faces.

area = 300 nm^2 + 1200 nm^2 + 1600 nm^2 = 3100 nm^2

Step 3: Multiply the answer to Step 2 by 2 because the opposite sides of a cuboid have equal areas.

total surface area = 2 × 3100 nm^2 = 6200 nm^2

You need three steps to calculate the surface area of a cylinder:

Step 1: Calculate the total area of the two circular ends using the equation

area = 2 × π × radius^2

Step 2: Calculate the area of the curved surface using the equation

area = 2 × π × radius × length

Step 3: Add the answers to Steps 1 and 2 together.

15.2 Volumes

Use this expression to calculate the volume of a cuboid:

volume of cuboid = length × width × height

You can calculate the volume in different units depending on the units of length, width, and height.

Calculating the volume of a cuboid

Calculate the volume of a glass block of length 15 cm, width 6 cm, and depth 1.5 cm in cm^3 and m^3 (Figure 24).

Step 1: Calculate the volume using the equation

volume = length × width × height

= 15 cm × 6 cm × 1.5 cm

= 135 cm^3 = 140 cm^3 (to 2 sig. fig.)

Step 2: Convert the measurements to metres

length = 0.15 m

width = 0.06 m

height = 0.015 m

Step 3: Use the equation to calculate the volume.

volume = length × width × height

= 0.15 m × 0.06 m × 0.015 m

= 0.000135 m^3 = 0.00014 m^3 (to 2 sig. fig.)

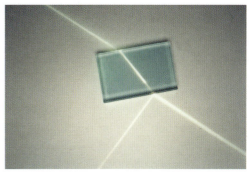

Figure 24 *Glass blocks are used in science experiments.*

16 2D representations of 3D objects

3D objects have width, height, and depth. 2D objects only have width and height. You cannot accurately show all aspects of a 3D object in a diagram. As such, it is easier to draw laboratory apparatus in 2D. Figures 25 and 26 show the same apparatus drawn in different ways.

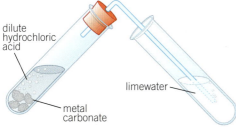

Figure 25 *A perspective drawing to give a sense of three dimensions.*

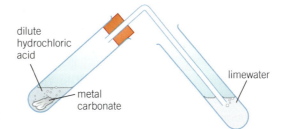

Figure 26 *A cross-section drawing with no attempt to represent three dimensions.*

Other 2D representations of 3D objects include:

- nets (2D shapes that can be cut out and folded to make 3D shapes)
- elevation views (showing objects from a side)
- plan views (showing objects from above or below).

Glossary

accurate A measure of how close a result is to the true value. An accurate result is very close.

acid A substance that ionises in water, releasing hydrogen ions.

acid rain Rain containing acidic gases from the air that make an acidic solution, which falls as rain.

acidic Having a pH value less than 7.

activation energy The minimum amount of energy needed for a reaction to start.

actual yield The mass of a product obtained in a chemical reaction.

addition polymer Substance consisting of molecules made from many repeat units, formed by addition reactions.

addition reaction Reaction in which an atom or group of atoms combines with a molecule to form a larger molecule, with no other product.

alcohol Member of a homologous series of organic compounds containing the hydroxyl group –OH.

algae Organisms that can carry out photosynthesis but that do not have roots, stems, or leaves.

alkali A soluble base.

alkali metal An element in Group 1 of the Periodic Table.

alkaline Having a pH value greater than 7.

alkane Saturated hydrocarbon with only single covalent bonds.

alkene Unsaturated hydrocarbon containing a carbon–carbon double bond.

allotropes Forms of an element in the same state but with different atomic arrangements.

alloy A mixture of a metal and at least one other element.

amino acid Monomer found in proteins.

anhydrous Completely dry or having no water of crystallisation.

anion An ion discharged at the anode during electrolysis.

anode The positive electrode used in electrolysis.

anthropogenic Changes produced as a result of human activities.

aqueous solution A solution in which the solvent is water.

aquifer Layer of rock that stores water underground.

atom Smallest particle of an element. Atoms have a nucleus containing protons and (usually) neutrons, surrounded by electrons.

atom economy Proportion of atoms in the reactants that end up in the desired product. It is given by (sum of M_r of desired product ÷ sum of M_r of all products) × 100.

atomic number Number of protons in the nucleus of an atom.

atomic radius Half the diameter of an atom.

Avogadro constant The number of entities in one mole, equal to the number of atoms in 12.0 g of ^{12}C atoms, 6.02×10^{23}.

backward reaction Reaction identified as occurring from right to left in the balanced equation for a reversible reaction.

balanced chemical equation A model for a reaction showing formulae and number of units for all substances involved.

balanced equation A model for a reaction showing formulae and number of units for all substances involved. The numbers of each atom are the same on both sides of the equation.

ball-and-stick model A representation of ionic compounds and molecules in which balls represent the atoms or ions, and springs or plastic links represent chemical bonds.

bar chart A way of presenting data when one variable is discrete or categoric and the other is continuous. Numerical values of variables are represented by the height or length of lines or rectangles of equal width.

base A substance that neutralises acids forming a salt and water only.

batch process Chemical process that makes products in limited amounts at different times.

bauxite Aluminium ore containing aluminium oxide.

binary ionic compound An ionic compound of two elements.

biofuel Fuel obtained from renewable materials produced by living organisms. Bioethanol and biodiesel are biofuels.

bioleaching Biological method of metal extraction in which bacteria speed up reactions that release soluble metal compounds from metal sulfides.

blast furnace Industrial reaction vessel for iron production.

block diagram Diagram modelling substances in which the general shape of the molecule is shown.

boil To change state from the liquid state to the gas state.

boiling point The temperature at which a substance changes from the liquid state to the gas state.

bond Attractive force between two particles.

bond energy The amount of energy that must be transferred to break a mole of a particular covalent bond.

bond length Distance between the centres of two atoms that are chemically joined together.

brittle When a substance is likely to crack or shatter when hit.

buckyball A type of fullerene molecule with a spherical or squashed spherical shape.

buffer A solution that resists changes in pH.

burette A graduated tube with a tap, used in titrations.

by-product A substance formed in a reaction in addition to the desired product. By-products may or may not be useful.

calibrate To adjust the reading on an instrument to match the observed value with the value of the quantity being measured.

carbohydrate Compounds containing carbon, hydrogen, and oxygen.

carbonates Ionic compounds that contain the carbonate ion.

carbon capture Process by which carbon dioxide emissions are collected and stored, and not allowed into the atmosphere.

carbon-neutral Refers to a process or activity in which there is no overall release of carbon dioxide to the atmosphere.

carboxyl group Functional group in carboxylic acids.

carboxylic acid Member of a homologous series of organic compounds containing the carboxyl group –COOH.

carrier gas An unreactive or inert gas used as the mobile phase in gas chromatography.

catalyse To speed up a reaction using a catalyst.

catalyst A substance that increases the rate of reaction, without being used up, by lowering the activation energy needed.

categoric variable A variable that can take on one of a limited, and usually fixed, number of possible values.

cathode The negative electrode used in electrolysis.

cation An ion discharged at the cathode during electrolysis.

ceramic Hard, non-metallic material such as brick or china.

chemical cell Device that produces a potential difference from chemical reactions.

chemical change Change that produces new substances.

chemical formula A description of a compound or an element that uses symbols to represent the atoms present. Numbers show if more than one atom of an element is present.

chemical property A feature of a substance that can only be determined by studying its chemical reactions.

chemical reaction Process in which substances react to form different substances.

chemical symbol Letter or letters used by chemists to represent an element, e.g. the symbol for mercury is Hg.

chromatogram The pattern produced when separating a mixture using chromatography.

chromatography A separation method that relies on the distribution of a substance between a mobile phase and a stationary phase.

climate change Long-term change in weather patterns.

closed system Apparatus in which substances cannot enter or leave the reaction mixture during a reaction.

collide To come into contact with something.

coke Product, mainly carbon, made by heating coal in the absence of air.

combustion Reaction in which a substance burns in oxygen.

complete combustion Burning in a plentiful supply of oxygen.

complex carbohydrate Polymer made from sugar monomers.

composite material Material made from two or more different materials, with different properties to those of the materials.

compound ion An ion formed when a group of atoms loses or gains electrons.

compressive strength Measure of how strong a material is when squashed.

concentrated When a solution contains a relatively high amount of solute.

concentration A measure of the amount of dissolved solute present in a solution.

concordant titres Results from a titration that are within $0.10\,cm^3$ of each other.

condensation polymer Substance with molecules made from many repeat units, formed by condensation reactions.

condensation reaction Reaction in which two molecules react together to form one larger molecule and one smaller molecule, usually water or hydrogen chloride.

condense To change from the gas state to the liquid state.

condensed Turned from the gas state to the liquid state.

condenser Apparatus that can cool and condense a substance.

conductor Able to carry electrical or thermal energy.

Contact process Industrial process to make sulfuric acid.

continuous process Chemical process that makes products in large amounts all the time.

continuous variable A variable that has values that can be any number between a maximum and a minimum.

control variable A variable that you have to keep the same in an investigation.

convection Movement of particles in a fluid owing to differences in density caused by temperature differences.

correlation A relationship where there is a link between two variables.

corrosion Reaction of a metal with substances in its surroundings, such as air or water.

covalent bond A strong force of attraction between the nuclei of two atoms that are sharing one or more pairs of electrons.

cracking Process of converting large alkanes into smaller alkanes and alkenes using high temperatures and a catalyst.

crude oil Complex mixture of hydrocarbons, mostly alkanes, formed from the remains of dead organisms over millions of years.

cryolite Compound in which aluminium oxide dissolves; it allows the electrolysis of aluminium oxide to happen at lower temperatures than if pure aluminium oxide were used.

crystalline The structure of substances in the solid state that have particles arranged in a giant lattice.

crystallisation The process by which crystals are formed during evaporation of a solvent from a solution.

cuboid A solid that has six rectangular faces at right angles to each other.

data Sets of values for variables.

decimal form Numbers that are between whole numbers can be written in decimal form, for example, 5.1 or 6.72.

delocalised electron A free electron in a molecule, ion, or solid metal that is not part of an individual atom or bond.

denatured When an enzyme's active site is damaged by high temperatures or extremes of pH, and catalytic activity stops.

density Mass per unit volume, measured in g/cm^3 or kg/m^3. An object with a high density is dense and feels heavy for its size.

dependent variable A variable that changes when you change the independent variable.

desalination Process of removing dissolved salts from water.

diamond A form of carbon in which each atom is covalently bonded to four other carbon atoms in a giant covalent structure.

diatomic molecule Molecule containing two atoms.

diffusion Net movement of particles from a place where they are in a high concentration, to one at a low concentration.

dilute A solution with a relatively low amount of solute.

directly proportional When the rate of increase of one variable is the same as that of another variable.

discharged In electrolysis, the gain or loss of electrons by an ion to become an atom.

discrete variable A variable that can only have whole-number values.

displace To replace or push out an element in a compound.

displacement Reaction in which a more reactive element displaces a less reactive element from its compounds.

displayed formula A diagram in which atoms are represented by chemical symbols, and covalent bonds by lines.

dissolve The process in which a solute and solvent mix completely to form a solution.

DNA Biological polymer made from nucleotide monomers. The sequence contains all the information needed to make an organism.

dot-and-cross diagram Model in which electron shells are shown as circles, with electrons as dots or crosses.

ductile Able to be drawn into a wire without breaking.

dynamic In terms of an equilibrium, this means that the reactions are still occurring.

economic impacts In science, the effects of an application of science that are to do with money.

effervescence Bubbling or fizzing observed in a chemical reaction.

electrode An electrical conductor used in electrolysis.

electrolysis A process in which an electric current is passed through a compound, causing a chemical change.

electrolyte A compound in its liquid state or in solution that contains ions and conducts electricity.

electron Subatomic particle surrounding the nucleus of an atom. It has a relative charge of -1, and a very small mass.

electron diagram A model of an atom or ion in which shells are represented by circles, and electrons by dots or crosses.

electronic structure The arrangement of the electrons in an atom.

electroplating Coating an object with a metal by electrolysis.

electrostatic forces Forces of attraction or repulsion between electrically charged particles.

element Substance whose atoms have the same atomic number.

empirical formula Formula showing the simplest whole-number ratio of the atoms of each element in a compound.

endothermic Describes a process in which there is an overall transfer of energy from the surroundings.

end-point The point at which the two reactants in a titration have exactly reacted together.

energy change The difference between the energy transferred from the surroundings to break bonds in the reactants, and the energy transferred from products when bonds form.

environmental Relating to the natural world and the impact of human activity on its condition.

enzyme Protein that acts as a catalyst in biological systems.

equilibrium Situation in a reversible reaction in a closed system in which the rates of the forward reaction and backward reaction are equal, and the concentrations of the products and reactants remain constant.

equilibrium position Description of the relative amounts of reactants and products in a reaction mixture at equilibrium.

equilibrium yield Amount of the desired product at equilibrium.

essential element One of the elements required for plant growth, including nitrogen, phosphorus, and potassium.

ethical issues Relating to moral principles or the branch of knowledge dealing with these.

evaporate When a substance turns from the liquid state to the gas state at a temperature below the boiling point.

excess When a reactant is present in an amount greater than that needed to react with the other reactant.

exothermic Describes a process in which there is an overall transfer of energy to the surroundings.

fair test An investigation in which all the variables are kept constant except the variable that the investigator changes and the variable that is measured.

fermentation Reaction in which ethanol and carbon dioxide are produced from glucose solution using enzymes from yeast.

fertiliser Substance added to soil to replace nutrients or minerals used by growing plants.

filtrate Liquid that passes through the filter during filtration.

filtration The process by which insoluble substances are separated from soluble substances using a filter.

finite resources Substances that are no longer being made or that are being made extremely slowly.

flame test A test that helps to identify metal ions in compounds from the colour they produce in a flame.

forward reaction Reaction identified as occuring from left to right in the balanced equation for a reversible reaction.

fossil fuel Fuel formed from the remains of dead organisms over millions of years, e.g., crude oil, coal, natural gas.

fraction In chemistry, substance separated during fractional distillation.

fractional distillation Method for separating a mixture of liquids with different boiling points into different fractions.

fractionating column A piece of apparatus used to improve the separation of solvents during fractional distillation.

freeze To change state from the liquid state to the solid state.

frequency table A frequency table shows the number of times that each categoric variable appears in a sample.

fuel cell Device that produces electricity by the reaction of a fuel with oxygen, without combustion.

fullerenes Family of carbon allotropes with tubes or balls of atoms.

functional group Atom, group of atoms, or type of bond that determines the chemical reactions of an organic compound.

galvanising Coating iron or steel with a thin layer of zinc.

gas chromatography A type of chromatography that uses silica or alumina packed into a metal tube as the stationary phase, and an unreactive or inert carrier gas as the mobile phase.

gas syringe Apparatus for measuring gas volumes.

general formula Chemical formula showing the relative number of atoms of each element in a compound.

giant covalent structure An arrangement of non-metal atoms joined by covalent bonds in a regular arrangement.

giant ionic lattice A structure of ionic compounds in which oppositely charged ions are held in a regular and repeating arrangement by strong electrostatic forces of attraction.

giant metallic lattice The repeating regular arrangement of metal atoms or ions in a metal in the solid state.

global warming Worldwide increase in temperature over time.

gradient The degree of steepness of a graph at any point.

graphene Carbon allotrope resembling a single layer of graphite.

graphite An form of carbon in which each atom is covalently bonded to three other carbon atoms to form a giant covalent structure in layers.

greenhouse effect Interaction of infrared radiation with molecules in the atmosphere, reducing the transfer of energy to space.

greenhouse gas Gas in the atmosphere that reduces the transfer of energy to space by infrared radiation.

group In the Periodic Table, a column of elements with similar chemical properties and the same number of electrons in the outer shells of their atoms.

Group 0 A vertical column of elements on the far right-hand side of the Periodic Table, containing unreactive non-metals.

Group 1 (IUPAC Group 17) A vertical column of elements on the far left-hand side of the Periodic Table, containing reactive metals.

Group 7 (IUPAC Group 17) A vertical column of elements on the right-hand side of the Periodic Table, containing reactive non-metals.

Haber process Industrial process to make ammonia.

haematite Iron ore that contains iron(III) oxide.

half equation A type of chemical equation that models the change that happens to one reactant in a reaction.

halide A compound containing a Group 7 (IUPAC Group 17) element and one other element, usually hydrogen or a metal.

halide ion A negative ion formed by a Group 7 (IUPAC Group 17) element.

halogen An element in Group 7 (IUPAC Group 17) of the Periodic Table.

hazard A possible source of danger.

high-grade ore Rock with a high concentration of a metal.

homologous series Family of organic compounds with the same functional group and general formula, similar chemical reactions, and similar trends in their physical properties.

hydrated Containing water or having water of crystallisation.

hydration In chemistry, a reaction in which water is added.

hydrocarbon Compound of hydrogen and carbon only.

hydrogen bond Type of intermolecular force. Hydrogen bonds are stronger than most intermolecular forces.

hydroxyl group Functional group, –OH, found in alcohols.

hypothesis An idea that is a way of explaining scientists' observations.

impure substance Material consisting of two or more different elements and/or compounds.

incomplete combustion Burning in a limited oxygen supply.

independent variable A variable you change that changes the dependent variable.

inert Very unreactive and unable to form compounds.

inert electrode Unreactive electrodes that do not change during electrolysis.

infrared radiation Electromagnetic radiation with a longer wavelength than visible light, used in cooking.

ingot Metal cast into a shape suitable for storage, transport, or further processing.

insoluble Describes a substance that will not dissolve.

instrumental method of analysis A technique for analysing a substance that depends upon a machine.

insulator Unable to carry electrical or thermal energy well.

intermolecular forces Weak forces of attraction that exist between simple molecules.

international system of units (SI units) The Système International d'Unités, or SI units, is the agreed set of units that we use in science.

inversely proportional When the rate of increase of one variable is the same as the rate of decrease of another variable.

ion Charged particle formed when an atom, or group of atoms, loses or gains electrons.

ionic bond Strong electrostatic force of attraction between oppositely charged ions.

ionic compound A compound containing oppositely charged ions from different elements.

ionic equation A type of chemical equation that models how oppositely charged ions form an ionic compound.

isotope Atoms with the same number of protons and electrons, but different numbers of neutrons, are isotopes of the same element.

IUPAC Group Numbering system, 1 to 18, for groups in the Periodic Table.

landfill Waste disposal where waste is put into the ground.

lattice Regular arrangement of atoms, ions, or molecules.

law of conservation of mass The principle that states that the total mass stays the same during a chemical reaction, because atoms are not created or destroyed during a chemical reaction.

Le Chatelier's principle Rule for determining the effect on the equilibrium position of a change in conditions.

life-cycle assessment 'Cradle-to-grave' analysis of the impact of making, using, and disposing of a manufactured product.

limewater Alkaline calcium hydroxide solution. It turns cloudy white in the presence of carbon dioxide.

limiting reactant A reactant present in an amount less than that needed to react completely with the other reactant in a chemical reaction.

line graph A way of presenting results when there are two numerical values.

line of best fit A smooth line on a graph that travels through or very close to as many of the points plotted as possible.

liquefied Turned into the liquid state, usually from the gas state.

litmus paper Blue litmus paper changes to red in an acidic solution, and red litmus paper changes to blue in an alkaline solution.

low-grade ore Rock with a low concentration of a metal.

malleable Able to be bent, hammered, or pressed into shape without cracking or shattering.

mass number Number of protons and neutrons in a nucleus.

mass spectrometer A machine that can measure masses of atoms and molecules.

matter Anything that has mass.

measuring cylinder A piece of apparatus used to measure different volumes of solution.

melt To change state from the solid state to the liquid state.

melting point The temperature at which a substance changes from the solid state to the liquid state.

meniscus The curve in the surface of a liquid.

metal A substance that is usually shiny, malleable, ductile, and a good electrical or thermal conductor.

metallic bond The strong electrostatic force of attraction between delocalised electrons and metal ions.

metric prefixes A symbol used to show multiples of a unit, such as the k in km.

mineral deficiency Situation in which an animal or plant does not receive enough of a mineral for healthy growth.

mixture Material consisting of two or more different substances, not chemically joined together.

mobile phase A substance in the liquid or gas state that moves during chromatography.

model Description, analogy, or equation that helps you to explain the physical world.

molar mass The mass in grams of one mole of a substance.

molar volume The volume occupied by one mole of a gas ($24\,dm^3$ at room temperature and pressure).

mole Amount of substance that contains the same number of particles (6.02×10^{23}) as there are atoms in $12.0\,g$ of ^{12}C.

molecular formula A description of a compound or an element that uses symbols for atoms, and numbers to show the actual number of atoms of each element in a molecule.

molecular ion Positive ion formed in a mass spectrometer when an electron is removed from an atom or molecule.

molecular ion peak The peak in a mass spectrum on the far right, corresponding to the molecular ion. Its mass-to-charge ratio is equal to the relative formula mass of the molecule.

molecule Particle consisting of two or more non-metal atoms chemically joined together by covalent bonds.

molten In the liquid state.

monatomic Existing as single atoms, such as the noble gases.

monomer Small molecules that can join together to form polymer molecules.

nanoparticle A particle between 1 nm and 100 nm across, containing just a few hundred atoms.

nanoparticulate Composed of nanoparticles.

nanotube A type of fullerene molecule in which carbon atoms form a hollow tube.

neutral A solution, not acidic or alkaline, with a pH of 7.

neutralisation A reaction between an acid and a base or alkali, producing a salt plus water.

neutron Subatomic particle found in the nucleus of an atom. It has no charge and a relative mass of 1.

noble gas An element in Group 0 (IUPAC Group 18) of the Periodic Table.

non-enclosed system Apparatus in which substances can enter or leave the reaction mixture during a reaction.

non-inert electrode An electrode that changes during electrolysis.

non-metal A substance that is usually dull, brittle, and a poor electrical or thermal conductor.

non-renewable Used faster than it can be replaced.

normal distribution A function that represents the distribution of many random variables as a symmetrical bell-shaped graph.

nucleotide Monomer found in DNA, consisting of an organic base, ribose sugar, and a phosphate group.

nucleus Positively charged central part of an atom, made up of protons and neutrons.

order of magnitude A number to the nearest power of ten.

ore Rock or mineral that contains enough metal to make it economical to extract the metal.

organic compound Substance consisting of molecules that contain carbon bound to a hydrogen, and possibly other elements.

outer shell The outermost occupied shell in an atom.

outlier A result that that is very different from the other measurements in a data set.

oxidation A gain of oxygen, or loss of electrons, by a substance in a chemical reaction.

oxide A type of compound formed when an element or compound gains oxygen.

oxidised Describes a substance that has gained oxygen, or lost electrons, in a chemical reaction.

oxidising agent A substance that oxidises another substance by donating oxygen or by accepting electrons.

paper chromatography Separation method that uses paper as the stationary phase and a solvent in the liquid state as the mobile phase.

particle Tiny piece of matter such as an atom, ion, or molecule.

particle model Scientific idea used to explain the properties of solids, liquids, and gases.

particulate Small particles, mostly carbon (soot) formed by incomplete combustion.

peer review Peer review is the checking and evaluation of a scientific paper by other expert scientists in order to help decide whether or not the paper should be published.

percentage A rate, number, or amount in each hundred.

percentage yield The proportion of the theoretical yield actually obtained during a reaction, expressed as a percentage. It is given by (actual yield ÷ theoretical yield) × 100.

period In the Periodic Table, a row in which the atomic number increases by 1 going from one element to the next.

Periodic Table A table in which the elements are arranged in rows (periods) and columns (groups), in order of increasing atomic number.

pH The relative acidity or alkalinity of an aqueous solution.

pH scale A scale that measures relative acidity or alkalinity of an aqueous solution, from 0 (strongly acidic) through 7 (neutral), to 14 (strongly alkaline).

pH titration curve Graph in which pH is plotted against volume of acid added to an alkali, or volume of alkali added to an acid.

phase In chemistry, a substance in the solid, liquid, or gas state.

photosynthesis Reaction in the cells of plants and algae that produces glucose and oxygen from carbon dioxide and water.

physical change Change, such as changes of state, that does not result in new substances being made.

physical property Feature of a substance that can be observed or measured, such as its melting point or colour.

phytoextraction Biological method of metal extraction in which plants absorb metals through their roots and concentrate them in their cells.

pipette filler Device to draw liquids into a pipette safely.

plum-pudding model Outdated model of the atom in which electrons are embedded in a sphere of positive charge.

pollutant Substance released into the environment that may cause harm to living things.

polyamide Condensation polymer containing many –CONH– (amide) groups.

polyester Condensation polymer containing many –COO– (ester) groups.

polymer Substance with molecules made from many repeat units.

polymerisation reaction Reaction in which small monomer molecules join together to make larger polymer molecules.

polythene Common name for poly(ethene), an addition polymer used for plastic bags.

potable water Water that is safe for drinking.

potential difference Difference in electrical potential produced by the separation of charge.

precipitate An insoluble product in the solid state, formed during a reaction involving solutions.

precipitation A type of reaction in which a precipitate forms.

precise This describes a set of repeat measurements that are close together.

precision A measure of how close the agreement is between measured values.

prediction A statement that says what you think will happen.

prefix In science, placing a letter before a unit.

pressure Force per unit area. Air pressure increases when you pump air into a tyre.

product The substance(s) formed in a chemical reaction.

protein Biological polymer made from amino acid monomers.

proton Subatomic particle found in the nucleus of an atom. It has a relative charge of +1 and a relative mass of 1.

pure substance Consisting of just one element or compound.

purity A measure of how pure a substance is. A 100% pure substance consists entirely of one element or compound.

qualitative Data that are descriptive or difficult to measure.

quantitative Data that are obtained by making measurements.

random error An error that causes there to be a random difference between measurement and true value each time you measure it.

rate of change The ratio between two related quantities.

ratio The quantitative relation between two amounts showing the number of times one value contains or is contained within the other.

raw material Substance obtained from the ground, air, or sea.

reactant A substance that takes part in a chemical reaction.

reaction pathway A reaction, or series of reactions, chosen for making a particular substance.

reaction profile A chart that shows the energy of reactants and products in a chemical reaction.

reaction rate The reaction rate for a given chemical reaction is the measure of the change in concentration of the reactants or the change in concentration of the products per unit time.

reaction time Duration of a reaction from the start to the end.

reactivity The tendency of a substance to take part in chemical reactions.

reactivity series A chart showing elements, usually metals, in order of how vigorously they react in chemical reactions.

reciprocal Related to another so that their product is 1.

recycling Processing a used material or object so that its substances can be reused.

redox A chemical reaction in which one substance is reduced while another is oxidised.

reduced Describes a substance that has lost oxygen, or gained electrons, in a chemical reaction.

reducing agent A substance that reduces another substance by accepting oxygen or by donating electrons.

reduction A loss of oxygen, or gain of electrons, by a substance in a chemical reaction.

relationship The way in which two or more things are connected.

relative atomic mass The mean mass of an atom of an element compared to 1/12 the mass of a ^{12}C atom, defined as 12 exactly. Its symbol is A_r.

relative formula mass M_r, the mean mass of a unit of a substance compared to 1/12 the mass of a ^{12}C atom, defined as 12 exactly. It is calculated by adding together the relative atomic masses for the atoms in the formula of a substance.

relative molecular mass Relative formula mass applied to a molecular substance.

renewable Can be replaced as it is used.

repeatable A measure of how close values are to each other when an experiment is repeated with the same equipment.

repeating unit A section of a polymer molecule that is repeated many times in the molecule.

reproducible When other people carry out an investigation and get similar results to the original investigation the results are reproducible.

residue Insoluble material left behind during filtration.

resin Substance that changes from the liquid state to the solid state because of chemical reactions.

resolution A measure of the smallest object that can be seen using an instrument.

retention time The time taken for a substance to travel through a gas chromatography column.

reversible reaction Chemical reaction that can proceed in either direction.

R_f value Relative distance travelled by a substance during chromatography, calculated as: distance travelled by substance ÷ distance travelled by solvent.

risk The chance of damage or injury from a hazard.

rusting Corrosion in which iron or steel reacts with oxygen and water to form hydrated iron(III) oxide or rust.

sacrificial protection Rust prevention in which a more reactive metal than iron, such as magnesium or zinc, corrodes in preference to iron or steel.

salt A compound formed when hydrogen ions in an acid are replaced by metal ions or ammonium ions. Salts are formed when acids react with bases, carbonates, or metals.

saturated In the context of hydrocarbon molecules, having only single carbon–carbon bonds.

saturated solution A solution containing the maximum mass of solute at a given temperature.

scientific questions Scientific questions are questions that can be answered by collecting and considering evidence.

shell Region of space in an atom that can hold a certain number of electrons.

significant figures (sig. fig.) Each of the digits of a number that are used to express it to the required degree of accuracy, starting from the first non-zero digit.

simple distillation Method to separate a solvent from a solution.

simple molecule A particle consisting of a few non-metal atoms joined together by covalent bonds.

slag Waste formed by impurities when extracting iron.

solubility A measure of how much solute can dissolve in a given solvent at a certain temperature.

soluble A substance that will dissolve in a given solvent.

solute Substance that dissolves in a solvent.

solution A mixture formed when one substance dissolves in another.

solvent A substance that can dissolve a solute to form a solution.

space-filling model A representation of ionic compounds and molecules in which balls represent the atoms or ions.

spectator ion A charged particle present in a reaction mixture, but that does not take part in the reaction.

spread The difference between the highest and lowest measurements of a set of repeat measurements.

standard form A way of writing down very large or very small numbers easily.

standard solution A solution with a known, accurate concentration.

starch–iodide paper Test paper used to detect chlorine. When damp, it turns blue-black in the presence of chlorine.

state The form that a substance has (solid, liquid, or gas) under given conditions of temperature and pressure.

state symbol Letters used to represent the physical state of a substance (g, gas; l, liquid; s, solid; aq, dissolved in water).

stationary phase A substance in the solid or liquid state that does not move during chromatography.

stoichiometry Describes the relative amounts of each substance involved in a chemical reaction.

strong acid An acid that is fully ionised in aqueous solution.

subatomic particle Particles from which atoms are made, including protons, neutrons, and electrons.

sublime To change state directly from the solid to the gas state.

successful collision In reactions, a collision between two particles with enough energy for a reaction to happen.

sugar Carbohydrate, a compound of carbon, hydrogen, and oxygen.

surroundings In a chemical reaction, anything that is not a reacting particle.

systematic error An error that causes there to be the same difference between a measurement and the true value each time you measure it.

tangent Straight line that touches a curve but does not cut through the curve when extended.

tare Set a balance to zero.

technology The application of scientific knowledge for practical purposes, especially in industry.

tensile strength Measure of how strong a material is when stretched.

theoretical yield The maximum possible mass of product that could be formed in a reaction, calculated using the law of conservation of mass.

thin-layer chromatography A type of chromatography that uses silica or alumina spread on a plate as the stationary phase, and a solvent in the liquid state as the mobile phase.

titration A method in which a neutralisation reaction is used to determine the concentration of an acid or an alkali.

titre A measured volume of solution added from a burette during a titration to reach the end-point.

transition metal An element in the block of the Periodic Table between Groups 2 and 3 (IUPAC Group 13).

trend A general pattern or direction of change.

uncertainty The doubt in the result because of the way that the measurement is made.

unsaturated In the context of hydrocarbon molecules, having a carbon–carbon double bond.

vapour A substance in its gas state and below its boiling point.

variable A quantity that can change, for example, time, temperature, length, mass.

volcanic activity Eruptions of volcanoes, releasing large volumes of gases into the atmosphere.

volumetric flask A piece of glassware used to make an accurate volume of solution, such as a standard solution.

volumetric pipette A piece of glassware used to measure an accurate volume of solution for a titration.

weak acid Acid only partially ionised in aqueous solution.

word equation A model of a reaction that describes reactants and products using their chemical names.

yield The mass of a particular product formed in a reaction.

y-intercept The y-intercept of a straight line graph is where the line crosses the y-axis.

Index

Reference material

Table 1 *SI base units.*

Physical quantity	Unit	Unit
length	metre	m
mass	kilogram	kg
time	second	s
temperature	kelvin	K
current	ampere	A
amount of a substance	mole	mol

Table 2 *SI derived units.*

Physical quantity	Unit(s)	Unit(s)
area	squared metre	m^2
volume	cubic metre; litre; cubic decimetre	m^3; l; dm^3
density	kilogram per cubic metre	kg/m^3
temperature	degree Celsius	°C
pressure	pascal	Pa
specific heat capacity	joule per kilogram per degree Celsius	J/kg/°C
specific latent heat	joule per kilogram	J/kg
speed	metre per second	m/s
force	newton	N
gravitational field strength	newton per kilogram	N/kg
acceleration	metre per squared second	m/s^2
frequency	hertz	Hz
energy	joule	J
power	watt	W
electric charge	coulomb	C
electric potential difference	volt	V
electric resistance	ohm	Ω
magnetic flux density	tesla	T

Table 3 *There are many different types of scientific apparatus. This table shows what they look like, how to draw them, and what you can use them for.*

Apparatus	What it looks like	Diagram	What you can use it for
test tube			• heating solids and liquids • mixing substances • small-scale chemical reactions
boiling tube			• A boiling tube is a big test tube. You can use it for the same things as a test tube.
beaker			• heating liquids and solutions • mixing substances
conical flask			• heating liquids and solutions • mixing substances
filter funnel			• to separate solids from liquids, using filter paper
evaporating dish			• to evaporate a liquid from a solution
condenser			• to cool a substance in the gas state, so that it condenses to the liquid state
tripod			• to support apparatus above a Bunsen burner
gauze			• to spread out thermal energy from a Bunsen burner • to support apparatus such as beakers, over a Bunsen burner
syringe			• to measure volumes of gases

Table 4 *The names and formulae of some common ions, with compound ions shaded.*

Positive ions	
ammonium	NH_4^+
hydrogen	H^+
lithium	Li^+
sodium	Na^+
potassium	K^+
silver	Ag^+
barium	Ba^{2+}
calcium	Ca^{2+}
copper(II)	Cu^{2+}
iron(II)	Fe^{2+}
lead(II)	Pb^{2+}
magnesium	Mg^{2+}
zinc	Zn^{2+}
aluminium	Al^{3+}
iron(III)	Fe^{3+}

Negative ions	
chloride	Cl^-
bromide	Br^-
iodide	I^-
hydroxide	OH^-
nitrate	NO_3^-
oxide	O^{2-}
carbonate	CO_3^{2-}
sulfate	SO_4^{2-}

Table 5 *Useful constants.*

Constant	Value	
Avogadro constant	6.02×10^{23} /mol	**H**
Molar gas volume at RTP	24 dm³	**H** **S**

Table 6 *Approximate sizes.*

Item	Approximate size (m)	
bond length	10^{-10}	
atomic radius	10^{-10}	
small molecule	10^{-9}	
nanoparticle	10^{-9} to 10^{-7} m	**S**

Table 7 *Properties of subatomic particles.*

Subatomic particle	Relative mass	Relative charge
proton	1	+1
neutron	1	0
electron	0.0005	−1

Table 8 *Flame test colours.*

Metal	Cation	Flame test colour
lithium	Li^+	red
sodium	Na^+	yellow
potassium	K^+	lilac
calcium	Ca^{2+}	orange-red
copper	Cu^{2+}	green-blue

Table 9 *Tests for gases.*

Gas	Test	Observation
oxygen	hold a glowing splint in the gas	splint relights
hydrogen	hold a lighted splint in the gas	gas ignites with a pop
carbon dioxide	bubble the gas through limewater	limewater turns milky or cloudy white
chlorine	hold damp blue litmus paper in the gas	litmus paper turns red then bleaches white

Table 10 *Colours of precipitates formed when dilute sodium hydroxide solution is added.* [S]

Name of cation	Cation	Colour of metal hydroxide precipitate
iron(II)	Fe^{2+}	green
iron(III)	Fe^{3+}	orange-brown
copper(II)	Cu^{2+}	blue
calcium	Ca^{2+}	white
zinc	Zn^{2+}	white (redissolves in excess sodium hydroxide)

Table 11 *Tests for negatively charged ions.* [S]

Name of anion	Anion	Test	Observation
carbonate	CO_3^{2-}	add dilute acid	bubbles produced (confirm they are carbon dioxide using limewater)
sulfate	SO_4^{2-}	acidify with a few drops of dilute hydrochloric acid, then add dilute barium chloride solution	white precipitate forms
chloride	Cl^-	acidify with a few drops of dilute nitric acid, then add dilute silver nitrate solution	white precipitate forms
bromide	Br^-		cream precipitate forms
iodide	I^-		yellow precipitate forms

Useful conversions

$dm^3 \rightarrow cm^3$ multiply by 1000

$cm^3 \rightarrow dm^3$ divide by 1000

Useful expressions

$$R_f = \frac{\text{distance travelled by substance}}{\text{distance travelled by solvent}}$$

[H]

$$\text{mass (g)} = \text{molar mass (g/mol)} \times \text{amount (mol)}$$

$$\text{concentration of solution (g/dm}^3) = \frac{\text{mass (g)}}{\text{volume of solution (dm}^3)}$$

[S]

$$\text{theoretical yield} = \frac{\text{mass of limiting reactant}}{\text{sum of } M_r \text{ for limiting reactant}} \times \text{sum of } M_r \text{ for product}$$

$$\text{percentage yield} = \frac{\text{actual yield}}{\text{theoretical yield}} \times 100$$

$$\text{atom economy} = \frac{\text{sum of } M_r \text{ of the desired product}}{\text{sum of } M_r \text{ of all products}} \times 100$$

[H] [S]

$$\text{volume of a gas (dm}^3) = \text{amount (mol)} \times \text{molar volume (dm}^3)$$

$$\text{concentration of solution (mol/dm}^3) = \frac{\text{amount (mol)}}{\text{volume of solution (dm}^3)}$$

Key
atomic number
Symbol
name
relative atomic mass

1	2											13	14	15	16	17	18
1 **H** hydrogen 1.0																	2 **He** helium 4.0
3 **Li** lithium 6.9	4 **Be** beryllium 9.0											5 **B** boron 10.8	6 **C** carbon 12.0	7 **N** nitrogen 14.0	8 **O** oxygen 16.0	9 **F** fluorine 19.0	10 **Ne** neon 20.2
11 **Na** sodium 23.0	12 **Mg** magnesium 24.3											13 **Al** aluminium 27.0	14 **Si** silicon 28.1	15 **P** phosphorus 31.0	16 **S** sulfur 32.1	17 **Cl** chlorine 35.5	18 **Ar** argon 39.9
19 **K** potassium 39.1	20 **Ca** calcium 40.1	21 **Sc** scandium 45.0	22 **Ti** titanium 47.9	23 **V** vanadium 50.9	24 **Cr** chromium 52.0	25 **Mn** manganese 54.9	26 **Fe** iron 55.8	27 **Co** cobalt 58.9	28 **Ni** nickel 58.7	29 **Cu** copper 63.5	30 **Zn** zinc 65.4	31 **Ga** gallium 69.7	32 **Ge** germanium 72.6	33 **As** arsenic 74.9	34 **Se** selenium 79.0	35 **Br** bromine 79.9	36 **Kr** krypton 83.8
37 **Rb** rubidium 85.5	38 **Sr** strontium 87.6	39 **Y** yttrium 88.9	40 **Zr** zirconium 91.2	41 **Nb** niobium 92.9	42 **Mo** molybdenum 95.9	43 **Tc** technetium	44 **Ru** ruthenium 101.1	45 **Rh** rhodium 102.9	46 **Pd** palladium 106.4	47 **Ag** silver 107.9	48 **Cd** cadmium 112.4	49 **In** indium 114.8	50 **Sn** tin 118.7	51 **Sb** antimony 121.8	52 **Te** tellurium 127.6	53 **I** iodine 126.9	54 **Xe** xenon 131.3
55 **Cs** caesium 132.9	56 **Ba** barium 137.3	57–71 lanthanoids	72 **Hf** hafnium 178.5	73 **Ta** tantalum 180.9	74 **W** tungsten 183.8	75 **Re** rhenium 186.2	76 **Os** osmium 190.2	77 **Ir** iridium 192.2	78 **Pt** platinum 195.1	79 **Au** gold 197.0	80 **Hg** mercury 200.6	81 **Tl** thallium 204.4	82 **Pb** lead 207.2	83 **Bi** bismuth 209.0	84 **Po** polonium	85 **At** astatine	86 **Rn** radon
87 **Fr** francium	88 **Ra** radium	89–103 actinoids	104 **Rf** rutherfordium	105 **Db** dubnium	106 **Sg** seaborgium	107 **Bh** bohrium	108 **Hs** hassium	109 **Mt** meitnerium	110 **Ds** darmstadtium	111 **Rg** roentgenium	112 **Cn** copernicium		114 **Fl** flerovium		116 **Lv** livermorium		

Figure 1 *The Periodic Table.*

Great Clarendon Street, Oxford, OX2 6DP, United Kingdom

Oxford University Press is a department of the University of Oxford.
It furthers the University's objective of excellence in research,
scholarship, and education by publishing worldwide. Oxford is a
registered trade mark of Oxford University Press in the UK and in
certain other countries

British Library Cataloguing in Publication Data
Data available

978 0 19 835982 1

10 9 8 7 6 5 4 3 2

Paper used in the production of this book is a natural, recyclable
product made from wood grown in sustainable forests.
The manufacturing process conforms to the environmental regulations
of the country of origin.

Printed in China by Golden Cup

This resource is endorsed by OCR for use with specification J248 OCR
GCSE (9–1) in Chemistry A (Gateway Science). In order to gain OCR
endorsement, this resource has undergone an independent quality
check. Any references to assessment and/or assessment preparation are
the publisher's interpretation of the specification requirements and are
not endorsed by OCR. OCR recommends that a range of teaching and
learning resources are used in preparing learners for assessment. OCR
has not paid for the production of this resource, nor does OCR receive
any royalties from its sale. For more information about the endorsement
process, please visit the OCR website, www.ocr.org.uk.

The authors and series editor would like to thank Sophie Ladden and
Margaret McGuire at OUP for their patience, encouragement, and
attention to all the small - but important - details.

Nigel Saunders would like to thank his family for their unending
patience, support, and supplies of chocolate.

Philippa Gardom Hulme would like to thank Barney, Catherine, and
Sarah for their never-ending support and patience, and for keeping
quietly out of the way in the early mornings. Thanks, too, to Claire
Gordon for her wise counsel over tea and scones, and for getting us all
going in the first place!